Advanced Carbon Materials and Technology

Scrivener Publishing
100 Cummings Center, Suite 541J
Beverly, MA 01915-6106

Advance Materials Series

The Advance Materials Series is intended to provide recent advancements of the fascinating field of advanced materials science and technology, particularly in the area of structure, synthesis and processing, characterization, advanced-state properties, and applications. The volumes will cover theoretical and experimental approaches of molecular device materials, biomimetic materials, hybrid-type composite materials, functionalized polymers, superamolecular systems, information- and energy-transfer materials, biobased and biodegradable or environmental friendly materials. Each volume will be devoted to one broad subject and the multidisciplinary aspects will be drawn out in full.

Series Editor: Dr. Ashutosh Tiwari
Biosensors and Bioelectronics Centre
Linkoping University
SE-581 83 Linkoping
Sweden
E-mail: ashutosh.tiwari@liu.se
Managing Editors: Swapneel Despande and Sudheesh K. Shukla

Publishers at Scrivener
Martin Scrivener(martin@scrivenerpublishing.com)
Phillip Carmical (pcarmical@scrivenerpublishing.com)

Advanced Carbon Materials and Technology

Edited by

Ashutosh Tiwari and S.K. Shukla

Scrivener
Publishing

WILEY

For general information on our other products and services or for technical support, please contact our Customer Care Department within the United States at (800) 762-2974, outside the United States at (317) 572-3993 or fax (317) 572-4002.

Wiley also publishes its books in a variety of electronic formats. Some content that appears in print may not be available in electronic formats. For more information about Wiley products, visit our web site at www.wiley.com.

For more information about Scrivener products please visit www.scrivenerpublishing.com.

Cover design by Russell Richardson

Library of Congress Cataloging-in-Publication Data:

ISBN 978-1-118-68623-2

Printed in the United States of America

10 9 8 7 6 5 4 3 2 1

Contents

Preface

The expansion of carbon materials is the focal point of materials research and technology which is mostly related to physics, chemistry, biology, applied sciences and engineering. Research on carbon materials has mainly focused on the aspects of fundamental physics that have unique electrical, thermal and mechanical properties applicable for a range of applications. The electrons in graphene and other derived carbon materials behave as dirac fermions due to their interaction with the ions of the lattice. This direction has led to the discovery of new phenomena such as Klein tunneling in carbon-based solid state systems, and the so-called half-integer quantum Hall effect due to a special type of Berry phase. In pursuit of the same goal, Advanced Carbon Materials and Technology offers detailed, up-to-date chapters on the processing, properties and technological developments of graphene, carbon nanotubes, carbon fibers, carbon particles and other carbon-based structures, including multifunctional graphene sheets, graphene quantum dots, bulky balls, carbon balls, and their polymer composites.

Nanoscaled materials have properties which make them useful for enhancing surface-to-volume ratio, reactivity, strength and durability. The chapter entitled, "Synthesis, Characterization and Functionalization of Carbon Nanotubes and Graphene: A Glimpse of Their Application," encompasses the principles of nanotubes and graphene production, new routes of preparation and numerous methods of modification essential for various potential applications. The chapter on, "Surface Modification of Graphene," covers a range of covalent and non-covalent approaches. In the chapter, "Graphene and Carbon Nanotube-Based Electrochemical Biosensors for Environmental Monitoring," the use of carbon nanotubes and numerous graphene-based affinity electrodes for the development of novel tools for monitoring environmental pollution are described. The chapter on, "Catalytic Application of

Carbon-Based Nanostructured Materials on Hydrogen Sorption Behavior of Light Metal Hydrides," describes the state-of-the-art of carbon nanotubes, carbon nanofibers and graphene as a catalyst for the aforesaid hydrogen storage materials. An informal presentation about recent progress in the advances in synthetic techniques for large-scale production of carbon nanotubes, their purification and chemical modification, and the emerging technologies they enable are presented in the chapter, "Carbon Nanotubes and Their Applications." Moreover, a chapter dedicated to the, "Bioimpact of Carbon Nanomaterials," discusses graphene, nanotubes and fullerenes, along with their nanotoxicity, nanoecotoxicity, and various biomedical applications.

Carbon nano-objects including fullerenes, carbon nanotubes, carbon quantum dots, shungites and graphenes, show unique photorefractive characteristics. The chapter on, "Advanced Optical Materials Modified with Carbon Nano-Objects," illustrates the spectral, photoconductive, photorefractive and dynamic properties of the optical carbon objects-based nanomaterials. "Covalent and Non-Covalent Functionalization of Carbon Nanotube: Applications," deals with the photocatalytic nature of carbon nanotube-based composites. Illustrated in, "Metal Matrix Nanocomposites Reinforced with Carbon Nanotubes," are the preparation and properties of nanocomposites based on aluminium, copper, magnesium, nickel and titanium with reinforced matrix of nanofiller carbon materials (e.g., nanoplatelets, nanoparticles, nanofibers and carbon nanotubes) using various processing techniques. The chapter also discusses reinforcement using carbon nanotubes, interfacial bonding, thermal, mechanical, and tribological properties and tne challenges related to the synthesis of composites.

Fly ash, a waste by-product of coal thermal power plants, is a carbon-based lightweight material. Fly ash is generally inexpensive and is considered to be an environmental hazard, thus utilization of fly ash in composites proves to be both economically and environmentally beneficial. In this way, use of fly ash in developing advanced composites is very encouraging for the next generation of advanced lightweight composites. The discussion in, "Aluminum/Fly Ash Syntactic Foams: Synthesis, Microstructure and Properties," is focused on the methods of synthesis for fly ash-filled aluminum matrix composites along with their microstructure and mechanical properties, and the tribological properties of Al/fly ash syntactic foams. The chapter entitled, "Engineering Behavior of Ash Fills," covers the extensive characterization, hardening, bearing capacity and settlement of ash fill technology. The chapter on, "Carbon-Doped Cryogel Thin Films Derived from Resorcinol Formaldehyde," presents results of the structural and optical properties of carbon-doped cryogel thin films derived from resorcinol formaldehyde.

This book is written for a large readership, including university students and researchers from diverse backgrounds such as chemistry, materials science, physics, pharmacology, medical science and engineering, with specializations in the civil, environmental and biomedical fields. It can be used not only as a textbook for both undergraduate and graduate students, but also as a review and reference book for researchers in materials science, bioengineering, medicine, pharmacology, biotechnology and nanotechnology. We hope that the chapters of this book will provide the readers with valuable insight into state-of-the-art advanced and functional carbon materials and cutting-edge technologies.

Editors
Ashutosh Tiwari, PhD
S.K. Shukla, PhD

Managing Editors
Swapneel Despande
Sudheesh K. Shukla

Part 1
GRAPHENE, CARBON NANOTUBES AND FULLERENES

Synthesis, Characterization and Functionalization of Carbon Nanotubes and Graphene: A Glimpse of Their Application

Mahe Talat and O.N. Srivastava*

Nanoscience and Nanotechnology Unit, Department of Physics, Banaras Hindu University, Varanasi, India

Abstract

Since the discovery of nanomaterials, carbon nanotubes structures have attracted great interest in most areas of science and engineering due to their unique physical and chemical properties and are supposed to be a key component of nanotechnology. The most recent addition to the family of carbon nanostructures is graphene. Graphene is a one-atom-thick material consisting of sp^2-bonded carbon with a honeycomb structure. It resembles a large polyaromatic molecule of semi-infinite size. In the past five years, graphene-based nanomaterials have been the focus of not only material scientists but also engineers and medical scientists. The interesting and exciting properties of single-layer graphene sheets have excited the scientific community especially in the areas of materials, physics, chemistry and medical science. The state-of-the-art CNT production encompasses numerous methods and new routes are continuously being developed. The most common synthesis techniques are arc discharge, laser ablation, high pressure carbon monoxide (HiPCO) and chemical vapor deposition (CVD) with many variants. Most of these processes take place in vacuum or with process gases. By choosing appropriate experimental parameters, large quantities of nanotubes can be synthesized by these methods. It is possible to control some properties of the final product, such as type of CNTs synthesized (MWNTs vs. SWNTs), the quality of the nanotubes,

Corresponding author: heponsphy@gmail.com

Ashutosh Tiwari and S.K. Shukla (eds.) Advanced Carbon Materials and Technology, (1–34)
2014 © Scrivener Publishing LLC

the amount and type of impurities, and some structural CNT features. In this chapter we discuss some of the methods employed in our lab for the synthesis and characterization of the CNTs and graphene. For application in biomedical and targeted drug delivery, the major limitation of these nanomaterials is their poor solubility, agglomeration and processibility. Functionalization of CNTs and GS is, therefore, necessary to attach any desired compounds including drug and also to enhance the solubility and biocompatibility of these nanomaterials. Two types of functionalization methods, i.e., covalent and non-covalent methods are generally being adopted. We deliberate these two procedures of functionalization of CNTs and GS. The merits of these two modes of functionalization will also be discussed.

Keywords: Synthesis, characterization, application, CNT, graphene

1.1   Introduction

Carbon is the base element of all organic materials, one of the most abundant elements on earth and also the only element of the periodic table that occurs in allotropic forms from 0 dimensions to 3 dimensions due to its different hybridization capabilities. It has the unique ability to form allotropes, which can be described by valence bond hybridization, sp^n. The well-known forms are diamond, where n = 3, and graphite, having n = 2. It was shown that all other 1<n<3 are possible. The valence bond hybridization determines the physical and chemical properties of the allotropes of carbon. Fullerene (zero-dimensional), carbon nanotubes (one-dimensional) and graphene (single layer of graphite, two-dimensional) are all made of sp^2-hybridized carbon atoms, whereas diamond (three-dimensional) is sp^3 hybridized. The discovery of "fullerenes" added a new dimension to the knowledge of carbon science, which led to the Nobel Prize being awarded to R.F. Curl, H. Kroto and R.E. Smalley in 1996. Stimulated by fullerene discoveries, the Japanese scientist Ijima [1] discovered another new form of carbon-graphitic tubules and the CNT was born. He also suggested the possible structure of these tubes, which was later proven right. The subsequent discovery of "carbon nanotubes" (CNTs) added a new dimension to the science and technology of new carbons.

The most recent addition to the family of carbon nanostructures is graphene. Graphene is a one-atom-thick material consisting of sp^2-bonded carbon with a honeycomb structure. It resembles a large polyaromatic molecule of semi-infinite size. In the past five years, graphene-based nanomaterials have been the focus of not only material scientists but also engineers and medical scientists. The interesting and exciting properties of single-layer graphene sheets, such as high mechanical strength, high elasticity and thermal conductivity, demonstration of the room-temperature quantum Hall effect, very high room temperature electron mobility, tunable optical properties, and a tunable band gap have excited the scientific community especially in the areas of materials, physics, chemistry and medical science. Following the series of Nobel Prizes awarded after the discovery of fullerenes, another nano star—graphene— received the 2010 Nobel Prize in Physics "for groundbreaking experiments regarding the two-dimensional material graphene" performed by Andre Geim and Konstantin Novoselov [2].

Therefore, the discovery and subsequent applications of these carbon nanomaterials have allowed the development of an entire branch of nanotechnology based on these versatile materials.

1.2   Synthesis and Characterization of Carbon Nanotubes

The first experimental evidence of carbon nanotubes (CNTs) came in 1991 [3] in the form of multi-wall nanotubes (MWNT), which motivated a sudden increase in nanotubes synthesis research. In 1993, the first experimental evidence of single-wall nanotubes (SWNT) was introduced [4]. Since then, the synthesis methods for CNTs have been developed tremendously. Production methods for carbon nanotubes (CNTs) can be broadly divided into two categories, chemical and physical, depending upon the process used to extract atomic carbon from the carbon-carrying precursor. Chemical methods rely upon the extraction of carbon solely through catalytic decomposition of precursors on the transition metal nanoparticles, whereas physical methods also use high energy sources, such as plasma or laser ablation to extract the atomic carbon. However, the most common synthesis techniques are arc discharge, laser ablation, high pressure carbon monoxide (HiPCO) and chemical vapor deposition (CVD) with many variants [5]. Most of these processes take place in vacuum or with process gases. By choosing appropriate experimental parameters, large quantities of nanotubes can be synthesized by these methods. It is possible to control some properties of the final product, such as type of CNTs synthesized (MWNTs vs SWNTs), the quality of the nanotubes, the amount and type of impurities, and some structural CNT features [6]. Other reported methods include plastic pyrolysis [7], diffusion flame synthesis and electrolysis using graphite electrodes immersed in molten ionic salts [8] and ball-milling of graphite [9].

In this chapter, we will discuss the method of synthesis of CNTs employed in our lab such as spray pyrolysis, arc discharge method and synthesis of CNTs by low pressure chemical vapor deposition (LPCVD) method, and catalytic decomposition of hydrocarbon gases, e.g., methane and ethylene, onto the Ferritin /Fe- SiO_2- Si substrates.

Synthesis of CNTs was carried out using spray pyrolysis-assisted CVD method, where ferrocene ($C_{10}H_{10}Fe$) was used as a source of iron (Fe) which acts as a catalyst for the growth of CNTs. Castor oil was used as the carbon source; castor oil contains carbon, hydrogen and lower amount of oxygen. The spray pyrolysis setup consisted of a nozzle (inner diameter ~ 0.5 mm) attached to a ferrocene-castor oil supply used for releasing the solution into a quartz tube (700 mm long and inner diameter 25 mm), which was mounted inside a reaction furnace [10].

Spray pyrolysis of castor oil-ferrocene solution at ~ 850°C in Ar atmosphere leads to a uniform thick black deposition on the inner wall of the quartz tube at the reaction hot zone (~ 850°C). Figure 1.1(a) shows the SEM morphology of the as-grown CNTs. The length of CNTs was ~ 5–10 µm. Structural details of the as-grown CNTs sample were further investigated by TEM. Typical TEM image of the as-grown CNTs is shown in Figure 1.1(b). The TEM investigation of the as-grown CNTs confirms that the CNTs are multi-walled in nature. These nanotubes have varying diameters ranging from ~ 20–60 nm. In the spray pyrolysis reaction, the castor oil-ferrocene solution was atomized via spray nozzle and sprayed through carrier gas (Ar). The Fe particles (liberated by the decomposition of ferrocene) were deposited on the inner walls of the quartz tube. The carbon species released from decomposition of castor oil and also from ferrocene got adsorbed on the Fe particles and diffused rapidly along the axial direction leading to the formation of CNTs. A study was also done using castor oil-ferrocene with ammonia solution so as to develop CNTs containing nitrogen, i.e., C-N nanotubes. This was done keeping in view the fact that nitrogen-doped CNTs are considered as one of the important ingredients of CNT-based electronics. Figure 1.2(a) shows

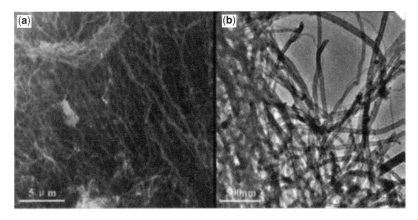

Figure 1.1 SEM (a) and TEM (b) micrographs of the as-grown CNTs obtained by spray pyrolysis of castor oil-ferrocene solution at ~ 850°C [11].

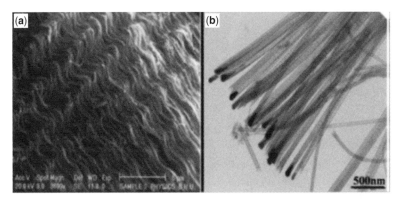

Figure 1.2 SEM (a) and TEM (b) micrographs of the as-grown C-N nanotubes obtained by spray pyrolysis of castor oil-ferrocene with ammonia solution at ~ 850°C [11].

a typical SEM micrograph of as-grown C-N nanotubes, which reveals the wavy morphology of nanotubes. These wavy nanotubes are most likely due to pentagonal and heptagonal defects that are introduced in the hexagonal sheets. TEM images of the as-grown C-N nanotubes are shown in Figure 1.2(b). These CNTs have bamboo-shaped structures. The TEM image in Figure 1.2(b) shows that the nanotubes have a range of diameters varying from ~ 50–80 nm. It is suggested that the bamboo-shaped morphologies arise from the incorporation of pyridine-like N atoms within the carbon framework [11]. Also, no encapsulated metal particle was found inside the nanotubes.

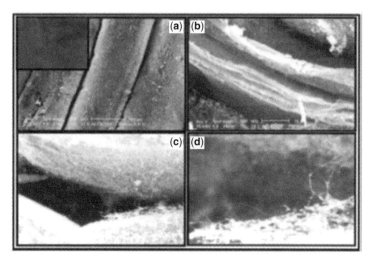

Figure 1.3 (a–d) SEM micrographs of as synthesized SWNTs webs, Inset of Fig. 1.3(a) shows the optical photograph of SWNTs web, some SWNTs which are coming out of the bundle.

Another method used is arc discharge method where SWNTs webs have been synthesized using Fe as catalysts by this method in argon atmosphere. SWNTs has been synthesized by using electric arc discharge of graphite cathode (3cm×1cm×1cm) and Fe as well as Ni-Y filled anode (5cm×0.8cm×0.8) at 200 torr pressure of argon gas. The SWNTs webs have been synthesized using Fe as catalysts by arc discharge method in argon atmosphere. Figure 1.3 shows the low magnification SEM micrographs of as-synthesized SWNTs web. The inset of Figure 1.3(a) shows the optical photograph of the as-deposited web on the chamber walls. The length of these webs is around 4 to 6 cm. The webs are ~75 to 100 µm thick, which abundantly contains the SWNTs, as is clear from Figure 1.3(b). Figure 1.3(c) shows the SEM image from the inner region between the two webs, dominantly containing SWNTs. A few SWNTs are also visible in Figure 1.3(d), which are coming out from these bundles.

Figure 1.4 shows the transmission electron micrographs of SWNTs bundles. Figure 1.4(a) reveals the bundles containing large amount of SWNTs along with the catalyst particle which have been used to synthesize these SWNTs. Figure 1.4(b) shows the HRTEM image of as synthesized SWNTs as shown in Figure 1.4(a). Figure 1.4(c,d) are from the different regions of the samples containing SWNTs bundles

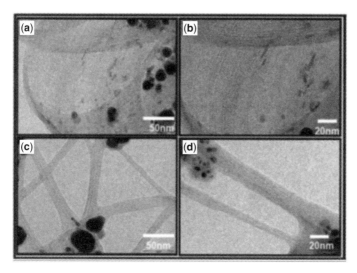

Figure 1.4 (a–d) HRTEM micrographs of SWNTs web from different regions.

and also catalysts particles and some amorphous carbon which have been coated onto the SWNTs bundles. The diameter of individual SWNTs is found to be approximately 1.6 nm. *(Synthesis of SWNTs and graphene by electric arc discharge method in Ar atmosphere, S. Awasthi, K. Awasthi and O.N. Srivastava, J Nanosci Nanotech [under submission].)*

Raman analysis of as-synthesized SWNTs webs was also done. In Figure 1.5, the presence of RBM at 132 cm^{-1} confirms the formation of SWNTs. The diameter of SWNTs as calculated from RBM is found to be 1.91 nm, which is in good agreement with the HRTEM results. Since G peak is related to the sp^2 bonding and is common to all the graphitic materials, and D peak is the defect-induced peak, showing the mixed state of sp^2-sp3 bonding, the intensity ratio of the D peak to the G peak can be used to quantify the quality of the graphitic samples. In the present case, this ratio was found to be ~0.08, which confirms that the as-synthesized nanotubes are highly crystalline and defect free in nature.

Through synthesis of CNTs by low pressure chemical vapor deposition (LPCVD) method we have grown CNTs (single-walled) by LPCVD method on Ferritin-based Fe catalyst. The SWNTs were grown using LPCVD unit (Automate, USA) (Fig.1.6). The substrate, i.e., ferritin-SiO$_2$-Si (supplied by Automate USA), was placed into the center of a 50 mm quartz tube reactor. First, the iron oxide particles were reduced under a H$_2$ pressure of 200 mbar at 800°C for

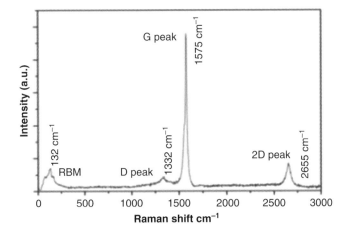

Figure 1.5 Raman spectrum of as-synthesized SWNTs webs.

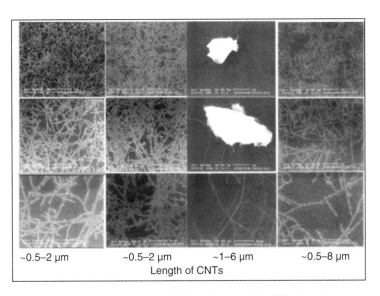

Figure 1.6 SEM images of as-grown CNTs using ferritin-SiO$_2$-Si substrate at different temperatures.

10 min. CNTs growth was then performed on the substrate at 800°C for 15 and 30 min under flows of 1400 sccm of methane, 100 sccm of ethylene and 500 sccm of H$_2$. After CVD reaction, the furnace was switched off and allowed to cool down to room temperature under Ar gas flow of 500 sccm. *(Synthesis of SWNTs and graphene by electric arc discharge method in Ar atmosphere, S. Awasthi, K. Awasthi and O.N. Srivastava, J Nanosci Nanotech [under submission].)*

1.3   Synthesis and Characterization of Graphene

Graphite is stacked layers of many graphene sheets, bonded together by week van der Waals force. Thus, in principle, it is possible to produce graphene from a high purity graphite sheet, if these bonds can be broken. Exfoliation and cleavage use mechanical and chemical energy, respectively, for breaking these weak bonds and separate out individual graphene sheets. To scale up the production, various synthetic methods are being developed. Some of the methods listed are presented below.

1.3.1   Micromechanical Cleavage of Highly Oriented Pyrolytic Graphite

The remarkably simple yet efficient method developed by Novoselov and Geim consists in using common adhesive tape to repeat the stick and peel process a dozen times which statistically brings a 1mm-thick graphite flake to a monolayer thin sample. The first piece of graphene sheet was obtained via manual mechanical cleavage of graphite with a Scotch tape [12], which seems to break the rule that no 2D crystals can exist under ambient conditions and shows us many unusual properties [12]. The exfoliated graphene manifests a unique structure and superior properties, although this production method is not applicable on a large scale. Inspired by this pioneering work, several alternative techniques have been developed for fabricating graphene materials.

Graphene has been made by four different methods. This approach, which is also known as the "Scotch tape" or peel-off method, was based on earlier work on micromechanical exfoliation from patterned graphite [13], and the fourth was the creation of colloidal suspensions. Micromechanical exfoliation has yielded small samples of graphene that are useful for fundamental study, although large-area graphene films (up to ~1cm^2) of single- to few-layer graphene have been generated by CVD growth on metal.

1.3.2   Chemical Vapor Deposition Growth of Graphene either as Stand Alone or on Substrate

Another feasible method is by chemical vapor deposition (CVD) and epitaxial growth, such as decomposition of ethylene on nickel surfaces [14]. These early efforts (which started in 1970) were followed by a large body of work by the surface-science community on "monolayer graphite" [15]. Epitaxial growth on electrically insulating surfaces such as SiC has also been used for the growth of graphene [16, 17]. One of the highly popular techniques of graphene growth is thermal decomposition of Si on the (0001) surface plane of single crystal of 6H-SiC [18]. Graphene sheets are found to be formed when H_2-etched surface of 6H-SiC was heated to temperatures of 1250 to 14500°C for a short time (1–20 minutes). Graphene epitaxially grown on this surface typically has 1 to 3 graphene layers; the number of layers being dependent on the decomposition temperature. In a similar process, Rollings *et al.* have produced graphene films as low as one atom thick [19]. The first report on planar few-layer graphene (PFLG) synthesized by CVD was in 2006 [20]. In this work, a natural, eco-friendly, low-cost precursor, camphor, was used to synthesize graphene on Ni foils. Camphor was first evaporated at 1800°C and then pyrolyzed in another chamber of the CVD furnace at 700 to 850°C using argon as the carrier gas. Large-area, high quality graphene can also be grown by thermal CVD on catalytic transition metal surfaces such as nickel and copper [21, 22]. Reina *et al.* prepared single- to few-layer graphene on polycrystalline Ni film of 1-2 cm² [23]. The Ni film (500 nm thick) was evaporated on a SiO2/Si substrate and was annealed in Ar+H_2 atmosphere at 900 to 10000°C, for 10 to 20 minutes. This annealing step created Ni grains of 5 to 20 μm in size. After CVD at 900 to 10000°C for 5 to 10 minutes, using 5 to 25 sccm CH4 and 1500 sccm H_2, graphene was found to form on the Ni—the size of each graphene being restricted by the Ni grain size. For Ni, mixed mono- and bi-layer graphene coverage of 87% has been reported [24, 25], while for Cu foils, an average of 95% of surfaces were covered by mono-layer graphene [26]. The graphene was later transferred to any substrate, keeping its electrical properties unchanged, thus making them suitable for various electronic applications. Typical CVD graphene growth uses gaseous hydrocarbons at elevated temperatures as the carbon source, such as methane, ethylene [27–29] and acetylene [3]. Single-layer graphene were synthesized from ethanol on Ni foils in an Ar atmosphere under atmospheric pressure by flash cooling after CVD, but a wide variation in graphene layer number was observed over the metal surface [31]. Single- and few-layer graphene films were grown employing a vacuum-assisted CVD technique on Cu foils using n-hexane as a liquid precursor [32]. Copper appears to have a small affinity for oxygen that allows for graphene growth even if the source of carbon is a solid, such as the sugar as reported recently [33]. Graphene, thus synthesized and transferred onto a glass substrate, has shown 90% optical transmittance [34].

1.3.3   Chemical and Thermal Exfoliation of Graphite Oxide

Some recent success in regards to graphene includes chemical exfoliation through the formation of derivatized graphene sheets such as GO [35, 36], r-GO [37], or halogenated graphene, solvent-assisted ultrasonic exfoliation [38]. Graphite oxide was first prepared in the nineteenth century [39], and since then it has been mainly produced by the following methods pronounced by Brodie, Staudenmaier [40] and Hummers [41]. All three methods involve oxidation of graphite in the presence of strong acids and oxidants. The level of the oxidation can be varied on the basis of the method, the reaction conditions and the precursor graphite used. Although extensive research has been done to reveal the chemical structure of graphite oxide, several models are still being worked out. Graphite oxide consists of a layered structure of "graphene oxide" sheets that are strongly hydrophilic such that intercalation of water molecules between the layers readily occurs [42]. The interlayer distance between the graphene oxide sheets increases reversibly from 6 to 12 Å with increasing relative humidity [43]. Notably, graphite oxide can be completely exfoliated to produce aqueous colloidal suspensions of graphene oxide sheets by simple sonication [44] and by stirring the water/graphite oxide mixture for a long enough time [45]. The second approach is the oxidation-exfoliation-reduction of graphite powder [46]. Severe oxidation treatment converts graphite to hydrophilic graphite oxide which can be exfoliated into single-layer graphite oxide (graphene oxide) via stirring or mild sonication in water. Graphene oxide can be regarded as a functionalized graphene containing hydroxyl, epoxy and carboxylic groups, providing reaction sites for chemical modifications [47]. Reducing graphene oxide can partly restore its graphitic structure as well as conductivity [48]. Although reduced graphene oxide (r-GO), (also called chemically modified graphene [CMG], chemically converted graphene [CCG] or graphene), has considerable defects, it is one of the most widely used graphene-based renewable energy materials due to its low cost, facile preparation process, large productivity, and potential for functionalization [49].

1.3.4   Arc-Discharge Method

The arc-discharge method has also been used to prepare graphene sheets. Rao *et al.* reported for the first time that the arc-discharge method can also be used for the synthesis of graphene sheets [50]. By using graphite rods as electrodes, they have synthesized pure graphene with mainly 2–4 layers in the inner wall region of the arc chamber under relatively high pressure of hydrogen without any catalyst. Moreover, through this method, nitrogen-doped and boron-doped graphene sheets can also be easily synthesized with boron sources (B_2H_6) or nitrogen sources (pyridine) mixed into the hydrogen gas. However, the size and thickness of the pure graphene sheets still have room to improve, and the properties of the N- and B-doped graphene sheets synthesized also need further studies.

1.4   Methods Used in Our Lab: CVD, Thermal Exfoliation, Arc Discharge and Chemical Reduction

To scale up the production, various synthetic methods are being developed. One of the methods is micromechanical cleavage of highly oriented pyrolytic graphite, however, the yield is very low. Another feasible method is by chemical vapor deposition (CVD) and epitaxial growth, chemical and thermal exfoliation of graphite oxide. Graphite oxide was first prepared in the nineteenth century [51], and since then it has been mainly produced by the following methods pronounced by Brodie, Staudenmaier and Hummers. All three methods involve oxidation of graphite in the presence of strong acids and oxidants. Graphite oxide consists of graphene sheets decorated mostly with epoxide and hydroxyl groups. Graphite oxide resulting from the deployment of these methods when subjected to thermal exfoliation leads to the formation of graphene oxide, and that method is called thermal exfoliation. Here the dried graphite oxide powder (~ 200 m) was placed in a quartz tube (diameter ~ 25 mm and length ~ 1.3 m). The sample was flushed with Ar for 15 min and the quartz tube was quickly inserted into a furnace preheated

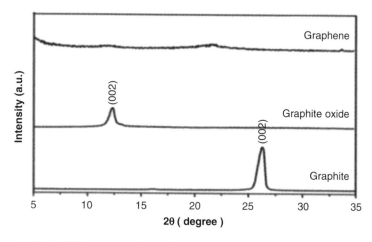

Figure 1.7 **The** XRD patterns of graphite, graphite oxide and graphene samples.

to 1050°C and held in the furnace for 30 s. The as-prepared GO was a brownish powder while the exfoliated version was of light consistency and shiny black. The XRD pattern of graphite, graphite oxide and graphene are shown in Figure 1.7. The XRD pattern of graphite shows an intense peak $2\theta = 26.4°$. This peak corresponds to 002 plane of graphite with interlayer spacing of 0.34 nm. In the XRD pattern of graphite oxide a new peak appears at $2\theta = 13.2°$, corresponding to the 002 plane of graphite oxide [52]. The interlayer spacing of GO is ~ 0.75 nm, which is significantly larger than that of graphite, due to intercalating oxide functional groups. The mechanism of exfoliation is mainly the expansion of CO_2 evolved into the interstices between the graphene sheets during rapid heating. The disappearance of native graphite XRD peaks in the XRD pattern of as-prepared graphene sample supports the formation of graphene sheets.

Few-layer graphene (FLG) has been synthesized by using electric-arc discharge of graphite electrodes in argon ambience at different pressure. The arc was maintained by continuously translating the anode to keep a constant distance of ~1 mm from the cathode. The deposits collected from the inner walls of the arc chamber have been characterized and Figure 1.8(a,b) shows the TEM images of as-synthesized FLG at 350 torr pressure of argon. Distinct features of the micrographs are the large-area graphene sheet-like structures as can clearly be seen in the Figure 1.8. Few-layer graphene nanosheets are clearly visible in the images. The

(a)   (b)

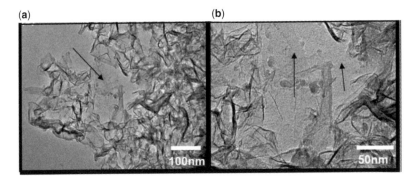

Figure 1.8 TEM micrographs of as-synthesized graphene nanosheets at 350 torr argon pressure.

Figure 1.8(b) shows the HRTEM image of the large-area graphene with minimum number of layers equal to four. The width of these graphene nanosheets is ~100-200 nm.

1.4.1   Raman Spectra

It is worth mentioning that an extensive analysis of graphene has been reported by Ferrari et al. [53], who have demonstrated that the second order Raman peak centered at 2700 cm−1 (2D peak) is the characteristic graphene feature and can be very useful in identifying the number of layers in a few-layer graphene sample. The number of layers in a graphene film can be estimated from the intensity, shape and position of the G and 2D bands. While the 2D band changes its shape, width and position with an increasing number of layers, the G-band peak position shows a down-shift with the number of layers. From XRD and TEM analysis it has been found that at 350 torr the minimum numbers of graphene layer are formed. By monitoring the width and position of this 2D peak one can deduce the number of layers from graphene samples. In Figure 1.9 at 350 torr, the 2D peak is at 2687cm^{-1}, confirming the formation of FLG.

Synthesis of large-area, high-quality and uniform graphene films on metal substrate by chemical vapor deposition (CVD) is shown in Figure 1.10. Chemical vapor deposition (CVD) is one of the main interesting synthetic procedures because it employs hydrocarbon decomposition over substrates, where metal nanoparticles have been placed. A one meter long quartz tube is used in the CVD system in our laboratory. The CVD system has a two zone furnace and the

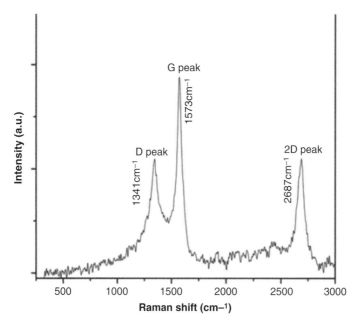

Figure 1.9 Raman Spectra at 350 torr. *(Synthesis of SWNTs and graphene by electric arc discharge method in Ar atmosphere, S. Awasthi, K. Awasthi and O.N. Srivastava, J Nanosci Nanotech [under submission].)*

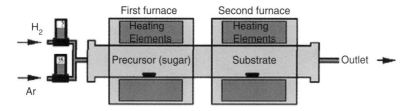

Figure 1.10 CVD set-up.

diameter of the quartz tube is 2 inches. The quartz tube is adjusted in the two zone CVD furnace. The solid carbon precursor (sugar) is placed in an alumina boat and placed on the middle portion in the first zone furnace. Alloy pieces are placed in a second alumina boat and it is placed in the second zone furnace. The distance between substrate and precursor is nearly 22 cm. In the first few (15) minutes only high purity gas flows. After that, the heating starts in the second zone (substrate) furnace set at 850°C in the presence of 200 sccm flow rate of argon gas. In the metal substrate we used copper and gold foil, and copper-gold alloy (Cu_3Au_2) as a substrate. Next,

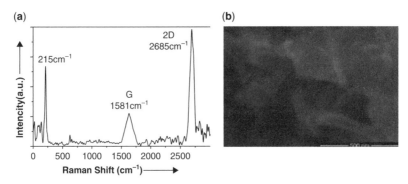

Figure 1.11 (a) Raman spectra of graphene deposited on Cu_3Au_2 substrate; (b) image captured by Raman spectrophotometer.

CH_4 was introduced at a flow rate of 30 sccm and H_2 at 50 sccm for annealing. After this the solid carbon precursor was vaporized and deposited on the substrate in the form of graphene.

In Figure 1.11(a) given below, Raman spectra of graphene was deposited on Cu_3Au_2 substrate. The appearance of a well-pronounced 2D peak at $2685cm^{-1}$ reveals the formation of high quality (probably single-layer) graphene. While in Figure 1.11(b), the growth of a large graphene island on Cu_3Au_2 substrate is visible in the image captured by Raman spectrophotometer.

1.4.2   Electrochemical Exfoliation

The electrochemical methods of preparing GN flakes involve the application of cathodic or anodic potentials or currents in either aqueous (acidic or other media) or non-aqueous electrolytes. One of the most important parameters for consideration of scaling-up the electrochemical technology is the yield of GN flakes. Single- or multi-layered GN flakes can easily be produced in short periods of time. It is a simple and fast method to exfoliate graphite into thin graphene sheets, mainly AB-stacked bilayered graphene with large lateral size (several to several tens of micrometers). In this method we take a chemical mixed with deionized water; graphite is used as an anode and platinum is used as cathode, and voltage is applied. Within a few minutes whole graphite foil is exfoliated into thin graphene sheets and dispersed into whole electrolyte (Figure 1.12). The electrical properties of these exfoliated sheets are radially superior to commonly used reduced graphene oxide, whose preparation

(a)   (b)

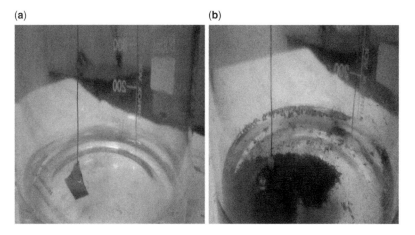

Figure 1.12 A set up in our lab showing (a) before exfoliation; (b) after exfoliation.

typically requires many steps including oxidation of graphite and high temperature reduction.

1.5   Functionalization of Carbon Nanotubes and Graphene

Beyond synthesis by different techniques, it appears immediately clear that CNTs and graphene need processing after their synthesis. To address this issue, purification methods and, above all, functionalization approaches, are essential to allow manipulation and further application of this material. Usually, metal nanoparticles and amorphous carbon are present as a synthetic residue. In general, these carbon nanomaterials are a fluffy powder difficult to manage, while chemical functionalization contributes to the preparation of more homogenous and soluble material. SWNTs are highly polarizable smooth-sided carbon compounds with attractive interaction of 0.5 eV per nanometer of inter-tube contact. This extreme cohesive force makes it difficult to disperse SWNTs into individual state. Pristine SWNTs tend to agglomerate in the polymer matrix and form bundles. Similarly, pristine graphene is also hydrophobic, so producing stable suspension of graphene in water or organic solvents is an important issue for the fabrication of many graphene-based devices [54, 55]. Prevention of aggregation was of

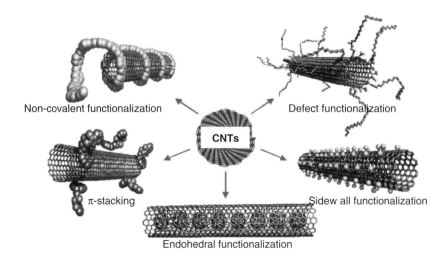

Non-covalent functionalization

Defect functionalization

CNTs

π-stacking

Sidew all functionalization

Endohedral functionalization

Scheme 1.1 A wide variety of functionalizations adapted from ref. [61].

particular importance for graphene sheets because most of their unique properties were only associated with individual sheets and keeping them well separated was required. Strategies for functionalizing these carbon nanomaterials are important for the pursuit of these applications so these materials should be dispersed uniformly and form stable suspension. In recent years, new functionalization methods have been developed to disperse these carbon nanomaterials which allow their application [56, 57] A wide variety of functionalizations have been reported in the literature [58–60], the most important of which are summarized in Scheme 1.1. We can categorize mainly covalent and non-covalent approaches.

1.5.1 Covalent Functionalization

In this method CNTs are functionalized by nonreversible attachment of appendage on the sidewalls and/or on the tips. Also in this case, many different approaches are reported [13]. Briefly, reactions can be performed at the sidewall site (sidewall functionalization) or at the defect sites (defect functionalization), usually localized on the tips. In the first case, fluorination with elemental fluorine at high temperature (400–600°C) has been explored, accomplishing further substitutions with alkyl groups. Furthermore, radical addition via diazonium salt has been proposed by Tour's group [62]. On the other hand, cycloadditions have found wide interest. Cycloadditions have been reported, such as carbene [2+1] cycloadditions or Diels-Alder via microwave (MW) irradiation cycloaddition. Side defects functionalization occurs via amidation or esterification reactions of carboxylic residues obtained on CNTs. Moreover, it is feasible that in general caps (i.e., tips when they are not cut) are more reactive than sidewalls because of their mixed pentagonal-hexagonal structure. The f-MWCNTs are not modified in their electronic structure and new properties can be added by means of functionalization. Instead, electronic properties of SWCNTs are perturbed by covalent functionalization and double bonds are irreversibly lost. This may affect conductive property, preventing further CNT applications.

1.5.2   Non-Covalent Functionalization

Recently, non-covalent functionalization has been preferred as it has several advantages over covalent functionalization. Noncovalent functionalization preserves the structural and electrical properties of CNTs and graphene which may be advantageous for future application. Noncovalent functionalization of these carbon nanonmaterials is basically through van der Waals, electrostatic and π-stacking interaction, etc. [63]. CNTs wrapping by polymers, including DNA, have been studied [64]; also, proteins are able to non-convalently interact with CNTs and these are often used for biosensor applications [65]. Furthermore, these procedures are usually quite simple and quick, and involve simple steps like ultrasonication, filteration, magnetic stirring and centrifugation, etc., and also do not perturb the electronic structure of CNTs, graphene and SWCNTs in particular.

These functionalization methods generally involve the conjugation of these carbon nanomaterials with the biological species like protein, carbohydrates and nucleic acid, etc.

In our lab we have functionalized CNT and graphene both by covalent and non-covalent methods. Covalent functionalization with amine groups of CNTs was achieved after such steps as carboxylation, acylation and amidation [66]. These CNTs were treated with a concentrated H_2SO_4/HNO_3 mixture to form a stable aqueous suspension containing individual oxidized CNTs with carboxyl groups (Figure 1.13). Then carboxylated CNTs were treated with ethylenediamine [$NH_2(CH_2)2NH_2$], forming an active amine group on the nanotubes surface. The optical image of as-synthesized CNTs and amino-functionalized CNTs is also shown in Figure 1.14(a).

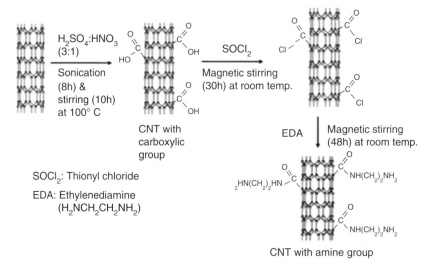

Figure 1.13 Schematic of the reaction scheme to form carbon nanotube (CNT) with amino functionalization [66].

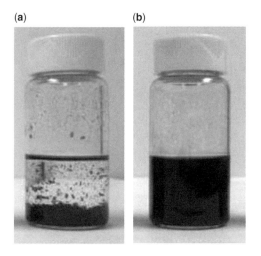

Figure 1.14 (a) As synthesized CNTs; (b) Amino functionalized CNTs.

Several reports are available for functionalizing graphene and CNTs by biomolecules. The biological molecules which possess hydrophobic and hydrophilic moieties provide a more efficient means to solubilize the nanomaterial in water than the use of surfactants and polymers. Molecules containing aromatic groups or electron-rich environments have also been reported to modify nano-tubes/nanosheets via -π stacking. The non-covalent approaches are based on interactions of the hydrophobic part of the adsorbed mol-ecules with nanotube sidewalls through van der Waals, -π, CH-, and other interactions, and aqueous solubility is provided by the hydrophilic part of the molecules. If any, the charging of the nano-tube surface by adsorbed ionic molecules additionally prevents nanotube aggregation by the coulombic repulsion forces between modified CNTs. In the last few years, the non-covalent treatment of CNTs with surfactants and polymers has been widely used in the preparation. On the basis of these observations, we explored the use of a novel amino-acid-based non-covalent functionalization of graphene and CNTs using amino acid L-cysteine.

Purified CNTs and graphene were dispersed in double distilled water and sonicated for one hour at room temperature for CNTs and one and a half hours for graphene to obtain homogeneous solu-tion. Then, 0.1 M of L-cysteine was added and sonicated for 30 min followed by 2 h for CNTs and 3 h for graphene of constant stirring to each solution. The CNTs and graphene so obtained were thor-oughly washed with double-distilled water in centrifuge at 10,000 rpm for 10 min and the solution phase was discarded. This washing was repeated five times in order to remove any unbound *L-cysteine*.

1.5.3   FTIR Analysis of CNTs and FCNTs

The presence of functional groups after the functionalization of CNTs and functionalized CNTs by FTIR spectroscopy are shown in Figure 1.15. Figure 1.15(a) shows the FTIR spectra of purified CNTs, the band at $3450 cm^{-1}$ is attributed to the presence of –OH group on the surface of CNTs and is believed to be due to the oxidation during the purification of nanotubes. The peak present at $1585 cm^{-1}$ corresponds to the C=C stretching vibration of CNTs. In the FTIR spectra of amino FCNTs, Figure 1.15(b), the peak at $3430 cm^{-1}$ is due to the NH_2 stretch of the amine group overlapped with –OH stretching vibration. The presence of peaks at 1472 and $1385 cm^{-1}$ correspond to the N-H and C-N bond stretching of amine group, respectively. The peaks at 2955 and $2835 cm^{-1}$ are due to the –CH stretching of CH_2 group.

The TEM image of GS reveals a wrinkled paper-like structure (Figure 1.16a). The inset in Figure 1.16(a) is the selected area elec-tron diffraction pattern (SAED) of GS, showing a clear diffraction

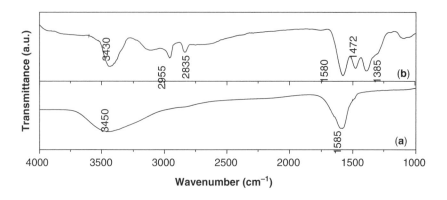

Figure 1.15 FTIR spectra of (a) CNTs and (b) FCNTs FTIR analysis of CNTs and FCNTs.

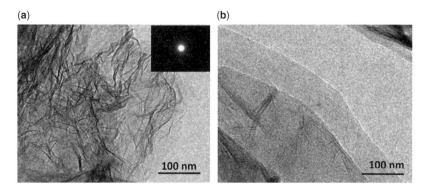

Figure 1.16 TEM images of (a) GS and (b) FGS.

spot. The diffraction spots were indexed to hexagonal graphite crystal structure, whereas the TEM images of FGS shows a smoothened surface which is due to the functionalization by L-cysteine Figure 1.16(b).

1.6   Applications

Nanomaterials have become the focus of scientific research in the past few years because of their unique electronic, chemical, optical, thermal and mechanical properties. Among the numerous active nanomaterials, carbon nanomaterials are at the forefront of research in a variety of physical and chemical disciplines. Both multi- and single-walled carbon nanotubes and graphene sheets (GS) show great promise for advancing the fields of biology [67], medicine [68], electronics [69], composite material [70], energy technology [71], etc. Currently there is a flurry of activity amongst scientists to exploit the unique properties of these carbon nanomaterials for potential application. We have also used these carbon nanomaterials for various applications such as drug delivery, biosensing, protein immobilization, protein and DNA immobilization, hydrogen storage, etc. Some of the applications used in our lab are discussed briefly below.

a) Targeted killing of Leishmania donovani in vivo and in vitro with amphotericin B attached to functionalized carbon nanotubes and graphene

The development of new drug delivery systems is attractive as it allows optimization of the pharmacological profile and the therapeutic properties of existing drugs. Within the family of nanomaterials, carbon nanotubes and graphene have emerged as a new and efficient tool for transporting therapeutic molecules, due to their unique physical and chemical properties. By employing the drug delivery system of f-CNTs to the known antileishmanial drug AmB, we found that antileishmanial efficiency is significantly increased in both *in vivo* and *in vitro* settings. This, together with low cytotoxicity of f-CNT–AmB, means that it is a viable compound for further drug development [72]. The synthesis of f-CNT–AmB was performed on a large scale and it was stored at room temperature for at least six months without any loss of efficacy.

Encouraging *in vitro/in vivo* results have prompted us to carry out further research on graphene-based drug delivery for *Leishmaniasis* treatment. The results obtained by targeted drug delivery were more promising as compared to CNT-based drug delivery. The probable reason could be due to the large surface area of graphene enabled to load more drug, thereby improving the bioavailability of drug to the cells. Figure 1.17 below shows the drug-attached graphene and Figure 1.18 shows the improved efficacy of graphene-based drug delivery as compared to CNT.

b) Attachment of biomolecules (protein and DNA) with amino f-CNTs

The biomolecules (e.g., bovine serum albumin [BSA] protein and DNA) have been attached to the multi-walled CNTs through

(a) (b)

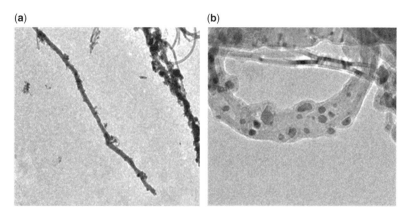

Figure 1.17 The presence of black particles on the surface of nanotube represents the attachment of AmB with CNT [72].

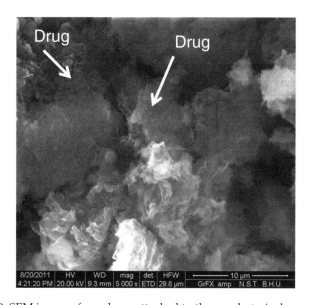

Figure 1.18 SEM image of graphene attached to the amphoteric drug.

interaction between amino f-CNTs and biomolecules. The as-synthesized CNTs, f-CNTs and amino f-CNTs with BSA protein and DNA samples have been characterized by TEM and FTIR spectroscopy. The TEM observations clearly confirm the attachment of BSA protein and DNA to the amino f-CNTs. The FTIR (Figure 1.20) results show the presence of carboxylic (at 1720 cm^{-1} C=O) and amino

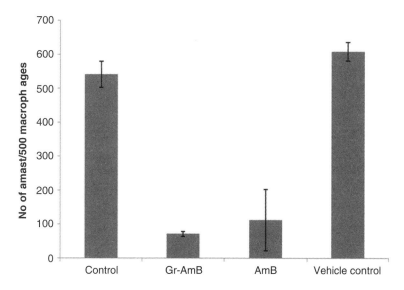

Figure 1.19 *In vivo* Efficacy of Gr-AmB.

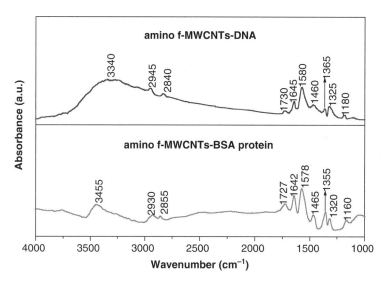

Figure 1.20 FTIR spectrum of the attachment of biomolecules (BSA protein and DNA) [73].

groups (at 3540 cm⁻¹ N-H) in the f-CNTs. The attachment of biomolecules (BSA protein and DNA) to amino f- CNTs is confirmed by the shift of the C=O (amide bond) peak in the amino f-CNTs- BSA protein/ DNA samples [73].

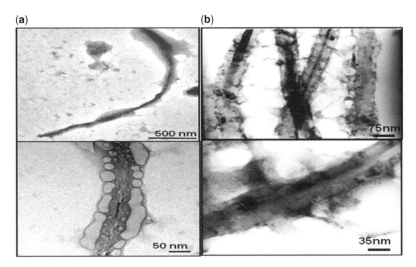

Figure 1.21 (a) TEM images of as prepared amino f-MWCNT-BSA protein sample. (b) TEM images of as-prepared amino f-MWCNT-DNA sample [73].

c) Immobilization of beta-Galactosidase onto functionalized graphene and its analytical applications

Beta-galactosidase is a vital enzyme with diverse application in molecular biology and industries. It was covalently attached onto functionalized graphene nano-sheets for various analytical applications based on lactose reduction. The enzyme was coupled to the functionalized graphene with the help of a spacer arm (cysteamine) and a crosslinker (glutaraldehyde). The functionalized graphene was dissolved in phosphate buffer to make a final preparation. Functionalized graphene-coupled enzyme stored at 4°C showed excellent reusability with negligible loss up to three cycles, and a retention of more than 92% enzymatic activity after 10 cycles of repeated use [74].

d) Carbon nanomaterials and hydrogen storage

Our group has synthesized helical carbon nanofibers (HCNFs) by employing hydrogen storage intermetallic LaNi5 as the catalyst precursor. It was observed that oxidative dissociation of LaNi5 alloy occurred during synthesis. In order to explore the application potential of the present as-synthesized CNFs, they were used as a catalyst for enhancing the hydrogen desorption kinetics of sodium aluminum hydride ($NaAlH_4$). It was observed that the present as-synthesized HCNFs, with metallic impurities, indeed work as an

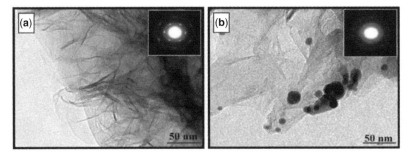

Figure 1.22 Transmission electron micrograph (TEM) images of functionalized (a) and coupled (b) graphene showing fine transparent sheets with insets showing the characteristic Selected-area electron diffraction pattern (SAD). Functionalized graphene sheets appear transparent, whereas islands of immobilized enzyme can be seen in dark shades.

effective catalyst [75]. Thus there is an enhancement of w5 times in kinetics when as-synthesized HCNFs are used as the catalyst.

1.7   Conclusion

In this chapter we have discussed some of the recent progress in CNT and graphene synthesis and its applications. The research in this area is still in its infancy and much more work is needed to realize the technological potential of graphene and CNTs. We should expect many other applications that exploit the unique properties of CNT and graphene to appear in the coming years. Keeping this in view we have synthesized these nanomaterials by applying different routes so as to explore and exploit its enormous potential, and also to obtain a high quality material for different applications from drug delivery, energy storage to electronics.

Acknowledgements

The authors are grateful to the Nano Science and Technology Initiative, Department of Science and Technology (DST), India, for financial support. We would also like to thank Prof. C.N.R. Rao and all the lab members who are engaged and contributed in this nanoscience research. The financial support from CSIR, and MNRE, New Delhi, are also gratefully acknowledged. M.T. thanks UGC-D.S. Kothari Fellowship for the financial support.

References

1. S. Iijima, and T. Ichihashi, Single-shell carbon nanotubes of 1-nm diameter, *Nature*, 1993. 363(6430): p. 603–605.3.
2. A.K. Geim, and K.S. Novoselov. The rise of graphene. *Nature Mater.* 6, 183–191 (2007).
3. S. Iijima, Helical microtubules, *Nature* 354, 56–58, 1991.
4. D.S. Bethune, et al., Cobalt-catalyzed growth of carbon nanotubes with single-atomic-layer walls, *Nature*, 1993. 363(6430): p. 605-607.
5. J. Prasek, J. Drbohlavova, J. Chomoucka, J. Hubalek, O. Jasek, V. Adam, and R. Kizek. Methods for carbon nanotubes synthesis—review, *J. Mater. Chem.* 21, 15872–15884, 2011.
6. K. Shen, H. Xu, Y. Jiang, T. Pietraß, The role of carbon nanotube structure in purification and hydrogen adsorption, *Carbon* 42, 2315-2322, 2004.
7. W.D. Zhang, L. Shen, I.Y. Phang, T. Liu, Surface energy components of a dye-ligand immobilized pHEMA membranes: Effects of their molecular attracting forces for non-covalent interactions with IgG and HSA in aqueous media, *Macromolecules* 37, 256-259, 2004.
8. S. Nakazawa, T. Yokomori, M. Mizomoto, Flame synthesis of carbon nanotubes in a wall stagnation flow, *Chem. Phys. Lett.* 403, 158-162, 2005.
9. X. Lu, M. Yu, H. Huang, and R.S. Ruoff, Tailoring graphite with the goal of achieving single sheets, *Nanotechnology* 10, 269–272, 1999.
10. A. Srivastava, C. Galande, L. Ci, L. Song, C. Rai, D. Jariwala, et al., Novel liquid precursor-based facile synthesis of large-area continuous, single, and few-layer graphene films, *Chem. Mater.* 22 (11), 3457–3461, 2010.
11. C.J. Lee, S.C. Lyu, H.W. Kim, J.H. Lee, and K.I. Cho, *Chem. Phys. Lett.* 359, 115, 2002.
12. K.S. Novoselov, A.K. Geim, S.V. Morozov, D. Jiang, Y. Zhang, S.V. Dubonos, I.V. Grigorieva, and A.A. Firsov, Electric field effect in atomically thin carbon films, *Science* 306, 666-669, 2004.
13. X. Lu, M. Yu, H. Huang, and R.S. Ruoff, Tailoring graphite with the goal of achieving single sheets, *Nanotechnology* 10, 269–272, 1999.
14. M. Eizenberg and J.M. Blakely. Carbon monolayer phase condensation on Ni(111), *Surf. Sci.* 82, 228–236, 1970.
15. T. Aizawa, R. Souda, S. Otani, Y. Ishizawa, and C. Oshima, Anomalous bond of monolayer graphite on transition-metal carbide surfaces, *Phys. Rev. Lett.* 64, 768–771, 1990.
16. C. Berger, Z. Song, X. Li, X. Wu, N. Brown, C. Naud, D. Mayou, T. Li, J. Hass, A.N. Marchenkov, E.H. Conrad, P.N. First, and W.A. Heer, Electronic confinement and coherence in patterned epitaxial graphene, *Science* 312, 1191–1196, 2006.

17. K.V. Emtsev, A. Bostwick, K. Horn, J. Jobst, et al., Towards wafer-size graphene layers by atmospheric pressure graphitization of silicon carbide, *Nature Mater.* 8, 203–207, 2009.

18. C. Berger, Z. Song, T. Li, X. Li, A.Y. Ogbazghi, R. Feng, Z. Dai, et al., Ultrathin epitaxial graphite: 2D electron gas properties and a route toward graphene-based nanoelectronics, *J. Phys. Chem. B* 108 (52), 19912–19916, 2004.

19. E. Rollings, G.H. Gweon, S.Y. Zhou, B.S. Mun, J.L. McChesney, B.S. Hussain, A.V. Fedorov, et al., Sythesis and characterization of atomically-thin graphite films on a silicon carbide substrate, *J. Phys. Chem. Solids* 67, 2172-2177, 2006.

20. P.R. Somani, S.P. Somani, and M. Umeno. Planer nanographenes from camphor by CVD, *Chemical Physics Letters* 430, 56-59, 2006.

21. S. Bae, H. Kim, Y. Lee, X. Xu, J.-S. Park, Y. Zheng, et al., Roll-to-roll production of 30-inch graphene films for transparent electrodes, *Nat. Nano.* 5(8), 574–578, 2010.

22. Q. Liu, W. Ren, D.W. Wang, Z.G. Chen, S. Pei, B. Liu, F. Li, H. Cong, C. Liu, and H.M. Cheng, *In situ* assembly of multi-sheeted buckybooks from single-walled carbon nanotubes, *ACS Nano* 3(3), 707–713, 2009.

23. A. Reina, S. Thiele, X.T. Jia, S. Bhaviripudi, M.S. Dresselhaus, J.A. Schaefer, et al., Growth of large-area single- and bi-layer graphene by controlled carbon precipitation on polycrystalline Ni surfaces, *Nano Res.* 2(6), 509–516, 2009.

24. A. Reina, X. Jia, J. Ho, D. Nezich, H. Son, V. Bulovic, M.S. Dresselhaus, and J. Kong, Large area, few-layer graphene films on arbitrary substrates by chemical vapor deposition chemical vapor deposition, *Nano Lett.* 9, 30–35, 2009.

25. K.S. Kim, Y. Zhao, H. Jang, S.Y. Lee, J.M. Kim, K.S. Kim, J.H. Ahn, P. Kim, J.Y. Choi, and B.H. Hong, Large-scale pattern growth of graphene films for stretchable transparent electrodes, *Nature* 457, 706-710, 2009

26. X. Li, W. Cai, J. An, S. Kim, et al., Large-area synthesis of high-quality and uniform graphene films on copper foils, *Science* 324, 1312-1314, 2009.

27. P.W. Sutter, J.I. Flege, E.A. Sutter, Epitaxial graphene on ruthenium, *Nat. Mater.* 7, 406–411, 2008.

28. H. Ueta, M. Saida, C. Nakai, Y. Yamada, M. Sasaki, S. Yamamoto, Highly oriented monolayer graphite formation on Pt (111) by a supersonic methane beam, *Surf Sci.* 560(1–3), 183–190, 2004.

29. J. Coraux, A.T. N'Diaye, C. Busse, T. Michely, Structural coherency of graphene on Ir(111), *Nano Lett.* 8(2), 565–570, 2008.

30. G. Nandamuri, S. Roumimov, and R. Solanki, Chemical vapor deposition of graphene films. *Nanotechnology* 21(14), 145604 (1-4), 2010.

31. Y. Miyata, K. Kamon, K. Ohashi, R. Kitaura, M. Yoshimura, H. Shinohara, A simple alcohol-chemical vapor deposition synthesis

of single-layer graphenes using flash cooling, *Appl. Phys. Lett.* 96, 263105–263107, 2010.

32. A. Srivastava, C. Galande, L. Ci, L. Song, C. Rai, D. Jariwala, et al., Novel liquid precursor-based facile synthesis of large-area continuous, single, and few-layer graphene films, *Chem. Mater.* 22(11), 3457–3461, 2010.

33. Z. Sun, Z. Yan, J. Yao, E. Beitler, Y. Zhu, and J.M. Tour, Growth of graphene from solid carbon sources, *Nature* 468, 549–552, 2010.

34. W. Choi, I. Lahiri, R. Seelaboyina, Y.S. Kang. Synthesis of graphene and its applications: A review, *Critical Reviews in Solid State and Materials Sciences*, 35(1), 52–71, 2010.

35. M. Hirata, T. Gotou, S. Horiuchi, M. Fujiwara, M. Ohba, Thin-film particles of graphite oxide high-yield synthesis and flexibility of the particles, *Carbon* 42, 2929–2937, 2004.

36. D.A. Dikin, S. Stankovich, E.J. Zimney, R.D. Piner, G.H.B. Dommett, G. Evmenenko, S.T. Nguyen, and R.S. Ruoff, Preparation and characterization of graphene oxide paper, *Nature* 448, 457–460, 2007.

37. V.C. Tung, M.J. Allen, Y. Yang, and R.B. Kaner, High-throughput solution processing of large-scale grapheme, *Nature Nanotech* 4, 25–29, 2009.

38. Y. Hernandez, et al., High-yield production of graphene by liquid-phase exfoliation of graphite, Nature Nanotechnology 3, 563-568, 2008.

39. B.C. Brodie, Sur le poids atomique du graphite. *Ann. Chim. Phys.* 59, 466 (1860).

40. L. Staudenmaier, Verfahren zur Darstellung der Graphitsaure. *Ber. Deut. Chem. Ges.* 31, 1481 (1898).

41. W.S. Hummers and R.E. Offeman, Preparation of graphitic oxide, *J. Am. Chem. Soc.* 80, 1339–1339, 1958.

42. A. Buchsteiner, A. Lerf, and J. Pieper, Water dynamics in graphite oxide investigated with neutron scattering, *J. Phys. Chem. B* 110, 22328–22338, 2006.

43. A.A. Balandin, S. Ghosh, W. Bao, I. Calizo, D. Teweldebrhan, F. Miao, and C.N. Lau, Superior thermal conductivity of single-layer graphene, *Nano Lett.* 8, 902–907, 2008.

44. S. Stankovich, D.A. Dikin, R.D. Piner, K.A. Kohlhaas, A. Kleinhammes, et al., Synthesis of graphene-based nanosheets via chemical reduction of exfoliated graphite oxide, *Carbon* 45, 1558–1565, 2007.

45. I. Jung, et al., Simple approach for high-contrast optical imaging and characterization of graphene-based sheets, *Nano Lett.* 7, 3569–3575, 2007.

46. S. Park and R.S. Ruoff, Chemical methods for the production of graphenes, *Nature Nanotechnology* 4, 217–224, 2009.

47. D.R. Dreyer, S. Park, C.W. Bielawski, and R.S. Ruoff, The chemistry of graphene oxide, *Chem. Soc. Rev.* 39, 228–240, 2010.

48. C. Zhu, S. Guo, Y. Fang, and S. Dong, Reducing sugar: New functional molecule for the green synthesis of graphene nanoshests, *ACS Nano* 4, 2429–2437, 2010.

49. Y. Sun, Q. Wu, and G. Shi.Graphene based new energy materials, *Energy Environ. Sci.* 4, 1113–1132, 2011.

50. C.N.R. Rao, A.K. Sood, K.S. Subrahmanyam, A. Govindaraj, Epitaxial graphite: 2D electron gas properties and a route toward graphene-based, *Angew. Chem. Int. Ed.* 48, 7752–7777, 2009.

51. C. Schafhaeutl, On the combination of carbon with silicon and iron, and other metals, forming the different species of cast iron, steel, and malleable iron, *Phil. Mag.* 16, 570–590, 1840.

52. H.C. Schniepp, J.L. Li, M.J. McAllister, H. Sai, M.H. Alonso, D.H. Adamson, R.K. Prud'homme, R. Car, D.A. Saville, and I.A. Aksay, Functionalized single graphene sheets derived from splitting graphite oxide, *J. Phys. Chem. B*, 110 (17), 8535–8539, 2006.

53. A.C. Ferrari, J.C. Meyer, V. Scardaci, C. Casiraghi, M. Lazzeri, F. Mauri, S. Piscanec, D. Jiang, K.S. Novoselov, S. Roth, and A.K. Geim, Raman spectrum of graphene and graphene layers, *Phys. Rev. Lett.* 97, 187401, 2006.

54. D. Li, M.B. Müller, S. Gilje, R.B. Kaner, G.G. Wallace, Processable aqueous dispersion of graphene nanosheets, *Nat. Nanotechnol.* 2008;3:101–5.

55. Z. Liu, J.T. Robinson, X. Sun, H. Dai, PEGylated nano-graphene oxide for delivery of water insoluble cancer drugs, *J. Am. Chem. Soc.* 2008;130:10876–7.

56. Y. Lin, S. Taylor, H. Li, K.A.S. Fernando, L. Qu, W. Wang, L. Gu, B. Zhou, and Y.P. Sun, Advances toward bioapplications of carbon nanotubes, *J. Mater. Chem.* 14(2004), pp. 527–541.

57. A. Carrillo, J.A. Swartz, J.M. Gamba, R.S. Kane, N. Chakrapani, B.Wei, and P.M. Ajayan, Noncovalent functionalization of graphite and carbon nanotubes with polymer multilayers and gold nanoparticles, *Nano Lett.* 3(2003), pp.1437–1440.

58. P. Singh, S. Campidelli, S. Giordani, D. Bonifazi, A. Bianco, M. Prato, Organic functionalisation and characterisation of single-walled carbon nanotubes, *Chemical Society Reviews* 2009, *38*, 2214–2230.

59. A. Hirsch, Functionalization of single-walled carbon nanotubes, *Angewandte Chemie - International Edition* 2002, 41, 1853–1859.

60. D. Tasis, N. Tagmatarchis, A. Bianco, M. Prato, Chemistry of carbon nanotubes. *Chemical Reviews* 2006, 106, 1105–1136.

61. A. Hirsch, Functionalization of single-walled carbon nanotubes, *Angewandte Chemie-International Edition* 41(11), 1853-1859 (2002).

62. B. Price, J. Hudson, J. Tour, Green chemical functionalization of single-walled carbon nanotubes in ionic liquids, *Journal of the American Chemical Society* 2005, 127, 14867–14870.

63. B. Long, M. Manning, M. Burke, B.N. Szafranek, G. Visimberga, G. Thompson, J.C. Greer, I.M. Povey, J. MacHale, G. Lejosne, D. Neumaier, and A.J. Quinn, Non-covalent functionalization of graphene using self-assembly of alkane-amines, *Adv. Funct. Mat.* 22(2012), 717–725.

64. D. Tasis, N. Tagmatarchis, A. Bianco, M. Prato, Chemistry of carbon nanotubes, *Chemical Reviews* 2006, 106, 1105–1136.

65. A. Jorio, G. Dresselhaus, M.S. Dresselhaus (Eds), *Carbon Nanotubes: Advanced Topics in the Synthesis, Structure, Properties and Applications; 1st ed.*; Springer, 2008.
66. K. Awasthi, D.P. Singh, S.K. Singh, et al., Attachment of biomolecules (protein and DNA) to amino-functionalized carbon nanotubes. *New Carbon Materials* 2009; 24: 301–6.
67. M. O'Connor, N.K. Sang, A.J. Killard, et al., Mediated amperometric immunosensing using single walled carbon nanotube forests, *Analyst*, 129(12), pp. 1176–1180, 2004.
68. L. Zhang and T.J. Webster, Nanotechnology and nanomaterials: Promises for improved tissue regeneration, *NanoToday*, 4(1), pp. 66–80, 2009.
69. J. Robertson, G. Zhong, S. Esconjauregui, C. Zhang, S. Hofmann, Synthesis of carbonnanotubes and graphene for VLSI interconnects, *Microelectronic Engineering* 107, 2013, 210–218.
70. H.X. Kong, Hybrids of carbon nanotubes and graphene/graphene oxide. *Curr. Opin. Solid State Mater. Sci.* (2013), http:// dx.doi. org/10.1016/j.cossms.2012.12.002
71. H. Wang, X. Yuan, Y. Wu, H. Huang, X. Peng, G. Zeng, H. Zhong, J. Liang, M.-M. Ren, Graphene-based materials: Fabrication, characterization and application for the decontamination of wastewater and wastegas and the hydrogen storage/generation, *Advances in Colloid and Interface Science* (2013).
72. V.K. Prajapati, K. Awasthi, S. Gautam, T.P. Yadav, M. Rai, O.N. Srivastava, and S. Sundar, Targeted killing of Leishmania donovani in vivo and in vitro with amphot ericin B attached to functionalized carbon nanotubes, *J. Antimicrob. Chemother.* 66(2011), pp.874–879.
73. K. Awasthi, D.P. Singh, Sunil Singh, D. Dash, and O.N. Srivastava. Attachment of biomolecules (Protein and DNA) to amino-functionalized carbon nanotubes, *New Carbon Materials* 24(4) Dec. 2009.
74. Immobilization of b-Galactosidase onto Functionalized Graphene Nano-sheets Using Response Surface Methodology and Its Analytical Applications (2012) Kishore D, Talat M, Srivastava ON and Kayastha AM *PLoS ONE* 7(7) : e40708. doi:10.1371/journal.pone.0040708.
75. Z. Qian, M. Sterlin, L. Hudson, H. Raghubanshi, R.H. Scheicher, B. Pathak, C.M. Araujo, A. Blomqvist, B. Johansson, O.N. Srivastava, and R. Ahuja. Excellent catalytic effects of graphene nanofibers on hydrogen release of sodium alanate, *J. Phys. Chem. C* 2012, 116, 10861–10866.

2

Surface Modification of Graphene

Tapas Kuila,* Priyabrata Banerjee and Naresh Chandra Murmu

CSIR-Central Mechanical Engineering Research Institute, Durgapur, India

Abstract

Surface modification of graphene is of crucial importance for their end applications. Surface treated graphene can be processed very easily by layer-by-layer assembly, spin-coating, vacuum filtration techniques, etc. The presence of functionalizing agent prevents restacking of graphene layers and forms stable dispersion either in water or in organic solvents. Surface modification of graphene can be done in several ways such as with chemical, electrochemical as well as sonochemical methods. Chemically functionalized graphene can be prepared by covalent and non-covalent surface modification techniques and it has been found that both of these aforesaid techniques (covalent and non-covalent) have been proven to be the most successful for the preparation of processable graphene. In chemical methods, first the surface treatment of graphene oxide (GO) can be carried out followed by the reduction of functionalized GO. However, the removal of excess modifier, consumption of organic solvent, use of toxic reducing agents, surface defects and loss of electrical conductivity of graphene are the major concerns to the materials scientists. In order to overcome these problems, several studies have been carried out in recent times for the preparation of surface-modified graphene directly from graphite. Sonication of graphite in presence of surface modifier dissolved in a suitable solvent leads to the formation of graphene. However, the yield is much less (~5%) as compared to chemical functionalization techniques. On the other hand, one-step electrochemical exfoliation of graphite into graphene is the most effective method for graphene preparation. Interestingly, this is a completely green route for graphene production that is less time consuming, uses water as solvent, consumes much less electricity, easily isolates product from the un-reacted graphite and has high production yield. Surface-modified graphene is widely used in polymer

---

*Corresponding author: t_kuila@cmeri.res.in

Ashutosh Tiwari and S.K. Shukla (eds.) Advanced Carbon Materials and Technology, (35–86) 2014 © Scrivener Publishing LLC

nanocomposites, super-capacitor devices, drug delivery systems, solar cells, memory devices, transistor devices, biosensors, etc.

Keywords: Graphene, surface-modified graphene, polymer nano-composites, super-capacitor devices, drug delivery system, solar cells, memory devices, transistor device, biosensor

2.1   Introduction

Graphene is a one-atom-thick, sp^2 carbon-based, two-dimensional material that has attracted significant research interest due to its remarkable electrical, mechanical, chemical, electrochemical, optical, sensing and thermal properties [1–12]. Due to its versatile extraordinary properties, graphene finds profound applications in the areas of polymer composites, energy storage electrode materials, detection of biomolecules, detection of hazardous element/chemicals, drug delivery systems, transistor devices, lubricants, hydrogen storage, nanofluids, photovoltaic cells, etc. [13–29]. In order to harness all these applications, large-scale production of graphene is very essential. There are several methods for the preparation of graphene such as micromechanical exfoliation of graphite, epitaxial growth of graphene on electrically insulating surfaces, chemical vapor deposition (CVD) or plasma enhanced CVD (PE-CVD) growth of graphene on electrically insulating or conducting substrates, electrical arc-discharge, electrochemical exfoliation of graphite and liquid phase intercalation-exfoliation of graphite [30–50]. Micromechanical exfoliation of graphite to graphene is suitable for defect free and large-area graphene synthesis but not suitable for mass production of graphene. Similarly, CVD, PE-CVD and electrical arc-discharge methods suffer from expensive experimental set-up. Although electrical arc-discharge may be suitable for mass production of graphene, detailed investigations are required before commercialization of this technique. One-step electrochemical exfoliation of graphite to graphene is a fruitful method for the preparation of graphene [45–47]. This method is environmentally friendly and inexpensive as compared to CVD, PE-CVD and electrical arc-discharge techniques. The main disadvantage of this method is the low production yield. In contrast, the easiest way for the large-scale production of graphene is chemical oxidation of graphite, conversion of graphite oxide to graphene oxide (GO) followed by reduction by chemical route or via electro-chemical or thermal techniques [51–54]. However, the single layer of graphene sheets always tend to agglomerate and usually forms a three-dimensional stack of graphite due to π-π interactions result-ing in disruption of the inherent properties of monolayer graphene. The processibility of graphene sheets after reduction of GO is ham-pered due to the removal of hydrophilic oxygen functional groups from the surface of graphene sheets. In such cases, the dispersion stability of the reduced GO (rGO) is hampered significantly due to the increase of interlayer cohesive energy (van der Waals interac-tion) with increasing sheet size [55–57]. Pristine graphene cannot be used as an electrode material for supercapacitor devices owing to its wettability problem with the electrolyte. Application of pris-tine graphene as biosensor is also hampered due to its inability to interact with the functional groups of the biomolecules. Moreover, the band gap in pristine graphene obtained by CVD or PE-CVD growth and electrical arc-discharge or by micromechanical exfo-liation of graphite is zero [58–61]. Thus, pristine graphene is not suitable for transistor device application due to the turned-off problem. Surface modification of graphene is a permanent solution to overcome all these problems. Many more useful properties can result from graphene after surface modification by different func-tionalities or additional molecules. One of the usual ways to obtain surface-modified graphene is the chemical modification of GO fol-lowed by reduction with suitable reducing agents. GO sheets are composed of planar, graphene-like aromatic domains by a network of cyclohexane-like units as shown in Figure 2.1 [62–64]. The sheets of GO bear different oxygen functional groups such as hydroxyl, epoxy, ether, diol, and ketone groups. Depending on these oxy-gen functionalities, the surface of GO can be covalently and non-covalently modified to obtain functionalized graphene sheets. An alternative approach to prepare functionalized graphene sheets is one-step sono-chemical and electrochemical exfoliation of graph-ite into functionalized graphene sheets. Figure 2.2 shows different surface modification techniques to obtain functionalized graphene [55]. The incorporation of polar functional groups onto graphene surfaces facilitates the dissolution of functionalized graphene sheets in water and polar solvents. The dispersion stability is dic-tated primarily by enthalpic interaction between the functionalized graphene and solvents. Grafting of small molecules or macromol-ecules prevents agglomeration of graphene sheets by a remarkable

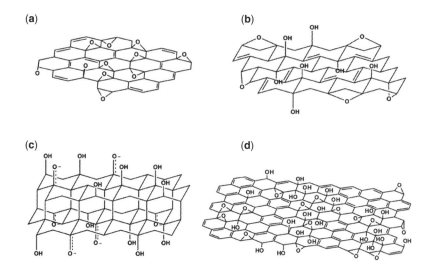

Figure 2.1 (a) Hofman, (b) Ruess, (c) Nakajima-Matsuo, and (d) Lerf-Klinowski model structures of GO showing the presence of oxygen functionalities above and below the basal plane. Reproduced from refs. [62–64].

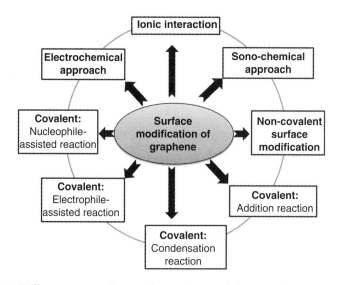

Figure 2.2 Different approaches for the surface modification of graphene.

decrease in enthalpic interaction primarily driven by the system entropy [65].

A general review on the synthesis, characterization, properties and application of surface-modified graphene is reported by Kuila *et al.* [55]. It describes a detailed overview on the application of surface-modified graphene. In the present discussion, we have high-lighted the synthesis of surface-modified graphene starting from GO by covalent and non-covalent surface modification techniques. One-step sono-chemical and electrochemical exfoliation of graphite to functionalized graphene has also been discussed. The reduction of GO to graphene and appearance of new functional groups in graphene are confirmed by morphological and structural character-ization of graphene. Extensive research has been carried out on the surface treatment of graphene. However, limitations with respect to dispersibility, electrical conductivity, and complexity in experimen-tal procedures have not been eliminated completely. The current discussion will highlight the advantages and disadvantages of the reported methods for the production of surface-modified graphene as well as their applications in the areas of polymer composites, biosensors, drug-delivery, transistor devices, hydrogen storage, lubricants, nanofluids, etc. [66–81].

2.2   Surface-Modified Graphene from GO

Reduction of GO causes stacking of graphene sheets due to the removal of oxygen containing functional groups from the sur-face of GO. The interlayer van der Waals force is increased by the removal of oxygen-containing groups (epoxy, hydroxyl and carboxyl groups) during the reduction process. Since most of the unique properties of graphene are associated with the existence of monolayer graphene sheets, the tendency to form graphene stack still remains a challenge in processing bulk-quantities. In order to overcome this problem and to improve its dispersibility and processability, many attempts have already been performed for the surface modification of GO prior to reduction. GO can be functionalized by covalent and non-covalent surface modification techniques. Different types of organic molecules, ionic liquids, bio-molecules, macromolecules, etc., have been used as surface modifying agents [55].

2.2.1   Covalent Surface Modification

Covalent surface modification of graphene can be performed in four different ways: nucleophilic substitution reaction, electrophilic addi-tion reaction, condensation reaction and cyclo-addition reaction.

2.2.1.1   Nucleophile-Assisted Reaction

The surface of GO consists of different oxygen functional groups such as hydroxyl, carboxyl, epoxy, etc. Due to the difference in elec-tronegativity, the carbon center of epoxy functional groups of GO has an inherent tendency to react with the electron donating func-tional groups. Thus, nucleophilic substitution reaction occurs very easily at room temperature (RT) and in an aqueous medium. All types of aromatic and aliphatic amines, amine terminated biomol-ecules and ionic liquids, amino acids, organic silanes, small molec-ular polymers, etc., bearing the lone pairs of electron can act as a nucleophile [13, 55, 82, 83].

Simultaneous reduction and surface modification of GO can be performed by using amino acids such as L-cysteine and l-glu-tathione [84, 85]. Chen *et al.* showed that L-cysteine can success-fully reduce GO to functionalized graphene sheets as confirmed by UV-vis spectroscopy analysis. Figure 2.3 shows that the π-π* plas-mon peak of GO appeared at ~230 nm and is shifted to ~270 nm after reduction, and the peak intensity increases with an increase in the reduction reaction time. Moreover, the n-π* transition peak of GO at ~300 nm is disappeared after reduction with L-cysteine. The shift of π-π* plasmon peak to higher wavelength and disappearance of

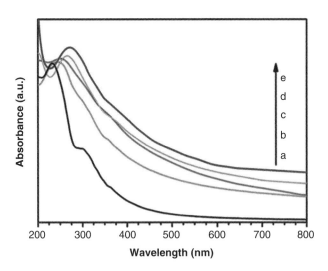

Figure 2.3 UV-vis spectra of aqueous dispersion of GO before (a) and after being reduced by L-cysteine for different reduction times 12 h (b), 24 h (c), 48 h (d), 72 h (e). Reproduced from ref. [84].

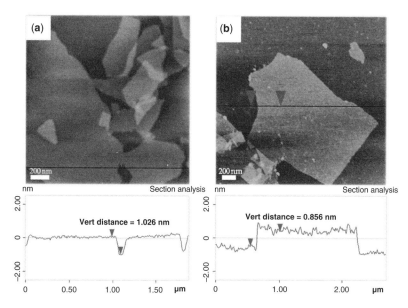

Figure 2.4 AFM images and section analysis of GO (a) and rGO (b) nanosheets absorbed on freshly cleaved mica. Reproduced from ref. [84].

n-π* transition peak in the reduced GO (rGO) is due to the restoration of π-electronic conjugated structure of graphene. The average thickness of rGO is reduced to 0.8 nm from 1.0 nm for GO (Figure 2.4). The decrease of thickness in rGO is due to the removal of oxygen functional groups after reduction. The rGO sheets form stable dispersion in N,N-Dimethylformamide (DMF). These results indicate that L-cysteine can reduce GO to functionalized graphene sheets without using any additional stabilizer. Similarly, Pham *et al.* showed that L-gluthathione reduced GO forms stable dispersion in tetrahydrofuran (THF), DMF, dimethylsulfoxide (DMSO), and water as solvent [85]. Removal of oxygen functionalities from the surface of GO has been confirmed by Fourier transform infrared (FT-IR) and X-ray photoelectron spectroscopy (XPS) analysis. All these observations suggest that L-gluthathione plays an important role as a capping agent in the stabilization of graphene simultaneously. Bose and his coworkers showed the dual role of glycine for the reduction and chemical modification of GO [86]. The-NH$_2$ functionalities of glycine reacts with the epoxide functionalities of GO and forms surface-modified graphene. Depending on the concentration of glycine used, rGO forms stable dispersion in water.

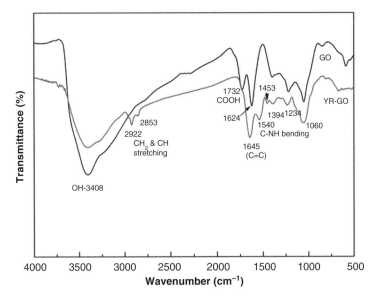

Figure 2.5 FT-IR spectra of GO and YR-GO. Reproduced from ref. [87].

Khanra *et al.* reported simultaneous biofunctionalization and reduction of GO using baker's yeast. The amine functional groups of nicotinamide adenine dinucleotide phosphate (NADPH) reacts with epoxide functionalities of GO and forms covalent bond. FT-IR spectroscopy shows that the carboxyl (-COOH) and epoxide (-C-O-C-) functionalities are removed completely in the yeast-reduced GO (YR-GO). Figure 2.5 also shows the appearance of new peaks in the region of 2922 and 2853 cm^{-1} [87]. The sharp peak at 1540 cm^{-1} in the YR-GO corresponds to the C-NH bending vibration confirming that the amine group of NADPH reacts with the epoxy groups of the GO. The electrical conductivity of YR-GO has been found to be 43 S m^{-1} at RT. All these findings confirm the removal of oxygen functionalities of GO and formation of functionalized graphene. Dopamine-functionalized graphene sheets have also been prepared successfully [88]. The amine functional groups of dopamine directly couple with the epoxide functional groups of GO. The functionalized graphene forms stable dispersion in water. Park *et al.* showed the chemical crosslinking of GO sheets by poly-allylamine (PAA) [89]. Polyallylamine is a long alkyl chain with a number of reactive –NH$_2$ functional groups that can easily react

ADVANCED CARBON MATERIALS AND TECHNOLOGY; ED. BY
ASHUTOSH TIWARI.

Cloth   492 P.

HOBOKEN: JOHN WILEY, 2014
SER: ADVANCE MATERIALS SERIES.

ED: LINKOPING UNIV. CO-PUB. W/ SCRIVENER PUB.
COLLECTION OF NEW ESSAYS.

ISBN 1118686233   **Library PO#** GENERAL APPROVAL

	List	195.00	USD
5461 UNIV OF TEXAS/SAN ANTONIO	**Disc**	17.0%	
App. Date 3/05/14 MEN.APR 6108-11	**Net**	161.85	USD

SUBJ: 1. CARBON. 2. CARBON COMPOSITES. 3.
NANOSTRUCTURED MATERIALS.

CLASS TA455   DEWEY# 620.193   LEVEL ADV-AC

YBP Library Services

ADVANCED CARBON MATERIALS AND TECHNOLOGY; ED. BY
ASHUTOSH TIWARI.

Cloth   492 P.

HOBOKEN: JOHN WILEY, 2014
SER: ADVANCE MATERIALS SERIES.

ED: LINKOPING UNIV. CO-PUB. W/ SCRIVENER PUB.
COLLECTION OF NEW ESSAYS.

ISBN 1118686233   **Library PO#** GENERAL APPROVAL

	List	195.00	USD
5461 UNIV OF TEXAS/SAN ANTONIO	**Disc**	17.0%	
App. Date 3/05/14 MEN.APR 6108-11	**Net**	161.85	USD

SUBJ: 1. CARBON. 2. CARBON COMPOSITES. 3.
NANOSTRUCTURED MATERIALS.

CLASS TA455   DEWEY# 620.193   LEVEL ADV-AC

with oxygen functional groups of GO sheets. The maximum dispersibility of PAA-modified graphene in water has been found to be 0.2 mg ml^{-1}.

Kang *et al.* showed the simultaneous reduction and surface modification of GO by a one-step poly(norepinephrine) functionalization [90]. During the polymerization of norepinephrine, GO is reduced to rGO and a homogeneous coating is formed on the surface of graphene sheets. The poly(norepinephrine)-coated rGO forms stable dispersion in water and can be considered as a multipurpose platform for graphene composite materials. The mussel-inspired chemical functionalization of graphene is non-toxic as it uses water as a solvent.

Yu *et al.* prepared stable colloidal suspensions of oleylamine-modified graphene [91]. The modified graphene exists as single layer with the maximum dispersibility in n-octane up to 3.82 mg ml^{-1}. Similarly, primary alkyl amines have extensively been used for the preparation of organomodified graphene. Primary alkyl amines play a dual role in the simultaneous reduction and surface modification of GO. The modified graphene disperses well in xylene, toluene, DMF, DMSO, acetone, $CHCl_3$, etc. Reduction of GO can be confirmed by the disappearance of the characteristics of the oxygen functional groups in the modified graphene as confirmed by FT-IR and X-ray photoelectron spectroscopy (XPS). Appearance of new peaks/band related to the stretching vibration of -CH_2- and C-N bonds confirm the formation of functionalized graphene.

Kim *et al.* and Liu *et al.* showed the ternary roles of ethylenediamine and polyethyleneimine (PEI), respectively [92, 93]. Ethylenediamine has been used for the simultaneous reduction, functionalization and stitching of GO as shown in Figure 2.6. The surface modification of GO using ethylenediamine can easily take place under mild conditions. The rGO shows good electrical conductivity (~1075 S m^{-1}) and forms stable dispersion in polar solvents. In comparison to unstitched graphene, ethylenediamine-modified graphene can efficiently improve the mechanical properties of linear low density polyethylene (LLDPE) composites suggesting that the sliding of graphene layers is prevented due to stitching by ethylenediamine. Similarly, depending on the concentration, PEI can act as a reducing agent, a surface modifier and as a polymer host [93]. Reduction of GO and formation of surface-modified graphene is shown schematically

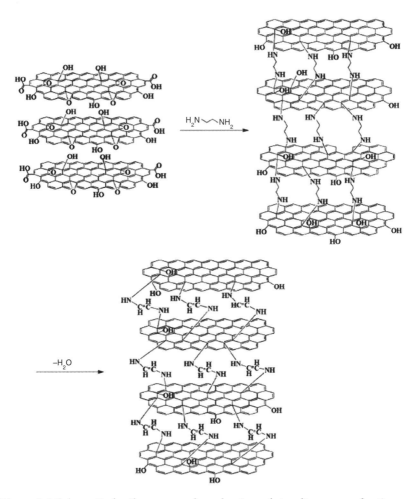

Figure 2.6 Schematic for the proposed mechanism of simultaneous reduction, functionalization and stitching of GO using ethylenediamine. Reproduced from ref. [92].

in Figure 2.7. The PEI-reduced GO (PEI-rGO) shows high electrical conductivity (~ 492 S m^{-1}) and does not form stable dispersion in water when the concentration of PEI is very low (PEI/GO= 0.02). Notably, the low content of PEI is not sufficient enough to maintain the polar-polar interaction with water. The electrical conductivity starts to deteriorate with increasing PEI concentration and PEI-rGO forms stable dispersion in water. The decrease of electrical conductivity is due to the attachment of insulating PEI molecules on the surface of rGO. The composite films act as a gas barrier film when the PEI/GO ratio reaches 7. The decrease in gas permeability is due

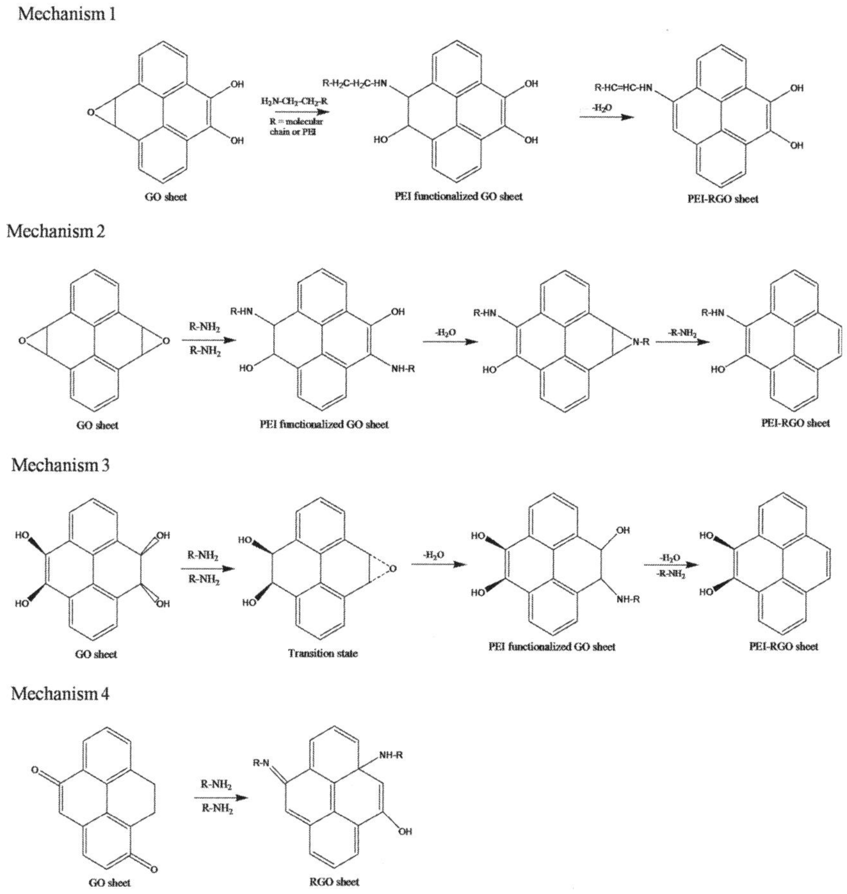

Mechanism 1

Mechanism 2

Mechanism 3

Mechanism 4

Figure 2.7 Mechanisms of reduction and surface modification of GO to PEI-rGO. Reproduced from ref. [93].

to the formation of highly-oriented brick and mortar structure which can increase the tortuous path length of the diffusing gas molecules.

Yang *et al.* showed the surface modification of graphene sheets using 3-aminopropyltriethoxysilane (APTS) [94]. The chemically converted graphene (f-CCG) disperses in water, ethanol, DMF, DMSO and APTS. The average thickness of f-CCG has been recorded as *ca.* 1.78 nm. The increased thickness is due to the grafting of silane chains on the surface of graphene. The thermogravimetric analysis shown in Figure 2.8 reveals that the weight loss of f-CCG below 200°C is much less, indicating the removal of oxygen functional groups of GO after reacting with APTS. About 10.6% weight loss has been recorded in the temperature range of 550–650°C that

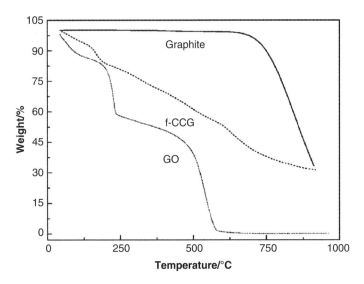

Figure 2.8 TGA of GO, graphite and f-CCG. Reproduced from ref. [94].

is caused by the thermal degradation of silane moieties attached with graphene. Wang *et al.* showed the chemical functionalization of GO using amino-terminated Kevlar oligomer [95]. The functionalized graphene disperses well in N-methyl pyrolidone (NMP) at a concentration of 0.2 mg ml⁻¹. The functionalized graphene also disperses well in polymethyl methacrylate (PMMA) matrix and the percolation in mechanical properties is achieved with very low loadings (≤ 0.2 wt%).

Kuila *et al.* and Wang *et al.* prepared surface-modified graphene by the reaction of primary alkyl amine with GO [14, 96–98]. A series of alkyl amine with different carbon chain lengths was used as surface modifier. It can be seen that the interlayer spacing in surface-modified graphene is dependent on the carbon chain length [99]. Li *et al.* demonstrated that the primary amine used for the surface modification of GO can simultaneously reduce to graphene sheets [100]. The surface-modified graphene can disperse in a variety of organic solvents such as toluene, xylene, THF, DMF, CHCl₃, DMSO, acetone, etc. The surface-modified graphene can be used as reinforcing filler for the preparation of LLDPE and polystyrene-based composites.

2.2.1.2 Electrophile-Assisted Reaction

Electrophile-assisted reaction of GO with surface modifying agents involves the displacement of a hydrogen atom by an electrophile.

Grafting of aryl diazonium salt to the surface of partially reduced GO is an example of an electrophile-assisted surface modification process. Si and Samulski prepared water-dispersible graphene using diazonium salt of sulfanilic acid [101]. Basically, the preparation of surface-modified graphene through electrophile-assisted reaction occurs in three steps: (1) preparation of partially reduced GO using mild reducing agents under controlled reaction time and temperature; (2) preparation of aryl diazonium salt of surface modifier at reduced temperature and coupling with partially reduced GO or rGO or graphene prepared by physical methods; (3) post-reduction of surface-treated, partially reduced GO with hydrazine monohydrate to remove the residual oxygen functional groups. Electrical conductivity of sulfanilic acid modified graphene has been found to be 1250 S m^{-1} at RT. Yu *et al.* have compared the nucleophilic and electrophilic surface modification of GO using 4-aminoazobenzene-4'-sulfonate (SAS) and its aryl diazonium salt (ADS), respectively [102]. The oxygen functional groups of GO are reduced significantly in the ADS-functionalized graphene (ADS-G) as compared to the SAS-functionalized graphene (SAS-G). This is attributed to the two-step reduction of GO in the previous case compared to the later. Figure 2.9 shows that the intensity of the C-N peak at 285.9 eV is relatively higher in ADS-G than that of the SAS-G. The better surface modification is further confirmed by the high dispersibility of ADS-G (2.9 mg ml^{-1}) compared to that of the SAS-G (1.4 mg ml^{-1}). The high electrical conductivity of ADS-G (1120 S m^{-1}) as in comparison to the SAS-G (149 S m^{-1}) is due to the good restoration of π-electronic conjugated structure during the two-step reduction process. Lomeda *et al.* have demonstrated the preparation of diazonium-functionalized graphene starting with sodium dodecyl benzene sulfonate (SDBS)-wrapped GO. The resulting functionalized graphene forms stable dispersion in DMF, NMP and N,N'-dimethylacetamide (DMAc) up to 1 mg ml^{-1} with minimal sedimentation [103].

Jin *et al.* synthesized solution-dispersed functionalized graphene by click chemistry [104]. The functionalization of graphene has been carried out by using 4-propargyloxybenzenediazoniumtetrafluoroborate. Avinash *et al.* showed the covalent modification of GO using ferrocene [105]. The functionalized graphene exists as monolayer (thickness ~ 0.91 nm) and shows excellent magnetic behavior at RT. Sun *et al.* have prepared soluble graphene by edge-selective functionalization of thermally expanded graphite with 4-bromo-aniline [106].

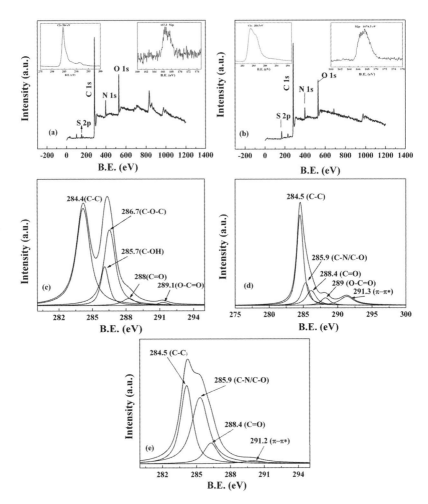

Figure 2.9 XPS survey of (a) SAS-G and (b) ADS-G. The corresponding S 2p bands are shown in the inset of Figure 2.7(a,b). The C1s of (c) GO, (d) SAS-G and (e) ADS-G. Reproduced from ref. [102].

Thin chemically-assisted exfoliated graphene (CEG) sheets can be prepared by the mild sonication of bulk functionalized graphite in DMF. The solubility of these CEG is higher as compared to that of the bulk graphite, and 70% of these soluble flakes contain less than 5 layers. Wu *et al.* prepared aryl radical functionalized SiO_2-nanoparticle-decorated graphene on the Si wafer substrate [107]. Micro Raman mapping analysis confirmed that the radicals can selectively react with the regions of graphene that are covered by SiO_2 nanoparticles.

Electrophile-assisted surface modification of graphene has been performed by solvothermal reduction of a GO suspension in NMP [108]. In this case, simultaneous reduction and surface modification of GO occurs in NMP at ~180°C. The solvothermal reduction of GO occurs at high temperature while the surface modification is carried out by *in-situ*-generated peroxy radicals during heating in aerial condition. It is assumed that the NMP is oxidized by dissolved oxygen at the -carbon position (or methyl group of NMP) to form peroxy radicals. These free radicals are supposed to attach on the surface of rGO. The functionalized graphene disperses well in DMF, NMP, polycarbonate, ethanol and THF. The electrical conductivity and dispersibility of NMP-reduced GO reaches 21600 S m^{-1} and 1.4 mg ml^{-1} (in NMP) at RT, respectively. Park *et al.* have demonstrated the *in-situ* thermochemical reduction and surface modification of GO using the diazonium salt of 4-iodoaniline as shown in Figure 2.10 [109]. It is seen that the electrical conductivity of the prepared functionalized graphene (42000 S m^{-1}) is almost five

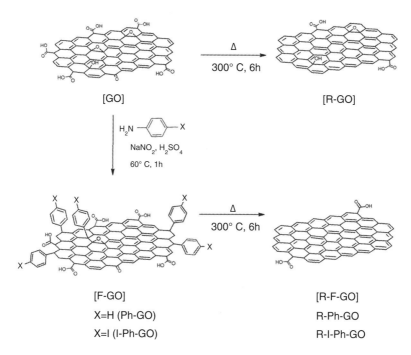

Figure 2.10 Schematic of functionalized graphene oxide (F-GO) and reduced graphene oxide (r-F-GO). Phenyl-functionalized GO (Ph-GO) and GO can be thermally reduced; 4-iodophenyl-functionalized GO (I-Ph-GO) can be thermochemically reduced at 300°C. [Reproduced from ref. 109].

times higher than that of the typical reduced GO. This is attributed to the reduction of GO by the thermally released HI from I-Ph-GO.

Covalent electron transfer chemistry of graphene with diazonium salts has been investigated in detail [110]. Bekyarova *et al.* showed the route for the surface modification of epitaxially grown graphene by aryl diazonium salt of 4-nitro aniline [111]. The reaction is spontaneous due to the electron transfer from graphene layer to the diazonium salt. Sharma *et al.* have investigated the electron transfer chemistry of graphene with 4-nitrobenzene diazonium tetrafluoroborate by using Raman spectroscopy [112]. It is seen that the monolayer graphene sheets are ~10 times more reactive than bi- or multilayers of graphene. They showed that the reactivity of graphene at the edges is at least two times higher than the reactivity of the bulk.

Huang *et al.* have reported for the first time that the conductivity of aryl groups functionalized graphene increases rather than decreases [113]. This is attributed to the presence of electron-withdrawing nitrophenyl groups which can create holes on the basal plane of graphene. The charge carrier density of functionalized graphene increases by increasing the extent of surface modification using nitrophenyl groups. Moreover, the covalent bonds between the nitrophenyl groups and graphene perform as defects so that the scattering of the carriers increases and the mean free path of the carriers decreases. It has been suggested that if the extent of reaction is relatively low, the nitrophenyl groups linking to the surface of graphene will be relatively few. In that situation, the scattering effect is obviously more than the charge transfer effect, leading to a decrease in conductivity of the functionalized graphene. However, if the extent of reaction exceeds a certain level, the effect of increasing charge carrier density will overrun the effect of reducing carrier mobility, resulting in an increase of conductivity.

2.2.1.3 Condensation-Assisted Reaction

Organic amines, isocyanate, diisocyanate and polymers will generally take part in condensation reaction with graphene via the formation of ester linkages. Stankovich *et al.* used a series of organic isocyanates for the surface modification of GO [15, 114]. The schematic for the formation of isocyanate-modified GO is shown in Figure 2.11. It is seen that the isocyanate-modified GO is highly dispersible in DMF at a concentration of 1 mg ml^{-1}. Organic diisocyanates can also be used for the preparation of

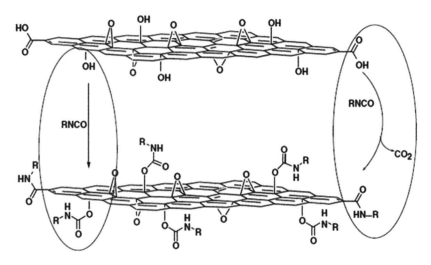

Figure 2.11 Proposed reactions during the isocyanate treatment of GO where organic isocyanates react with the hydroxyl (left oval) and carboxyl groups (right oval) of graphene oxide sheets to form carbamate and amide functionalities, respectively. Reproduced from ref. [114].

chemically-functionalized graphene [115]. Reaction is generally used to occur with the carboxyl and hydroxyl groups on both sides of the graphene sheets forming crosslinked structure. The graphene samples functionalized with 4,4′-diisocyanato-3,3′-dimethyl-biphenyl and 1,4-diisocyanato-benzene are designated as GHPM-1 and GHPM-2, respectively. Similarly, GHPM-3 and GHPM-4 indicate functionalized graphene obtained from GO dispersion in DMF treated by 4,4′-diisocyanato-3,3′-dimethyl-biphenyl and 1,4-diisocyanato-benzene. The chemical crosslinking of GO using diisocyanates can be confirmed by XRD pattern observation as shown in Figure 2.12. The basal reflection peak of pure GO appears at $2\theta =$ 11.62° corresponding to the interlayer distance of 0.76 nm. The XRD pattern of GHPM-1 and GHPM-3 show a sharp peak at $2\theta = 16.11°$ corresponding to an interlayer spacing of 0.55 nm. The decrease of interlayer spacing is attributed to the crosslinking of adjacent GO sheets by 4,4′-diisocyanato-3,3′-dimethyl-biphenyl. On the other hand, the interlayer spacing in GHPM-2 and GHPM-4 has been found to be 0.42 nm due to the modification of GO sheets by 1,4-diisocyanato-benzene.

Xu *et al.* prepared amine terminated porphyrin (TPP-NH₂)-functionalized graphene by using SOCl₂-activated GO in DMF

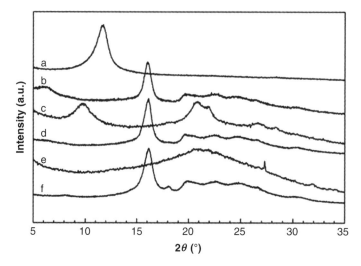

Figure 2.12 XRD patterns of (a) powdery GO, (b) GHPM-1, (c) GHPM-2, (d) GHPM-3, (e) GHPM-4, and (f) r-GHPM-3. Reproduced from ref. [115].

[116]. The formation of covalent bonds between graphene and TPP-NH$_2$ has been confirmed by UV-vis absorption and FT-IR study. Attachment of TPP-NH$_2$ remarkably improves the dispersibility of functionalized graphene in organic solvents. Figure 2.13 shows that the fluorescence spectra of TPP-NH$_2$ is significantly quenched in the functionalized graphene due to the possible electron-transfer process. Zhang *et al.* used amine-terminated dextran compound for the preparation of functionalized graphene [117]. The dispersion of the functionalized graphene shows excellent stability in different physiological solutions.

Shen *et al.* prepared functionalized graphene by the reaction of NaOH-treated GO with 1-bromobutane [118]. The reactive site of 1-bromobutane reacts with the carboxylate anions of the GO leading to the formation of functionalized graphene. The functionalized graphene sheets disperse well in organic solvents and the solutions obey Beer's law. The resulting organic dispersions are homogeneous, exhibiting long-term stability, and are made up of graphene sheets of a few hundred nanometers large length.

Goncalves *et al.* and Ren *et al.* synthesized PMMA and polystyrene-grafted graphene by atom transfer radical polymerization (ATRP) [119, 120]. The grafting of PMMA or polystyrene occurs on the surface of graphene. Liu *et al.* also prepared highly flexible films of polystyrene-modified graphene which is highly dispersible in

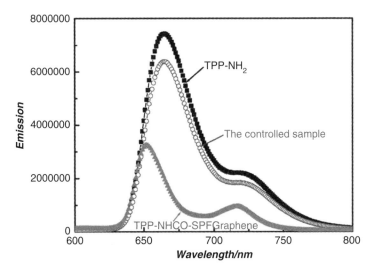

Figure 2.13 Fluorescence spectroscopic changes observed for TPP-NH$_2$, the control sample, and TPP-NHCO-SPF graphene in DMF, with the normalization of the absorbance of the Soret band excitation wavelength (419 nm) to the same value (0.24). Reproduced from ref. [116].

organic solvents [121]. The electrical conductivity of the functionalized graphene has been measured to be ~ 170 S m^{-1} at RT.

A soluble GO covalently functionalized with zinc phthalocyanine (PcZn), GO-PcZn, has been synthesized by an amidation type of reaction [122]. The amine functionality of PcZn reacts with the carboxyl functionality of GO and forms functionalized graphene. The PcZn unit of functionalized graphene acts as an electron donor and the GO as an electron acceptor. Therefore, this material is very suitable as p-n junction materials. The functionalized graphene exhibits much larger nonlinear optical extinction coefficients and broadband optical limiting performance than GO at both 532 and 1064 nm, indicating a remarkable accumulation effect via the covalent linkage of GO and PcZn. Amine-functionalized prophyrin (TPP-NH$_2$) and fullerene have been successfully used for the preparation of functionalized graphene [123]. The intramolecular donor-acceptor interaction between the two moieties of TPP-NH$_2$ and graphene in functionalized graphene may include a charge transfer from the photoexcited singlet TPP-NH$_2$ to the graphene moiety. Therefore, it has been assumed that the processable functionalized graphene materials can bring a competitive entry into the realm of optical limiting and optical switching materials for photonic and optoelectronic devices.

Salavagione *et al.* prepared polyvinyl chloride (PVC)- and polyvinyl alcohol (PVA)-modified graphene by the esterification reaction [124, 125]. A surface modification of PVC has been carried out prior to the surface modification of graphene. The condensation of modified PVC with GO has been carried out in presence of DCC/DMAP followed by NaBH$_4$ reduction. Surface-modified graphene can be obtained by the condensation reaction of PVA with SOCl$_2$-activated GO in presence of DCC/DMAP followed by reduction of modified GO using hydrazine monohydrate. Differential scanning calorimetric analysis (DSC) showed that the melting peak of PVA disappears in the modified product and the partially crystalline PVA became totally amorphous, suggesting the existence of covalent linkage between PVA and graphene sheets (Figure 2.14). Formation of hydrogen bonding between the oxygen functional groups of GO and PVA may be another reason for the shrinkage of crystalline peak in the modified products.

Water-dispersible graphene has been fabricated by the chemical reduction of GO with the assistance of hydroxypropyl cellulose or chitosan (CS) covalently grafted on the graphene [126]. The -COOH functionalities of GO are converted to –COCl to increase the

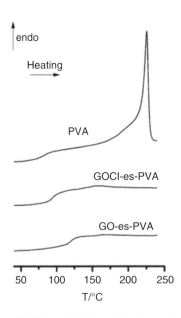

Figure 2.14 DSC curves of PVA and PVA modified graphene. Reproduced from ref. [125].

esterification reaction rate. The reaction between the -COCl functionalities of activated GO and hydroxyl functionalities of hydroxypropyl cellulose occurs for 48 hours at 120°C. On the contrary, the
reaction between the -COCl groups of GO and -NH$_2$ functionalities
of CS occurs within 2 h at 130°C. This is attributed to the difference in reactivity of –OH and –NH$_2$ with GO. Both the functionalized graphenes are highly dispersible in water and find application
in biological and medical fields, such as in drug delivery, transplant devices, and biosensors. Chitosan-modified graphene can
be prepared under microwave irradiation in DMF medium, which
involves the reaction between the –COOH functional groups of GO
and the amido groups of CS followed by reduction of CS-modified
GO using hydrazine monohydrate [127]. The immobilization of CS
on the graphene sheets can be easily identified by the observation
of TEM images as shown in Figure 2.15. It is seen that the unmodified graphene sheets are transparent, rippled silk waves and some
of them are inclined to overlap due to their hydrophobic character.
However, after CS modification, the surface becomes rough and the
transparency is decreased due to the attachment of surface modifier on the graphene sheets. The CS-modified graphene disperses
well in aqueous acetic acid solution and exhibits electrorheological
behavior under an applied electric field.

Zhan *et al.* prepared 4-aminophenoxyphthalonitrile functionalized graphene sheets using SOCl$_2$-treated GO in DMF medium

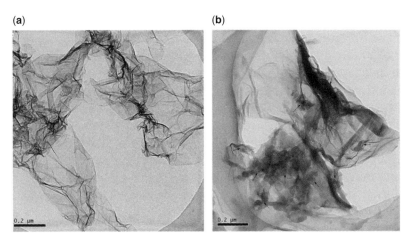

(a)   (b)

Figure 2.15 TEM images of (a) unmodified and (b) CS-modified graphene sheets.
Reproduced from ref. [127].

at RT [128]. The nitrile groups of surface-modified graphene react with ferric ions on the surface of Fe_3O_4 with the help of relatively high boiling point solvent ethylene glycol to form a functionalized graphene/Fe_3O_4 hybrid. It is seen that the Fe_3O_4 nanoparticles are homogeneously dispersed on the graphene sheets and exhibit excellent magnetic properties indicating their suitability for the preparation of polymer composite materials.

Hou and his coworkers demonstrated a method to prepare silane-functionalized graphene sheets [129]. The sheets of GO have been silanized with N-(trimethoxysilylpropyl) ethylenediamine triacetic acid (EDTA). Silanization occurs through the condensation reaction between the trialkoxy groups of silane and the hydroxyl group's of graphene. The EDTA-functionalized graphene forms stable dispersion in water at a concentration level of 1.0 mg ml^{-1}. Graphene oxide can be silylated by various alkylchlorosilanes in butylamine medium to obtain new intercalated compounds [130]. Butylamine serves a dual role for the exfoliation of graphite oxide as well as for scavenging HCl molecules which causes the decomposition of silylated graphite oxide. Interestingly, it is seen that the silylating reagents containing two or three chlorine atoms at silicon successfully reacts with the hydroxyl groups of GO and forms Si-O bonding. However, there is no reaction when the silylating reagents consist of only one chlorine atom.

Worsley et al. prepared dispersible graphene from graphite fluoride [131]. Reaction of Tetramethylethylenediamine (TMEDA) with graphite fluoride at low temperature produces soluble graphene sheets. Organic soluble graphene can also be prepared by the condensation reaction of $SOCl_2$-activated GO and octadecyl amine [132]. Pham et al. demonstrated an efficient strategy for the preparation of water-dispersible graphene using GO and polyglycerol [133]. Covalent grafting of polyglycerol onto the surface of GO can be carried out by in situ ring-opening polymerization of glycidol. Polyglycerol-grafted graphene forms stable dispersion in water and can be successfully attached with Fe-core/Au-shell nanoparticles. The resulting hybrid materials exhibit good water dispersion stability and excellent magnetic properties suggesting their utility as promising materials to meet the demands for potential applications. Liu et al. demonstrated the use of amine-terminated polyethylene glycol (PEG)-modified GO for the delivery of water-insoluble cancer drugs [134]. The oxygen functional groups of PEG increases the water dispersibility of GO. The water-insoluble drugs like SN38 attached on the surface of PEG-modified GO by π-π interaction and, subsequently, can be easily delivered to the affected cell. The unique 2D shape and ultrasmall size (down to 5 nm) of PEG-modified GO may offer interesting *in vitro* and *in vivo* behaviors. Shen *et al.* also showed the preparation of biocomposites by condensation reaction of GO with different biomolecules such as adenine, cystine, nicotinamide and ovalbumin [135]. For the preparation of composites, the -COOH groups of GO are activated by N-ethyl-N'-(3-dimethylaminopropyl) carbodiimide hydrochloride (EDAC) followed by the reaction with the amine groups of biomaterials. Zhang *et al.* prepared a hierarchical structure of functionalized graphene by the reaction of GO with fullerene [136]. The condensation reaction occurs between the -COOH from the fullerene surface and -OH from the graphene surface. The grafting of fullerene on the surface of graphene serves as a space impediment to prevent re-stacking of exfoliated graphene sheets.

2.2.1.4   Cycloaddition-Assisted Reaction

In addition reaction two or more molecules combine together to form a larger one. The molecules having carbon-carbon double or triple bonds or carbon-hetero atom double bond (C=O, C=N) can undergo addition reactions. Organic pericyclic reactions induced by heat or photoirradiation are the best example of addition reaction. Covalent functionalization of graphene can be performed by the cycloaddition reaction of azomethine ylide, Poly(oxyalkylene) amines, alkylazides, azidotrimethylsilane, etc. [137–146]. Choi *et al.* prepared chemically-functionalized graphene by the reaction of epitaxially grown graphene with azidotrimethylsilane (ATS) [137]. The reaction occurs via [2+1] cycloaddition of nitrene radical or biradical pathways with graphene. The adsorbed nitrene species are stable up to 200°C and the functionalized graphene could be useful for electronic device applications. Suggs *et al.* studied the electronic properties of perfluorophenylazide-functionalized graphene using first-principles density functional calculations [138]. The theoretical investigation shows that the sp^2 network structure of graphene is preserved after [2+1] cycloaddition reaction. However, it is seen that the π-electronic conjugation of graphene is hampered significantly after chemical functionalization. The band gap of graphene is increased after perfluorophenylazide functionalization and the semimetallic nature of graphene is converted to a semiconducting

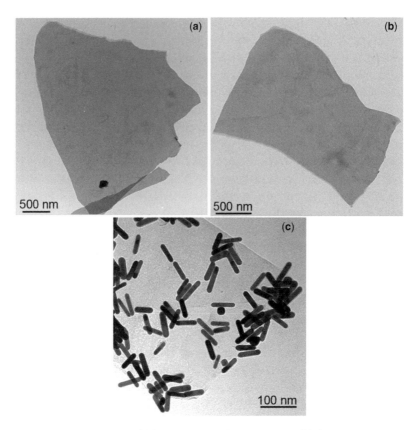

Figure 2.16 TEM images of (a) pristine graphene in NMP, (b) functionalized graphene, (c) gold-nanorod-decorated functionalized graphene sheets. Reproduced from ref. [139].

state. Quintana *et al.* and Wang *et al.* prepared functionalized graphene by the 1,3-dipolar cycloaddition of azomethine ylides with graphene dispersion in NMP [139, 140]. The amino functional groups of azomethine ylides directly attached with graphene sheets can selectively bind to gold nanorods. Figure 2.16 shows the TEG image of pristine graphene sheets in NMP dispersion, functionalized graphene sheets and gold-nanorod-decorated functionalized graphene sheets.

Georgakilas *et al.* prepared defect-free functionalized graphene by 1,3 dipolar cycloaddition of azomethine ylide with pristine graphene derived from graphite directly in pyridine solvent [141]. The chemically-functionalized graphene forms stable dispersions in organic solvents and water. The dispersibility of the functionalized graphene has been tested by UV-vis spectroscopy analysis. The absorption of the dispersion increases continuously at moderate concentrations and the estimated absorption coefficient is 1300. A unified approach to covalently functionalize graphene nanosheets based on nitrene chemistry is reported [142]. This strategy allows many kinds of functional moieties and polymers covalently bonded on graphene, resulting in functional graphene nanosheets and 2D macromolecular brushes, respectively. The resulting functionalized graphene can easily disperse in water, DMF, toluene, and chloroform. The electrical conductivities of the different functionalized graphene are in the range of 1×10^{-1} to 1×10^{3} S m^{-1}.

Zhang *et al.* synthesized tetraphenylporphyrin (TPP)- and palladium TPP (PdTPP)-functionalized graphene directly from pristine graphene through one-pot cycloaddition reactions [143]. Thermogravimetric analysis shows that the functionalized graphene contains ~18% of TPP and ~20% PdTPP. The energy/electron transfer between graphene sheets and covalently bonded TPP (or PdTPP) can be confirmed by fluorescence and phosphorescence quenching with concomitant decreases in excited state lifetimes. In comparison to other surface modifying agents, porphyrin modification can allow the retention of the inherent properties of graphene. Based on the excited state electronic properties of porphyrin, these kinds of hybrid materials may find applications in several areas of photovoltaic cell, sensors and catalysis.

Vadukumpully *et al.* showed a unique approach for the functionalization of surfactant-wrapped graphene sheets with various alkylazides [144]. It is seen that the dispersibility of functionalized graphene in common organic solvents can be improved with concomitant, increasing the alkyl chain length and polar functional groups. The average thicknesses of the functionalized graphene sheets have been found to be 1–1.3 nm. The free carboxylic acid groups of functionalized graphene sheets can bind to gold nanoparticles and the resulting hybrid forms stable dispersion in organic solvents. The electrical conductivity of the functionalized graphene is increased by threefold after the attachment of gold nanoparticles.

Quintana *et al.* have performed selective organic functionalization of graphene by the 1,3-dipolar cycloaddition of polyamidoamine dendron on the surface of graphene [145]. The functionalized graphene is highly dispersible in DMF and selectively recognizes gold nanoparticles. Xu *et al.* prepared a series of graphene-polyacetylene hybrid by utilizing the nitrene chemistry reaction [146].

The resulting hybrid materials are dispersible in different organic solvents and show good photoluminescence properties as compared to pristine graphene.

2.2.2   Non-covalent Surface Modification

Non-covalent surface modification of graphene does not require sharing of a pair of electrons, but rather involves various types of electromagnetic interactions including van der Waals forces, π-π interactions, hydrogen bonding and ionic interactions, etc. In general, non-covalent surface modification refers to intermolecular attractive forces that are entirely different from covalent bonding. These kinds of bonding are weak by nature and must therefore work together to have a significant effect. Non-covalent surface modification does not create extra defects in graphene and, thus, the electronic properties of surface-modified graphene does not alter significantly as compared to pristine graphene. Different types of surface modifying agents such as surfactants, polymers, ionic liquids, small organic molecules, DNA, peptides, etc., can be used for the non-covalent surface modification of graphene.

Utilizing the non-covalent interactions, the first stable aqueous dispersion of graphitic nanoplatelets were prepared by the reduction of exfoliated graphite oxide in the presence of poly(sodium 4-styrenesulfonate) (PSS) [147]. The PSS attached onto the surface of graphitic nanoplatelets by π-π interaction and the van der Walls interaction. The aqueous dispersion stability of the functionalized graphitic nanoplatelets is achieved due to the presence of hydrophilic functional groups of PSS. Wang *et al.* and Bose *et al.* prepared a stable aqueous dispersion of PSS-modified graphene sheets [98, 148]. The aqueous dispersion of PSS-functionalized graphene obeys Beer's law at moderate concentrations and the molar extinction coefficient has been found to be 126.8 ml mg^{-1} cm^{-1}. Yan *et al.* used PSS for the surface modification of thermally exfoliated and reduced graphite oxide (TEG) sheets for supercapacitor application [149]. The hydrophilic character of TEG can be improved by the PSS modification in the aqueous suspension of TEG. The original specific capacitance of TEG (~35 F g^{-1} at a current density of 1 A g^{-1}) can be increased to 278.8 F g^{-1} after PSS treatment. This is attributed to the improved wettability of TEG with the aqueous electrolyte and introduction of pseudocapacitance after surface modification. PSS-modified TEG shows about 97% retention in specific capacitance after 1000 charge-discharge cycles at a current density of 1 A g^{-1}. Hao *et al.* prepared 7,7,8,8-tetracyanoquinodimethane (TCNQ)-anion-stabilized graphene from expanded graphite [37]. The strong π-π interaction of TCNQ with graphene helps to form non-covalent bond, negative charges of adsorbed TCNQ prevents the restacking of individual graphene sheets and the hydrophilic functional groups maintains aqueous dispersibility of TCNQ-functionalized graphene. It is also dispersible in highly-polar organic solvents such as DMF and DMSO.

Water dispersible graphene can be prepared by the surface modification of GO with sulfonated polyaniline (SPANI) followed by the reduction of functionalized GO [150]. The maximum dispersibility of SPANI-modified-graphene in water is >1 mg ml^{-1}. The modified graphene shows high conductivity, good electrocatalytic activity and stability. Choi *et al.* prepared organic solvent dispersible graphene by non-covalent functionalization with amine-terminated polystyrene [151]. It is seen that the protonated form of the amine functionalities of polystyrene undergo non-covalent (ionic) interaction with the carboxylate groups (COO$^-$) of rGO sheets providing high dispersibility in various organic media. Chen *et al.* prepared water-dispersible graphene by the reaction of GO with thionine [152]. It is found that the thionine interacts with graphene sheets via π-π interaction as evident from the fluorescence and UV-vis spectra analysis. The electrical conductivity of the functionalized graphene is 10^7 times higher than that of pure GO.

A smart pH-responsive aqueous complex can be prepared from GO by the hydrophobic interaction of polyacrylamide as surface modifying agent [153]. The surface of graphene is covered by polyacrylamide as observed from the increased thickness of modified graphene compared with GO. The carboxylate functionalities of the modified GO become protonated at low pH resulting in the agglomeration of rGO sheets. Similarly, the aqueous complex again becomes unstable at higher pH (>7) due to the base-induced reduction of GO as reported earlier [42]. Qi *et al.* demonstrated the preparation of a novel amphiphilic graphene composite by using an amphiphilic coilrod-coil conjugated triblock copolymer (PEG-OPE) as surface-modified agents [154]. The maximum dispersibility of PEG-OPE-modified graphene in water, ethanol, methanol, toluene, chloroform, acetone, THF, DMSO and DMF has been found to be 5.2, 5.5, 5.7, 5.1, 4.8, 4.6, 5.3, 5.7 and 5.0 mg ml^{-1}, respectively. The PEG-OPE-modified graphene offers good biocompatibility and

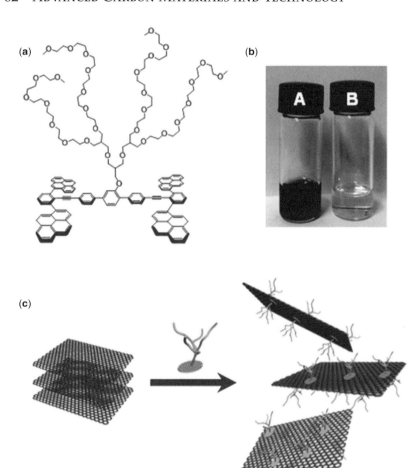

Figure 2.17 (a) Chemical structure of amphiphile 1. (b) Photograph of graphene (A) and SWNT (B) in aqueous solution of 1. (c) Schematic representation of graphene dispersion. Reproduced from ref. [152].

high stability in the physiological environment. Lee *et al.* showed the preparation of water-dispersible graphene by the aromatic amphiphile consisting of a hydrophilic dendron and an aromatic segment that interacts with graphene through π-π interaction as shown in Figure 2.17 [155]. Interestingly, it is seen that the amphiphilic pyrene sheet does not stabilize carbon nanotube, suggesting that the amphiphile selectively recognizes the 2D graphene sheets. Thus, the current strategy provides a new technique for the separation of graphene from other carbon allotropes and fabrication of graphene-based composite materials. The graphene sheets of the film are transparent and show a good electrical property as evidenced by transparent resistance measurements.

Wang *et al.* showed a simple approach for the preparation of highly biocompatible, stable and conductive graphene hybrid modified by hydrophilic poly-L-lysine (PIL) [156]. The PIL is attached with graphene surface via hydrophobic interaction between butyl chains of PLL and graphene surface. In addition, the hybrid material is stabilized by the π-π interaction (π-electrons of –NH$_2$ with the π-electrons of graphene) and electrostatic interaction (between the protonated –NH$_2$ functionalities of PIL and negatively charged graphene) of graphene with PIL. The resulting hybrid material bears positive charge on its surfaces and, thus, is very useful for the immobilization of negatively charged hemoglobin by electrostatic interaction. The hemoglobin immobilized hybrid material shows excellent catalytic activity towards the detection of hydrogen peroxide at a concentration range of 10–80 μm. Blood compatible graphene/heparin conjugate has been synthesized through non-covalent interaction of rGO with heparin [157]. Negatively charged heparin biomolecules on graphene sheets convert hydrophobic graphene into stable aqueous dispersion of graphene. It is noteworthy that the non-covalently attached heparin preserve their bioactivity after conjugation with graphene. Yang *et al.* prepared highly concentrated (0.6–2 mg ml^{-1}) and stable aqueous dispersion of graphene sheets by non-covalent functionalization with sodium lignosulfonate (SLS), sodium carboxymethyl cellulose (SCMC), and pyrene-containing hydroxypropyl cellulose (HPC-Py) [158]. The non-covalent interaction of graphene with the modifiers has been confirmed by UV-vis spectra and fluorescence quenching analysis. Zhang *et al.* prepared functionalized graphene with water-soluble iron (III) meso-tetrakis (Nmethylpyridinum-4-yl) porphyrin (FeTMPyP) via π-π non-covalent interaction on electrode surface [159]. The functionalized graphene exhibits excellent biosensing activity towards the amperometric detection of glucose ranging from 0.5 mM to 10 mM. Similarly, poly(styrenesulfonic acid-g-pyrrole)-functionalized graphene can be used as a biosensing platform for the detection of hypoxanthine [160]. The functionalized graphene shows excellent electrocatalytic activity towards the oxidation of hydrogen peroxide and uric acid. Gao *et al.* showed that methylene green (MG) can be adsorbed on the surface of graphene through π-π interaction and exhibits excellent electrochemical as well as electrocatalytic properties [161]. The modified graphene can be used as a biosensing platform for the detection of NADH. Congo red can also be used for the surface modification of graphene through π-π interaction [162]. Congo-red-modified graphene forms stable dispersion in methanol, ethanol, water and DMF. The bulk electrical conductivity of the modified graphene has been found to be ~ 6850 S m^{-1} at RT.

Reduced graphene oxide (rGO)-polymer composites was prepared via π-π stacking interactions of rGO with the perylene bisimide-containing poly(glyceryl acrylate) (PBIPGA) [163]. The π-π stacking interaction between PBI and the rGO surface has been evidenced by the fluorescence and UV-visible absorption spectroscopy results. Wu *et al.* described the preparation of polystyrene (PS)-functionalized graphene sheets through one-step *in situ* ball milling techniques [164]. The PS chains have been attached on the surface of graphene by π-π interaction and the grafted ratio of PS to graphene is ~ 85%. It has been found that chains of PS can effectively prevent the agglomeration of graphene sheets and exhibit extraordinary electrical conductivity. Geng *et al.* demonstrated the preparation of functionalized graphene sheets using porphyrin [165]. The surface modifying agents used were 5,10,15,20-tetraphenyl-21H, 23H-porphine-p,p′,p″,p′″-tetrasulfonic acid tetrasodium hydrate (TPP-SO$_3$Na) and 5,10,15,20-tetrakis(4-trimethylammoniophenyl) porphyrin tetra(p-toluenesulfonate).. The negatively charged porphyrin (TPP-SO$_3$Na) renders hydrobhobic graphene sheets to be highly dispersible in water. Karousis *et al.* prepared IL-modified GO by using porphyrin as a counter anion of the IL unit [166]. It has been anticipated that the functionalized graphene can be employed towards the development of novel optoelectronic devices. Guo *et al.* prepared water-dispersible graphene by the non-covalent functionalization of graphene with tryptophan [167]. The functionalized graphene can disperse homogeneously in the poly(vinyl alcohol) nanocomposites and improves 23% of the tensile strength of the nanocomposites with only 0.2 wt% loadings. Water-dispersible graphene has been prepared by using poly[3-(potassium-6-hexanoate)thiophene-2,5-diyl](P3KT) for solar cell application [168]. The functionalized graphene has been used to prepare an active layer for devices without further treatment. The power conversion efficiency of 0.6 wt% of functionalized graphene-loaded device is 0.027%, which is much better than that of the pure P3KT-containing devices. Kodali *et al.* performed controlled chemical modification of graphene sheets using pyrenebutanoic acid-succinimidyl ester (PYR-NHS) [169]. The hydrophobic pyrene part of PYR-NHS non-covalently adsorbed on the surface of graphene sheets and that of the succinimide ester group provide water dispersibility. Sodium dodecyl benzene sulphonate (SDBS) has been widely used for the non-covalent surface modification of graphene [170–172]. The modified graphene exhibit improved electrical conductivity and dispersibility as compared to sodium dodecyl sulphate-modified graphene. Ionic liquid polymer (imidazolium)-functionalized graphene forms stable dispersion in water [173]. The negatively charged graphene sheets are partly stabilized by positively charged imidazolium cation. However, there is a possibility for the cation-π or π-π interaction. Yang *et al.* showed the preparation of non-covalently modified graphene sheets by imidazolium ionic liquids [174]. The vinyl-benzyl reactive sites of the imidazolium ionic liquids attached with graphene can react with methyl methacrylate to fabricate graphene/PMMA composites. The functionalized graphene disperses homogeneously and randomly in the PMMA matrix. The percolation threshold in electrical conductivity is achieved at 0.25 vol% of graphene loading and the electrical conductivity is recorded as 2.55 S m^{-1} at graphene loading of 1 vol%.

In general, GO is the starting material for the covalent and non-covalent surface modification of graphene. In these techniques, surface modification of GO by the external modifying agents is followed by the chemical reduction. These two methods are very useful while considering large-scale production of surface-modified graphene. However, the major disadvantages of these techniques are: (i) surface area of the resulting graphene decreases remarkably due to the destructive oxidation of graphite and sonication of graphite oxide, (ii) use of toxic and harmful reducing agents for the reduction of modified graphene, (iii) multi-step processes, (iv) large amounts of chemical waste, and (iv) poor electrical conductivity. In order to overcome these difficulties, a green method for the one-step exfoliation of graphite to graphene has been introduced. In the one-step process, graphene can be prepared by two ways: sono-chemical and electrochemical exfoliation of graphite. However, in both the cases, the synthesized graphene is stabilized by the π-π interaction.

Stable aqueous dispersion of graphene can be prepared by the sonication of graphite in the presence of aromatic small molecules/macromolecules. Exfoliation of graphite occurs by using 1-pyrenecarboxylic acid (PCA) or 9-anthralene carboxylic acid (9-ACA)

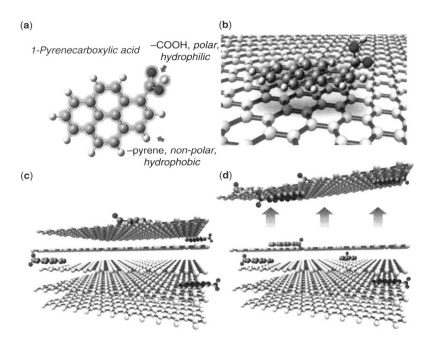

Figure 2.18 (a) Molecular structure of 1-pyrenecarboxylic acid with its polar (hydrophilic) and nonpolar (hydrophobic) parts indicated. (b) A PCA molecule can form a stable π-stacking bond with graphitic surfaces. (c) π-π stacking of graphitic layers and PCA. (d) π-π stacking of graphene and PCA. Reproduced from ref. [38].

in methanol-water solvent mixture [38, 175]. PCA or 9-ACA interacts with the graphite by π-π interaction when it is sonicated in methanol-water solution. The nonpolar pyrene or anthralene portions have fully conjugated π-network, and –COOH functionalities provide the hydrophilicity in the aqueous system. The exfoliation of graphite in the presence of PCA and their non-covalent interaction are shown in Figure 2.18. The PCA-modified graphene shows high specific capacitance (~120 F g⁻¹), power density (~105 kW kg⁻¹) and energy density (~9.2 Wh kg⁻¹). In contrast, the specific capacitance of 9-ACA-modified graphene has been recorded as 148 F g⁻¹. Economopoulos *et al.* prepared few-layer functionalized graphene using benzylamine as the surface modifying agent [176]. The exfoliation of graphite to functionalized graphene occurs under argon atmosphere with 2–10 h of continuous sonication. The schematic for the preparation of functionalized graphene is shown in Figure 2.19.

Lotya *et al.* demonstrated a method for the preparation of water-dispersible graphene in water-surfactant solutions [177].

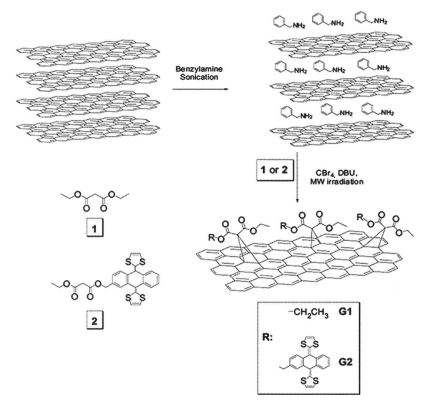

Figure 2.19 Schematic for the preparation of benzylamine modified graphene. Reproduced from ref. [176].

Transmission electron microscopy analysis shows that the yield for monolayer graphene production is only ~3% and that of few-layers (<5) is ~40%. Sim *et al.* prepared stable aqueous dispersion of graphene sheets by the ultrasonic agitation of graphite using sodium cholate and polyoxyethylene nonylphenyl ether aqueous solutions [178]. Smith *et al.* showed the one-step synthesis of graphene directly from graphite using 12 ionic and non-ionic surfactants [179]. Electron microscopy analysis shows that the average length of the prepared graphene sheets is typically 750 nm long and, on average, four layers thick. Zeta potential measurements reveal that the dispersed concentration of graphene in water is proportional to the magnitude of the electrostatic potential for the functionalized graphene while using ionic surfactants. On the contrary, for non-ionic surfactants, dispersed concentration increases linearly with the steric potential barrier stabilizing the flakes.

Liu *et al.* and Wajid *et al.* prepared polymer-stabilized graphene dispersions directly from graphite using macromolecule stabilizer [180, 181]. Polyvinylpyrrolidone (PVP)-functionalized graphene disperses well in a wide range of organic solvents. The dispersions are stable against lyophilization, pH changes, and temperatures greater than 100°C. Poly(styrene-co-butadiene-co-styrene) (SBS) interacts with graphene sheets through the π-π stacking and the resulting materials are highly dispersible in NMP. About 63% of SBS is adsorbed on the surface of graphene as evidenced from TGA. The percolation threshold in electrical conductivity of the composites is achieved with ~0.25 vol% of graphene loadings. Xu *et al.* demonstrated a convenient approach for the preparation of polystyrene surface-modified graphene starting from graphite flakes and a reactive monomer, styrene [182]. Polystyrene macromolecular chains are formed by the radical polymerization initiated by the ultrasonication of styrene and can make up to ~18 wt% of the functionalized graphene, as determined by thermal gravimetric analysis.

Liu *et al.* described the one-step green synthesis of graphene sheets directly from graphite [180]. They used a series of ionic liquid as electrolyte for the electrochemical exfoliation of graphite rod. The ionic liquids serve an additional role as surface modifying agents for graphene. In a typical procedure, two graphite rods are used as cathode and anode. The operating voltage is fixed at 15–20 V and electricity is passed for 10–12 h. Transparent ionic liquid solution turns to black with the progress of the reaction. The graphite anode is corroded and a stable dispersion of graphene is obtained in the solution. The full experimental procedures are shown schematically in Figure 2.20. The functionalized graphene disperses well in DMF and can be used for the preparation of polystyrene/graphene composites. The percolation threshold for electrical conductivity can be achieved with 0.1 vol% of graphene loadings and the nonconductive polystyrene exhibit electrical conductivity of 13.84 S m^{-1} at a graphene loading of 4.19 vol%.

Wang *et al.* used PSS as electrolyte and surface modifying agents for the electrochemical exfoliation of graphite to functionalized graphene sheets [184]. According to them, the electrochemical exfoliation occurs via adsorption of electrolyte/surface modifier on the surface of graphite electrode (anode) by π-π interaction and electrostatic forces followed by exfoliation of the graphene layers

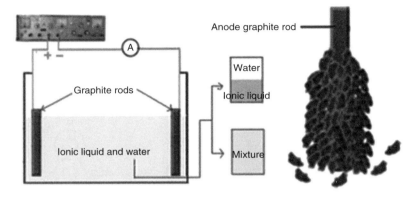

Figure 2.20 Schematic for the electrochemical exfoliation of graphite to graphene sheets. Reproduced from ref. [183].

along with the modifiers. Subsequently, the surface modifiers are again adsorbed on the graphite electrode via similar mechanism and functionalized graphene. The process was continued until the graphite anode corroded completely. Khanra *et al.* showed a similar type of exfoliation mechanism in their recent report using 9-ACA as surface modifier [45]. The specific capacitance of 9-ACA-functionalized graphene has been recorded as 577 F g^{-1} in 1 M H$_2$SO$_4$ solution. The very high specific capacitance is attributed to the appearance of electrochemical double layer capacitance (EDLC) along with the pseudocapacitance. The -COOH functionalities of the surface modifier and the oxygen functional groups of graphene takes part in pseudocapacitance reaction during the charge-discharge experiment. The oxidation-reduction mechanism is shown in Figure 2.21. The retention in specific capacitance is about 83.4% after 1000 charge-discharge cycles confirming its long-term electrochemical stability. The coulombic efficiency of the functionalized graphene-modified electrode is ~102% suggesting its suitability as an energy storage electrode material. Lee *et al.* also used PSS as surface modifying agent and electrolyte for the exfoliation of graphite [185]. However, they used graphite as anode and copper as cathode for the electrolysis. The 1-pyrene sulfonic acid sodium salt (1-PSA) can be used for the one-step electrochemical synthesis of functionalized graphene [186]. Raman spectral analysis suggests that the production yield for mono- and bi-layer graphene is increased to about 60% while using 1-PSA as electrolyte.

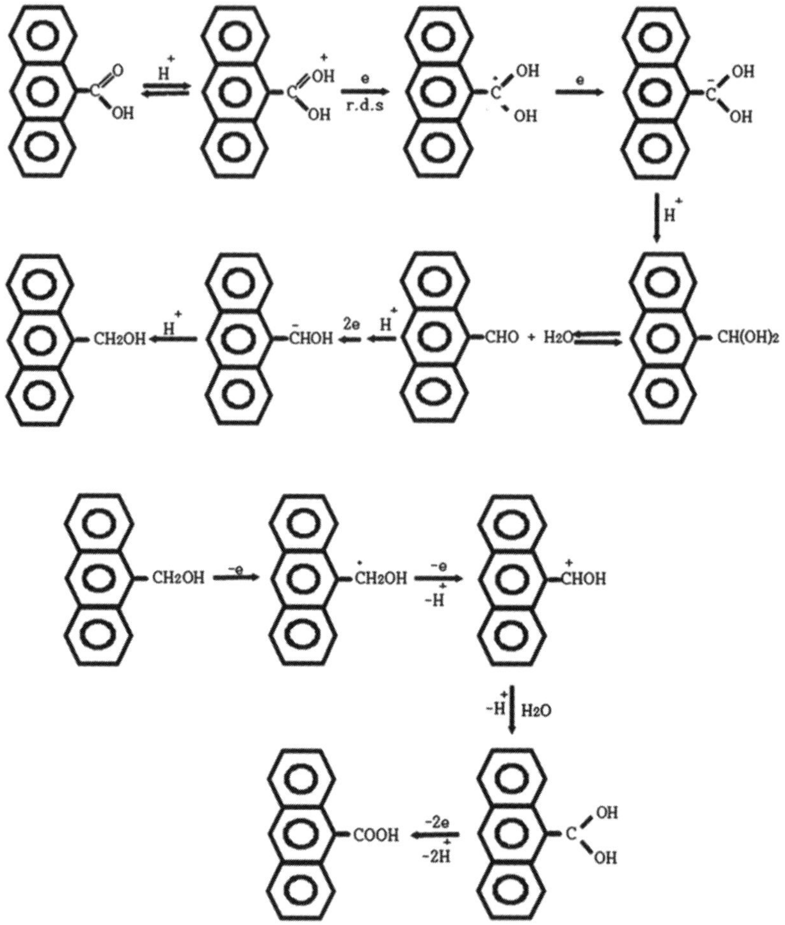

Figure 2.21 Proposed redox reaction of 9-ACA modified graphene during charge-discharge experiment. Reproduced from ref. [45].

2.3 Application of Surface-Modified Graphene

Processing of pristine graphene is very difficult due to its incompatibility with organic materials. In addition, the single layer of graphene always tends to be agglomerated resulting in loss of its exciting properties. Thus, surface modification of graphene is a prerequisite for preparing processable graphene. The surface-modified graphene not only forms stable dispersion in water and various organic solvents, but also in organic polymer matrix. It finds applications in the areas of polymer composites, biosensors, drug delivery systems, nanoelectronics, nanofluids, hydrogen storage, and lubricants, etc. [187–217].

2.3.1 Polymer Composites

Polymer composite is a biphasic material where one phase is dispersed in another phase. The surface-modified graphene can disperse in polymer matrix and can act as reinforcing filler for the enhancement of physiochemical properties. Stankovich *et al.* first showed the use of phenyl isocyanate-treated graphene as nanofiller while preparing polystyrene (PS)/graphene composites [15]. It is seen that the addition of only 2.4 vol% of surface-modified graphene almost entirely filled the composites. This is due to the large surface area of graphene. The percolation threshold in electrical conductivity is achieved with ~0.1 vol% of graphene loadings as shown in Figure 2.22. Eda *et al.* showed that the functionalized graphene-filled PS composites show electrical properties similar to individual monolayer of rGO sheets [187]. The PS composites exhibit p-type semiconductor behavior at higher temperatures.

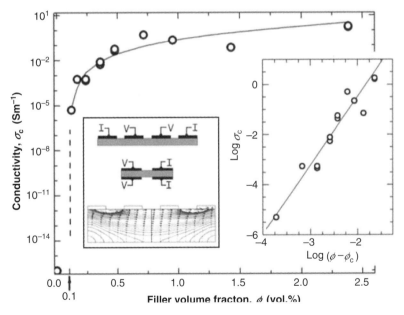

Figure 2.22 Electrical conductivity of phenyl isocyanate modified graphene filled polystyrene composites. Reproduced from ref. [15].

Kuila *et al.* demonstrated the use of dodecyl amine (DA)- and octa-decyl amine (ODA)-modified graphene (DA-G and ODA-G) as filler in the preparation of linear low density polyethylene (LLDPE)- and ethylene vinyl acetate (EVA)-based composites, respectively [14, 96, 97]. The functionalized graphene disperses well in the LLDPE and EVA matrix through hydrophobic interaction of alkyl chains of the modifier and polymer matrix. The tensile strength and storage modulus of the composites are increased with the addition of surface-modified graphene up to a certain content, and then decrease with further addition of fillers. Graphene-based epoxy composites have been investigated in detail [188–193]. It is seen that the addition of a very small amount of surface-modified graphene effectively enhances the mechanical and thermal stability of the composites as compared to the neat epoxy. This is attributed to the large specific surface area and high mechanical strength of graphene. Other commonly used polymers for the preparation of graphene-based composites are polyvinyl alcohol, polymethyl methacrylate, polypropylene, polyaniline, polypyrrole, liquid crystal polymers, chitosan, cellulose, nafion, polycarbonate, polyethylene terephthalate, polyimide, and polyvinyl chloride, etc. [194–203]. Graphene-based polymer composites find potential applications in the areas of the automobile and air-craft industry, high speed turbine blades, bone and tissue implantations, etc.

2.3.2   Sensors

The large surface area and excellent electrical conductivity of graphene allow it to act as an "electron wire" between the redox centers of an enzyme or protein and an electrode's surface [19]. Rapid electron transfer through surface-modified graphene facilitates accurate and selective detection of biomolecules, poisonous gases and heavy metal ions. The graphene-modified electrode can detect glucose concentration in blood sample very accurately and selectively. It has been found that the presence of ascorbic acid, uric acid, lactic acid, dopamine, etc., does not inhibit the detection of glucose in blood sample. In a few cases, the effect of gold nanoparticles on the glucose detection ability of graphene has been investigated. Surface-modified graphene can interact directly with functional groups of biomolecules in electrolyte solution. Hemoglobin, NADH, cholesterol, catechol, hydrogen peroxide, ascorbic acid, uric acid, etc., can be detected using surface-modified graphene as current collector [204, 205]. Heavy metal ions such as lead, cadmium, silver, mercury and arsenic have severe environmental and medical effects and, therefore, require careful monitoring [204–207]. It is seen that the sensor designed with surface-modified graphene selectively detects heavy metal ions in solution. Surface-modified graphene has the enormous potential to detect different gaseous molecules such as iodine, hydrogen, carbon monoxide, hydrogen, ammonia, chlorine, nitrogen dioxide, etc.

2.3.3 Drug Delivery System

Surface-modified graphene has played an important role in drug delivery as initiated by the Hongjie Dai group at Stanford University (California, USA) in 2008 [134]. They showed that the aromatic anticancer drugs (e.g., SN38 and doxorubicin) can be effectively loaded on the surface of PEG-treated graphene (PEGylated graphene) via π-π interactions. The extremely large surface area of graphene facilitates exceptionally high drug loading efficiency. The functionalized graphene has been successfully applied for *in vivo* cancer therapy.

2.3.4 Lubricants

Application of surface-modified graphene as additives in base lubricant oil is a rapidly progressing field of research due to its extremely high mechanical flexibility, large surface area, good friction reduction and anti-wear ability. Zhang *et al.* showed the utilization of oleic acid-modified graphene as lubricant [208]. The tribological properties of surface-modified graphene-contained oils were investigated using a four-ball tribometer. Figure 2.23 shows that the lubricant with optimized graphene contents (0.02–0.06 wt%) exhibits improved friction and anti-wear performance, with friction coefficient and wear scar diameter reduced by 17% and 14%, respectively. The improved friction behavior can be explained by the tribological mechanism proposed in Figure 2.23(c). A protective layer of graphene is formed on the surface of each steel ball at lower concentrations and introduces the enhanced anti-wear performance. On the contrary, the oil film becomes more discontinuous at higher concentrations which is responsible for the degradation of the antiwear properties. Lin *et al.* also confirmed that 0.075 wt% of stearic and oleic acid-treated graphene in oil improved the wear resistance and load-carrying capacity of the machine [209]. A

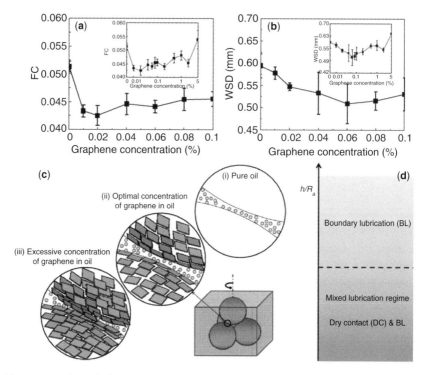

Figure 2.23 Four-ball test results: (a) FC versus graphene concentration; (b) WSD versus graphene concentration; (c) schematic diagram of the tribological mechanism of graphene sheets as oil additives; (d) lubrication regime transition. Reproduced from ref. [208].

recent report showed that alkylated graphene in organic solvents can also exhibit lubrication properties [27]. The alkylated graphene with variable alkyl chain lengths (Cn = 8, 12, 18) were prepared by the condensation reaction of alkylamine with $SOCl_2$-activated GO. It is seen that octadecyl amine-treated graphene dispersed in hexadecane reduces friction and wear by 26% and 9%, respectively, compared to hexadecane.

2.3.5 Nanofluids

The production of heat energy decreases the performance of many of the instruments and machines. In order to improve the performance of the instruments and machineries, most of the industries use fluids like DI water, ethylene glycol (EG), transformer oil, etc., as heat transfer fluids. However, the efficiency of heat transfer of these fluids is very low and deteriorates the productivity and lifetime of the equipment, electronic circuits, machines, etc. The efficiency of heat transfer can be increased by the addition of nanomaterials in the fluids. Baby *et al.* showed that the thermal conductivity can be increased by about 14% at 25°C with DI water as base fluid at a very low volume fraction of 0.056% [210, 211]. The thermal conductivity is enhanced by ~64% at 50°C at a similar content of functionalized graphene. Ghozatloo *et al.* showed that 0.05 wt% of functionalized graphene can enhance thermal conductivity by 14.1% when tested in water at 25°C [212]. The thermal conductivity is improved by 17% when the temperature is increased to 50°C.

2.3.6   Supercapacitor

The supercapacitor has been regarded as the next generation energy storage device. It can store electrical energy by electrochemical double layer capacitance and pseudocapacitance mechanism. The power density of the supercapacitor device is almost 10 times higher than that of conventional lead-acid battery or capacitor. Moreover, it can operate in a wide temperature range, is eco-friendly and can deliver energy very fast if needed. Graphene-based materials have the potential to be used in supercapacitor devices as energy storage electrode materials. However, pristine graphene finds electrode-electrolyte interface problematic due to the hydrophobic nature of graphene. In order to enhance the hydrophilicity of graphene, surface modification has been introduced [39, 175, 213–217]. The presence of functional groups on the surface of graphene not only overcomes the compatibility problem but also takes part in faradic reaction. It has been found that the specific capacitance of pristine graphene due to electrochemical double layer capacitance can be increased up to ~800–1000 F g^{-1} by introducing pseudocapacitance. Thus, surface modification plays an important role in determining the energy storage efficiency of graphene.

2.4   Conclusions and Future Directions of Research

Extensive research has been carried out on the surface modification of graphene to improve the processability of pristine graphene.

Different covalent and non-covalent techniques have been applied to achieve water/organic dispersible graphene. In both cases, the starting material for surface modification is graphene oxide, which can be obtained by the oxidation of graphite. However, oxidation of graphite is a destructive method and consumes lots of hazardous chemicals. Moreover, further reduction of surface-modified GO is essential to restore the π-electronic conjugated network structure. Different kinds of toxic reducing chemicals are generally used for the reduction of surface-modified GO. The use of these toxic chemicals should be minimized for the benefit of human health and protection of environmental damages. It is also noteworthy that the electrical conductivity of the surface-modified graphene obtained by covalent and non-covalent approaches decreases significantly due to the incomplete restoration of conjugated network. In addition, surface area of the graphene decreases remarkably due to the destructive oxidation-reduction methods. In order to overcome these problems associated with the surface modification of graphene, a one-step sono-chemical approach has been reported. This method is green and does not require the use of hazardous chemicals. However, the low production yield is of great concern in this technique. In contrast, one-step electrochemical exfoliation and stabilization of graphite to graphene in a suitable solvent is a fruitful method for obtaining surface-modified graphene. The presence of organic functional groups onto the surface of graphene sheets facilitates the formation of stable dispersion either in water or in the organic solvents. Surface-modified graphene can also be dispersed in organic polymer matrix to form high performance polymer composites for applications in the areas of the automobile and aircraft industry. The homogeneous dispersion of graphene in organic solvents is used extensively as nanofluids and lubricants. Surface-modified graphene also finds applications in the areas of biosensor, drug delivery and supercapacitor devices. However, there is not enough scope on the surface modification of graphene as the mass production of surface-modified graphene has not been reported so far in the literature. Selection of proper surface modifying agents can play an important role in determining the end properties of graphene. Thus, the preparation, characterization and application of highly conducting surface-modified graphene is in high demand for the development of next generation devices to be used in the areas of health, energy, mobility, security, amd economic and industrial growth.

Acknowledgement

This study was supported by the DST INSPIRE Faculty Programme (IFA12-CH-47), New Delhi, India. The authors are thankful to Prof. Gautam Biswas, the Director of CSIR-Central Mechanical Engineering Research Institute, Durgapur, India.

References

1. K.S. Novoselov, A.K. Geim, S.V. Morozov, D. Jiang, Y. Zhang, S.V. Dubonos, I.V. Grigorieva, and A.A. Firsov, *Science*, Vol. 306, p. 666, 2004.
2. A.K. Geim, and K.S. Novoselov, *Nat Mater*, Vol. 6, p. 183, 2007.
3. Y. Cui, S.N. Kim, S.E. Jones, L.L. Wissler, R.R. Naik, and M.C. McAlpine, *Nano Lett*, Vol. 10, p. 4559, 2010.
4. L. Liao, Y.C. Lin, M. Bao, R. Cheng, J. Bai, Y. Liu, Y. Qu, K. L. wang, Y. Huang, and X. Duan, *Nature,* Vol. 467, p. 305, 2010.
5. T. Mueller, F. Xia, and P. Avouris, *Nat Photonics*, Vol. 4, p. 297, 2010.
6. F. Bonaccoso, Z. Sun, T. Hasan, and A.C. Ferrari, *Nat Photonics*, Vol. 4, p. 611, 2010.
7. C. Chen, S. Rosenblatt, K.I. Bolotin, W. Kalb, P. Kim, I. Kymissis, H.L. Stormer, T.F. Heinz, and J. Hone, *Nat Nanotechnol* Vol. 4, p. 861, 2009.
8. G. Yu, L. Hu, N. Liu, H. wang, M. Vosgueritchian, Y. Yang, Y. Cui, and Z. Bao, *Nano Lett*, Vol. 11, p. 4438, 2011.
9. K.R. Ratinac, W. Yang, S.P. Ringer, and F. Braet, *Environ Sci Technol,* Vol. 44, p. 1167, 2010.
10. Y. Dan, Y. Liu, N.J. Kybert, Z. Luo, and A.T.C. Johnson, *Nano Lett*, Vol. 9, p. 1472, 2009.
11. F. Schedin, A.K. Geim, S.V. Morozov, E.W. Hill, P. Blake, M.I. Katsnelson, and K.S. Novoselov, *Nat Mater,* Vol. 6, p. 652, 2007.
12. Y. Liu, D. Yu, C. Zeng, Z. Miao, and L. Dai, *Langmuir*, Vol. 26, p. 6158, 2010.
13. T. Kuila, S. Bhadra, D. Yao, N.H. Kim, S. Bose, and J.H. Lee, *Prog Polym Sci*, Vol. 35, p. 1350, 2010.
14. T. Kuila, S. Bose, C.E. Hong, E. Uddin, N.H. Kim, P. Khnara, and J.H. Lee, *Carbon*, Vol. 49, p. 1033, 2011.
15. S. Stankovich, D.A. Dikin, G.H.B. Dommett, K.M. Kohlhaas, E.J. Zimney, E.A. Stach, R.D. Piner, S.T. Nguyen, and R.S. Ruoff, *Nat Mater* Vol. 442, p. 282, 2006.
16. Y. Zhu, S. Murali, M.D. Stoller, K.J. Ganesh, W. Cai, P.J. Ferreira, A. Pirkle, R.M. Wallace, K.A. Cychosz, M. Thommes, D. Su, E.A. Stach, and R.S. Ruoff, *Science*, Vol. 332, p. 1537, 2011.

17. Q. Wu, Y. Xu, Z. Yao, A. Liu, and G. Shi, *ACS Nano*, Vol. 4, p. 1963, 2010.
18. Z. Lei, L. Lu, and X. S. Zhao, *Energy Environ Sci*, Vol. 5, p. 6391, 2012.
19. T. Kuila, S. Bose, P. Khanra, A.K. Mishra, N.H. Kim, and J.H. Lee, *Biosens Bioelectron*, Vol. 26, p. 4637, 2011.
20. M.H. Yang, B.G. Choi, H. Park, W.H. Hong, S.Y. Lee, and T.J. Park, *Electroanalysis*, Vol. 22, p. 1223, 2010.
21. R.S. Dey and C.R. Raj, *J Phys Chem*, Vol. 114, p. 21427, 2010.
22. S. Lee, J. S. Yeo, Y. Ji, C. Cho, D.Y. Kim, S.I. Na, B.H. Lee and T. Lee, *Nanotechnology*, Vol. 23, 344013, 2012.
23. C.Y. Chu, J.T. Tsai, and C.L. Sun, *Inter J Hydrogen Energ*, Vol. 37, p. 13880, 2012.
24. Z. Jin, W. Lu, K.J. O'Neill, P.A. Parilla, L.J. Simpson, C. Kittrell, and J.M. Tour, *Chem Mater*, Vol. 23, p. 923, 2011.
25. Y. Chan, and J.M. Hill, *Nanotechnology*, Vol. 22, p. 305403, 2011.
26. S.H. Aboutalebi, S.A. Yamini, I. Nevirkovets, K. Konstantinov, and H.K. Liu, *Adv Energ Mater*, Vol. 2, p. 1439, 2012.
27. S. Choudhary, H.P. Mungse, and O.P. Khatri, *J Mater Chem*, Vol. 22, p. 21032, 2012.
28. Q. Liu, Z. Liu, X. Zhang, L. Yang, N. Zhang, G. Pan, S. Yin, Y. Chen, and J. Wei, *Adv Funct Mater*, Vol. 19, p. 894, 2009.
29. Y. Wu, Y. Lin, A.A. Bol, K.A. Jenkins, F. Xia, D.B. Farmer, Y. Zhu, and P. Avouris, *Nature*, Vol. 472, p. 74, 2011.
30. B. Guo, Q. Liu, E. Chen, H. Zhu, L. Fang, and J.R. Gong, *Nano Lett*, Vol. 10, p. 4975, 2010.
31. C.Y. Su, A.Y. Lu, C.Y. Wu, Y.T. Li, K.K. Liu, W. Zhang, S.Y. Lin, Z.Y. Juang, Y.L. Zhong, F.R. Chen, and L.J. Li, *Nano Lett*, Vol. 11, p. 3612, 2011.
32. K.S. Kim, H.J. Lee, C. Lee, S.K. Lee, H. Jang, J.H. Ahn, J.H. Kim, and H.J. Lee, *ACS Nano*, Vol. 5, p. 5107, 2011.
33. Z. Li, P. Wu, C. Wang, X. Fan, W. Zhang, W. Zhang, X. Zhai, C. Zeng, Z. Li, J. Yang, and J. Hou, *ACS Nano*, Vol. 5, p. 3385, 2011.
34. S.J. Byun, H. Lim, G.Y. Shin, T.H. Han, S.H. Oh, J.H. Ahn, H.C. Choi, and T.W. Lee, *J Phys Chem*, Vol. 2, p. 493, 2011.
35. Z. Peng, Z. Yan, Z. Sun, and J.M. Tour, *ACS Nano*, Vol. 5, p. 8241, 2011.
36. Z.L. Liu, C.Y. Kang, L.L. Fan, P.S. Xu, and C.W. Zou, *J Appl Phys*, Vol. 113, p. 014311, 2013.
37. R. Hao, W. Qian, L. Zhang, and Y. Hou, *Chem Commun*, p. 6576, 2008.
38. X. An, T. Simmons, R. Shah, C. Wolfe, K.M. Lewis, M. Washington, S.K. Nayak, S. Talapatra, and S. Kar, *Nano Lett*, Vol. 10, p. 4295, 2010.
39. T. Kuila, A.K. Mishra, P. Khanra, N.H. Kim, E. Uddin, and J.H. Lee, *Langmuir*, Vol. 28, p. 9825, 2012.
40. D.R. Dreyer, S. Murali, Y. Zhu, R.S. Ruoff, and C.W. Bielawski, *J Mater Chem*, Vol. 21, p. 3443, 2011.

41. Y. Chen, X. Zhang, P. Yu, and Y. Ma, *Chem Commun*, p. 4527, 2009.
42. X. Fan, W. Peng, Y. Li, X. Li, S. Wang, G. Zhang, and F. Zhang, *Adv Mater*, Vol. 20, p. 4490, 2008.
43. Z.S. Wu, W. Ren, L. Guo, J. Zhao, Z. Chen, B. Liu, D. Tang, B. Yu, C. Jiang, and H.M. Cheng, *ACS Nano*, Vol. 3, p. 411, 2009.
44. H.S.S. Ramakrishna Matte, K.S. Subrahmanyam, and C.N.R. Rao, *Nanomater Nanotechnol*, Vol. 1, p. 3, 2011.
45. P. Khanra, T. Kuila, S.H. Bae, N.H. Kim, and J.H. Lee, *J Mater Chem*, Vol. 22, p. 24403, 2012.
46. C.Y. Su, A.Y. Lu, Y. Xu, F.R. Chen, A.N. Khlobystov, and L.J. Li, *ACS Nano*, Vol. 5, p. 2332, 2011.
47. J. Wang, K.K. Manga, Q. Bao, and K.P. Loh, *J Am Chem Soc*, Vol. 133, p. 8888, 2011.
48. C.N.R. Rao, A.K. Sood, K.S. Subrahmanyam, and A. Govindaraj, *Angew Chem Int Ed*, Vol. 48, p. 7752, 2009.
49. K.S. Subrahmanyam, L.S. Panchakarla, A. Govindaraj and C.N.R. Rao, *J Phys Chem C*, Vol. 113, p. 4257, 2009.
50. L.S. Panchakarla, K.S. Subrahmanyam, S.K. Saha, A. Govindaraj, H.R. Krishnamurthy, U.V. Waghmare, and C.N.R. Rao, *Adv Mater*, Vol. 21, p. 4726, 2009.
51. T. Kuila, A.K. Mishra, P. Khanra, N.H. Kim, and J.H. Lee, *Nanoscale*, Vol. 5, p. 52, 2013.
52. S. Pei, and H.M. Cheng, *Carbon*, Vol. 50, p. 3210, 2012.
53. R. Larciprete, S. Fabris, T. Sun, P. Lacovig, A. Baraldi, and S. Lizzit, *J Am Chem Soc*, Vol. 133, p. 17315, 2011.
54. Y. Zhu, M.D. Stoller, W. Cai, A. Velamakanni, R.D. Piner, D. Chen, and R.S. Ruoff, *ACS Nano*, Vol. 4, p. 1227, 2010.
55. T. Kuila, S. Bose, A.K. Mishra, P. Khanra, N.H. Kim, and J.H. Lee, *Prog Mater Sci*, Vol. 57, p. 1061, 2012.
56. O.C. Compton, and S.T. Nguyen, *Small*, Vol. 6, p. 711, 2010.
57. Y. Zhu, S. Murali, W. Cai, X. Li, J.W. Suk, J.R. Potts, and R.S. Ruoff, *Adv Mater*, Vol. 22, p. 3906, 2010.
58. K.P. Loh, Q. Bao, P.K. Ang, and J. Yang, *J Mater Chem*, Vol. 20, p. 2277, 2010.
59. F. Xia, D.B. Farmer, Y.M. Lin, and P. Avouris, *Nano Lett*, Vol. 10, p. 715, 2010.
60. F. Yavari, C. Kritzinger, C. Gaire, L. Song, H. Gullapalli, T. Borca-Tasciuc, P.M. Ajayan, and N. Koratkar, *Small*, Vol. 22, p. 2535, 2010.
61. P. Avouris, *Nano Lett*, Vol. 10, p. 4285, 2010.
62. A. Lerf, H. He, M. Forster and J. Klinowski, *J Phys Chem B*, Vol. 102, p. 4477, 1998.
63. T. Szabo, O. Berkesi, P. Forgo, K. Josepovits, Y. Sanakis, D. Petridis and I. Dekany, *Chem Mater*, Vol. 18, p. 2740, 2006.
64. H. He, J. Klinowski, M. Forster and A. Lerf, *Chem Phys Lett*, Vol. 287, p. 53, 1998.

65. M. Fang, K. Wang, H. Lu, Y. Yang, and S. Nutt, *J Mater Chem*, Vol. 19, p. 7098, 2009.
66. M.M. Gudarzi, and F. Sharif, *J Colloid Interface Sci*, Vol. 365, p. 44, 2012.
67. H. Bao, Y. Pan, Y. Ping, N.G. Sahoo, T. Wu, L. Li, J. Li, and L.H. Gan, *Small*, Vol. 7, p. 1569, 2011.
68. X. Wang, H. Bai, Z. Yao, A. Liu, and G. Shi, *J Mater Chem*, Vol. 20, p. 9032, 2010.
69. Q. Wu, Y. Xu, Z. Yao, A. Liu, and G. Shi, *ACS Nano*, Vol. 4, p. 1963, 2010.
70. H. Pang, Y.C. Zhang, T. Chen, B.Q. Zeng, and Z.M. Li, *Appl Phys Lett*, Vol. 96, p. 251907, 2010.
71. H. Kim, and C.W. Macosko, *Polymer*, Vol. 50, p. 3797, 2009.
72. M.H. Yang, B.G. Choi, H. Park, S.Y. Lee, and T.J. Park, *Electroanalysis*, Vol. 22, p. 1223, 2010.
73. X. Kang, J. Wang, H. Wu, I. A. Aksay, J. Liu, and Y. Lin, *Biosens Bioelectron*, Vol. 25, p. 901, 2009.
74. J.D. Fowler, M.J. Allen, V.C. Tung, Y. Yang, R.B. Kaner, and B.H. Weiller, *ACS Nano*, Vol. 2, p. 301, 2009.
75. J.T. Robinson, F.K. Perkins, E.S. Snow, Z. Wei, and P.E. Sheehan, *Nano Lett*, Vol. 8, p. 3137, 2008.
76. M. He, J. Jung, F. Qiu, and Z. Lin, *J Mater Chem*, Vol. 22, p. 24254, 2012.
77. J. Wu, H.A. Becerril, Z. Bao, Z. Liu, Y. Chen, and P. Peumans, *Appl Phys Lett*, Vol. 92, p. 263302, 2008.
78. S.S. Gupta, V.M. Siva, S. Krishnan, T.S. Sreeprasad, P.K. Singh, T. Pradeep, and S.K. Das, *J Appl Phys*, Vol. 110, p. 084302, 2011.
79. W. Yu, H. Xie, X. Wang, and X. Wang, *Phys Lett A*, Vol. 375, p. 1323, 2011.
80. V. Eswaraiah, V. Sankaranarayanan, and S. Ramaprabhu, *ACS Appl Mater Interface*, Vol. 3, p. 4221, 2011.
81. A. Senatore, V. DÁgostino, V. Petrone, P. Ciambelli, and M. Sarno, *Hindawi Pub Corp*, Vol. 2013, ID. 425809.
82. W.R. Collins, E. Schmois, and T.M. Swager, *Chem Commun*, Vol. 47, p. 8790, 2011.
83. R. Ledezma, L. Arizmendi, J.A. Rodríguez, A. Castañeda, R.F. Ziolo, *Macromol Symp*, Vol. 325–326, p. 141, 2013.
84. D. Chen, L. Li, and L. Guo, *Nanotechnology*, Vol. 22, p. 325601, 2011.
85. T.A. Pham, J.S. Kim, J.S. Kim, and Y.T. Jeong, *Colloid Surf A; Physicochem Eng Aspect*, Vol. 384, p. 543, 2011.
86. S. Bose, T. Kuila, A.K. Mishra, N.H. Kim, and J.H. Lee, *J Mater Chem*, Vol. 22, p. 9696, 2012.
87. P. Khanra, T. Kuila, N.H. Kim, S.H. Bae, D.S. Yu, and J.H. Lee, *Chem Eng J*, Vol. 183, p. 526, 2012.
88. L.Q. Xu, W.J. Yang, K.G. Neoh, E.T. Kang, and G.D. Fu, *Macromolecules*, Vol. 43, p. 8336, 2010.

89. S. Park, D.A. Dikin, S.T. Nguyen, and R.S. Ruoff, *J Phys Chem C*, Vol. 113, p. 15801, 2009.
90. S.M. Kang, S. Park, D. Kim, S.Y. Park, R.S. Ruoff, and H. Lee, *Adv Funct Mater*, Vol. 21, p. 108, 2011.
91. W. Yu, H. Xie, X. Wang, and X. Wang, *Nanoscale Res Lett*, Vo. 6, p. 47, 2011.
92. N.H. Kim, T. Kuila, and J.H. Lee, *J. Mater Chem A*, Vol. 1, p. 1349, 2013.
93. H. Liu, T. Kuila, N.H. Kim, B.C. Ku, and J. H. Lee, *J Mater Chem A*, Vol. 1, p. 3739, 2013.
94. H. Yang, F. Li, C. Shan, D. Han, Q. Zhang, L. Niu, and A. Ivaska, *J Mater Chem*, Vol. 19, p. 4632, 2009.
95. Y. Wang, Z. Shi, and J. Yin, *Polymer*, Vol. 52, p. 3661, 2011.
96. T. Kuila, S. Bose, A.K. Mishra, P. Khanra, N.H. Kim, and J.H. Lee, *Polym Test*, Vol. 31, p. 31, 2012.
97. T. Kuila, P. Khanra, A.K. Mishra, N.H. Kim, and J.H. Lee, *Polym Test*, Vol. 31, p. 282, 2012.
98. G. Wang, B. Wang, J. Park, J. Yang, X. Shen, and J. Yao, *Carbon*, Vol. 47, p. 68, 2009.
99. A.B. Bourlinos, D. Gournis, D. Petridis, T. Szabo, A. Szeri, and I. Dekany, *Langmuir*, Vol. 19, p. 6050, 2003.
100. W. Li, X.Z. Tang, H.B. Zhang, Z.G. Jiang, Z.Z. Yu, X.S. Du, and Y.W. Mai, *Carbon*, Vol. 49, p. 4724, 2011.
101. Y. Si, and E.T. Samulski, *Nano Lett*, Vol. 8, p. 1679, 2008.
102. D.S. Yu, T. Kuila, N.H. Kim, P. Khanra, and J.H. Lee, *Carbon*, Vol. 54, p. 310, 2013.
103. J.R. Lomeda, C.D. Doyle, D.V. Kosynkin, W.F. Hwang, and J.M. Tour, *J Am Chem Soc*, Vol. 130, p. 16201, 2008.
104. Z. Jin, T.P. McNicholas, C.J. Shih, Q.H. Wang, G.L.C. Paulus, A.J. Hilmer, S. Shimizu, and M.S. Strano, *Chem Mater*, Vol. 23, p. 3362, 2011.
105. M.B. Avinash, K.S. Subrahmanyam, Y. Sundarayya, and T. Govindaraju, *Nanoscale*, Vol. 2, p. 1762, 2010.
106. Z. Sun, S.I. Kohama, Z. Zhang, J.R. Lomeda, and J.M. Tour, *Nano Res*, Vol. 3, p. 117, 2010.
107. Q. Wu, Y. Wu, Y. Hao, J. Geng, M. Charlton, S. Chen, Y. Ren, H. Ji, H. Li, D.W. Boukhvalov, R.D. Piner, C.W. Bielawski, and R.S. Ruoff, *Chem Commun*, Vol. 49, p. 677, 2013.
108. V.T. Pham, T.V. Cuong, S.H. Hur, E. Oh, E.J. Kim, E.W. Shin, and J.S. Chung, *J Mater Chem*, Vol. 21, p. 3371, 2011.
109. O.K. Park, M.G. Hahm, S. Lee, H.I. Joh, S.I. Na, R. Vajtai, J.H. Lee, B.C. Ku, and P.M. Ajayan, *Nano Lett*, Vol. 12, p. 1789, 2012.
110. G.L.C. Paulus, Q.H. Wang, and M.S. Strano, *Acc Chem Res*, Vol. 46, p. 160, 2013.

111. E. Bekyarova, M.E. Itkis, P. Ramesh, C. Berger, M. Sprinkle, W.A. de Heer, and R.C. Haddon, *J Am Chem Soc*, Vol. 131, p. 1336, 2009.

112. R. Smarma, J.H. Baik, C.J. Perera, and M.S. Strano, *Nano Lett*, Vol. 10, p. 398, 2010.

113. P. Huang, H. Zhu, L. Jing, Y. Zhao, and X. Gao, *ACS Nano*, Vol. 5, p. 7945, 2011.

114. S. Stankovich, R.D. Piner, S.T. Nguyen, and R.S. Ruoff, *Carbon*, Vol. 44, p. 3342, 2006.

115. D.D. Zhang, S.Z. Zu, and B.H. Han, *Carbon*, Vol. 47, p. 2993, 2009.

116. Y. Xu, Z. Liu, X. Zhang, Y. Wang, J. Tian, Y. Huang, Y. Ma, X. Zhang, and Y. Chen, *Adv Mater*, Vol. 21, p. 1275, 2009.

117. S. Zhang, K. Yang, L. Feng, and Z. Liu, *Carbon*, Vol. 49, p. 4040, 2011.

118. J. Shen, N. Li, M. Shi, Y. Hu, and M. Ye, *J Colloid Interface Sci*, Vol. 34, p. 377, 2010.

119. L. Ren, X. Wang, S. Guo, and T. Liu, *J Nanopart Res*, Vol. 13, p. 6389, 2011.

120. G. Goncalves, P.A.A.P. Marques, A. Barros-Timmons, I. Bdkin, M.K. Singh, N. Emami, and J. Gracio, *J Mater Chem*, Vol. 20, p. 9927, 2010.

121. Y.T. Liu, J.M. Yang, X.M. Xie, and X.Y. Ye, *Mater Chem Phys*, Vol. 130, p. 130, 794.

122. J. Zhu, Y. Li, Y. Chen, J. Wang, B. Zhang, J. Zhang, and W.J. Blau, *Carbon*, Vol. 49, p. 1900, 2011.

123. Z.B. Liu, Y.F. Xu, X.Y. Zhang, X.L. Zhang, Y.S. Chen, and J.G. Tian, *J Phys Chem B*, Vol. 113, p. 9681, 2009.

124. H.J. Salavagione, and G. Martinez, *Macromolecules*, Vol. 44, p. 2685, 2011.

125. H.J. Salavagione, M.A. Gomez, and G. Martinez, *Macromolecules*, Vol. 42, p. 6331, 2009.

126. Q. Yang, X. Pan, K. Clarke, and K. Li, *Ind Eng Chem Res*, Vol. 51, p. 310, 2012.

127. H. Hu, X. Wang, J. Wang, F. Liu, M. Zhang, and C. Xu, *Appl Surf Sci*, Vol. 257, p. 2637, 2011.

128. Y. Zhan, X. Yang, F. Meng, J. Wei, J. Zhao, and X. Liu, *J Colloid Interface Sci*, Vol. 363, p. 98, 2011.

129. S. Hou, S. Su, M.L. Kasner, P. Shah, K. Patel, and C.J. Madarang, *Chem Phys Lett*, Vol. 501, p. 68, 2010.

130. Y. Matsuo, T. Tabata, T. Fukunaga, T. Fukutsuka, and Y. Sugie, *Carbon*, Vol. 43, p. 2875, 2005.

131. K.A. Worsley, P. Ramesh, S.K. Mandal, S. Niyogi, M.E. Itkis, R.C. Haddon, *Chem Phys Lett*, Vol. 445, p. 51, 2007.

132. S. Niyogi, E. Bekyarova, M.E. Itkis, J.L. Mcwilliams, M.A. Hamon, and R.C. Haddon, *J Am Chem Soc*, Vol. 128, p. 7720, 2006.

133. T. A. Pham, N. A. Kumar, and Y. T. Jeong, *Syn Met*, Vol. 160, p. 2028, 2010.

134. Z. Liu, J.T. Robinson, X. Sun, and H. Dai, *J Am Chem Soc*, Vol. 130, p. 10876, 2008.
135. J. Shen, B. Yan, M. Shi, H. Ma, N. Li, and M. Ye, *J Colloid Interface Sci*, Vol. 356, p. 543, 2011.
136. Y. Zhang, L. Ren, S. Wang, A. Marathe, J. Chaudhuri, and G. Li, *J Mater Chem*, Vol. 21, p. 5386, 2011.
137. J. Choi, K.J. Kim, B. Kim, H. Lee, and S. Kim, *J Phys Chem C*, Vol. 113, p. 9433, 2009.
138. K. Suggs, D. Reuven, and X.Q. Wang, *J Phys Chem C*, Vol. 115, p. 3313, 2011.
139. M. Quintana, K. Spyrou, M. Grzelczak, W.R. Browne, P. Rudolf, and M. Prato, *ACS Nano*, Vol. 4, p. 3527, 2010.
140. X. Wang, D. Jiang, and S. Dai, *Chem Mater*, Vol. 20, p. 4800, 2008.
141. V. Georgakilas, A.B. Bourlinos, R. Zboril, T.A. Steriotis, P. Dallas, A.K. Stubos, and C. Trapalisa, *Chem Commun*, Vol. 46, p. 1766, 2010.
142. H. He, and C. Chao, *Chem Mater*, Vol. 22, p. 5054, 2010.
143. X. Zhang, L. Hou, A. Cnossen, A.C. Coleman, O. Ivashenko, P. Rudolf, B.J. van Wees, W.S. Browne, and B.L. Feringa, *Chem Eur J*, Vol. 17, p. 8957, 2011.
144. S. Vadukumpully, J. Gupta, Y. Zhang, G.Q. Xu, and S. Valiyaveettil, *Nanoscale*, Vol. 3, p. 303, 2011.
145. M. Quintana, A. Montellano, A.E. del Rio Castillo, G.V. Tendeloo, C. Bittencourt, M. Prato, *Chem Commun*, Vol. 47, p. 9330, 2011.
146. X. Xu, Q. Luo, W. Lv, Y. Dong, Y. Lin, Q. Yang, A. Shen, D. Pang, J. Hu, J. Qin, and Z. Li, *Macromol Chem Phys*, Vol. 212, p. 768, 2011.
147. S. Stankovich, R.D. Piner, X. Chen, N. Wu, S.T. Nguyen, and R.S. Ruoff, *J Mater Chem*, Vol. 16, p. 155, 2006.
148. S. Bose, N.H. Kim, T. Kuila, K.T. Lau, and J.H. Lee, *Nanotechnology*, Vol. 22, p. 295202, 2011.
149. Y. Yan, T. Kuila, N.H. Kim, B.C. Ku, and J.H. Lee, *J Mater Chem*, Vol. 1, p. 5892, 2013.
150. H. Bai, Y. Xu, L. Zhao, C. Li, and G. Shi, *Chem Commun*, p. 1667, 2009.
151. E.Y. Choi, T.H. Han, J. Hong, J.E. Kim, S.H. Lee, H.W. Kim, and S.O. Kim, *J Mater Chem*, Vol. 20, p. 1907, 2010.
152. C. Chen, W. Zhai, D. Lu, H. Zhang, and W. Zheng, *Mater Res Bull*, Vol. 46, p. 583, 2011.
153. L. Ren, T. Liu, J. Guo, S. Guo, X. Wang, and W. Wang, *Nanotechnology*, Vol. 21, p. 335701, 2010.
154. X. Qi, K. Y. Pu, H. Li, X. Zhou, S. Wu, Q. L. Fan, B. Liu, F. Boey, W. Huang, and H. Zhang, *Angew Chem Int Ed*, Vol. 49, p. 9426, 2010.
155. D.W. Lee, T. Kim, and M. Lee, *Chem Commun*, Vol. 47, p. 8259, 2011.
156. J. Wang, Y. Zhao, F.X. Ma, K. Wang, F.B. Wang, and X.H. Xia, *J Mater Chem B*, Vol. 1, p. 1406, 2013.
157. D.Y. Lee, Z. Khatun, J.H. Lee, Y.K. Lee, and I. In, *Biomacromolecules*, Vol. 12, p. 336, 2011.

158. Q. Yang, X. Pan, F. Huang, and K. Li, *J Phys Chem C*, Vol. 114, p. 3811, 2010.
159. S. Zhang, S. Tang, J. Lei, H. Dong, and H. Ju, *J Electroanal Chem*, Vol. 656, p. 285, 2011.
160. J. Zhang, J. Lei, R. Pan, Y. Xue, and H. Ju, *Biosens Bioelectron*, Vol. 26, p. 371, 2010.
161. H. Liu, J. Gao, M. Xue, M. Zhang, and T. Cao, *Langmuir*, Vol. 25, p. 12006, 2009.
162. F. Li, Y. Bao, J. Cai, Q. Zhang, D. Han, and L. Niu, *Langmuir*, Vol. 26, p. 12314, 2010.
163. L Q. Xu, L. Wang, B. Zhang, C.H. Lim, Y. Chen, K.G. Neoh, E.T. Kang, and G.D. Fu, *Polymer*, Vol. 52, p. 2376, 2011.
164. H. Wu, W. Zhao, H. Hu, and G. Chen, *J Mater Chem*, Vol. 21, p. 8626, 2011.
165. J. Geng, and H.T. Jung, *J Phys Chem C*, Vol. 114, p. 8227, 2010.
166. N. Karousis, S.P. Economopoulos, E. Sarantopoulou, and N. Tagmatarchis, *Carbon*, Vol. 48, p. 854, 2010.
167. J. Guo, L. Ren, R. Wang, C. Zhang, Y. Yang, and T. Liu, *Compos Part B: Eng*, Vol. 42, p. 2130, 2011.
168. Z. Liu, L. Liu, H. Li, Q. Dong, S. Yao, A.B. Kidd IV, X. Zhang, J. Li, and W. Tian, *Solar Energ Mater Solar Cell*, Vol. 97, p. 28, 2012.
169. V.K. Kodali, J. Scrimgeour, S. Kim, J.H. Hankinson, K.M. Carrol, W.A. de Heer, C. Berger, and J.E. Curtis, *Langmuir*, Vol. 27, p. 863, 2011.
170. Q. Zeng, J. Cheng, L. Tang, X. Liu, Y. Liu, J. Li, and J. Jiang, *Adv Funct Mater*, Vol. 20, p. 3366, 2010.
171. H. Chang, G. Wang, A. Yang, X. Tao, X. Liu, Y. Shen, and Z. Zheng, *Adv Funct Mater*, Vol. 20, p. 2893, 2010.
172. E. Uddin, T. Kuila, G.C. Nayak, N.H. Kim, B.C. Ku, and J.H. Lee, *J Alloy Compound*, Vol. 562, p. 134, 2013.
173. T.Y. Kim, H. Lee, J.E. Kim, and K.S. Suh, *ACS Nano*, Vol. 4, p. 1612, 2010.
174. Y.K. Yang, C.E. He, R.G. Peng, A. Baji, X.S. Du, Y.L. Huang, X.L. Xie, and Y.W. Mai, *J Mater Chem*, Vol. 22, p. 5666, 2012.
175. S. Bose, T. Kuila, A.K. Mishra, N.H. Kim, and J.H. Lee, *Nanotechnology*, Vol. 22, p. 405603, 2011.
176. S.P. Economopoulos, G. Rotas, Y. Miyata, H. Shinohara, and N. Tagmatarchis, *ACS Nano*, Vol. 4, p. 7499, 2010.
177. M. Lotya, Y. Hernandez, P.J. King, R.J. Smith, V. Nicolosi, L.S. Karlsson, F.M. Blighe, S. De, Z. Wang, I.T. McGovern, G.S. Duesberg, and J.N. Coleman, *J Am Chem Soc*, Vol. 131, p. 3611, 2009.
178. Y. Sim, J. Park, Y.J. Kim, and M.J. Seong, *J Korean Phys Soc*, Vol. 58, p. 938, 2011.
179. R.J. Smith, M. Lotya, and J.N. Coleman, *New J Phys*, Vol. 12, p. 125008, 2010.

180. Y.T. Liu, X.M. Xie, and X.Y. Ye, *Carbon*, Vol. 49, p. 3529, 2011.
181. A.S. Wajid, S. Das, F. Irin, H.S. Tanvir Ahmed, J.L. Shelburne, D. Parviz, R.J. Fullerton, A.F. Jankowski, R.C. Hedden, and M.J. Green, *Carbon*, Vol. 50, p. 526, 2012.
182. H. Xu, and K.S. Suslick, *J Am Chem Soc*, Vol. 133, p. 9148, 2011.
183. N. Liu, F. Luo, H. Wu, Y. Liu, C. Zhang, and J. Chen, *Adv Funct Mater*, Vol. 18, p. 1518, 2008.
184. G. Wang, B. Wang, J. Park, Y. Wang, B. Sun, and J. Yao, *Carbon*, Vol. 47, p. 3242, 2009.
185. S.H. Lee, S.D. Seo, Y.H. Jin, H.W. Shim, and D.W. Kim, *Electrochem Commun*, Vol. 12, p. 1419, 2010.
186. J.H. Jang, D. Rangappa, Y.U. Kwon, and I. Honma, *J Mater Chem*, Vol. 21, p. 3462, 2011.
187. G. Eda, and M. Chhowalla, *Nano Lett*, Vol. 9, p. 814, 2009.
188. X. Wang, W. Xing, P. Zhang, L. Song, H. Yang, and Y. Hu, *Compos Sci Technol*, Vol. 72, p. 737, 2012.
189. M.A. Rafiee, J. Rafiee, Z. Wang, H. Song, Z.Z. Yu, and N. Koratkar, *ACS Nano*, Vol. 3, p. 3884, 2009.
190. M.A. Rafiee, J. Rafiee, I. Srivastava, Z. Wang, H. Song, Z.Z. Yu, and N. Koratkar, *Small*, Vol. 6, p. 179, 2010.
191. J. Liang, Y. Wang, Y. Huang, Y. Ma, Z. Liu, J. Cai, C. Zhang, H. Gao, and Y. Chen, *Carbon*, Vol. 47, p. 922, 2009.
192. S. Ganguli, A.K. Roy, and D.P. Anderson, *Carbon*, Vol. 46, p. 806, 2008.
193. S. Wang, M. Tambraparni, J. Qiu, J. Tipton, and D. Dean, *Macromolecules*, Vol. 42, p. 5251, 2009.
194. H. Bai, C. Li, and G. Shi, *Adv Mater*, Vol. 23, p. 1089, 2011.
195. J.R. Potts, D.R. Dreyer, C.W. Bielawski, and R.S. Ruoff, *Polymer*, Vol. 52, p. 5, 2011.
196. R.K. Layek, S. Samanta, and A.K. Nandi, *Carbon*, Vol. 50, p. 815, 2012.
197. T. Ramanathan, A.A. Abdala, S. Stankovich, D.A. Dikin, M. Herrera-Alonso, R.D. Piner, D.H. Adamson, H.C. Schniepp, X. Chen, R.S. Ruoff, S.T. Nguyen, I.A. Ivaska, R.K. Prud'homme, and L.C. Brinson, *Nat Nanotechnol*, Vol. 3, p. 327, 2008.
198. K. Zhang, L.L. Zhang, X.S. Zhao, and J. Wu, *Macromolecules*, Vol. 22, p. 1392, 2010.
199. H. Bao, Y. Pan, Y. Ping, N.G. Sahoo, T. Wu, L. Li, J. Li, and L.H. Gan, *Small*, Vol. 7, p. 1569, 2011.
200. S. Biswas, H. Fukushima, and L.T. Drzal, *Compos Part A*, Vol. 42, p. 371, 2011.
201. Y.R. Lee, A.V. Raghu, H.M. Jeong, and B.K. Kim, *Macromol Chem Phys*, Vol. 210, p. 1247, 2009.
202. N.D. Luong, U. Hippi, J.T. Korhonen, A.J. Soninen, J. Ruokolainen, L.S. Johansson, J.D. Nam, L.H. Sinh, and J. Seppala, *Polymer*, Vol. 52, p. 5237, 2011.

203. D. Zhang, X. Zhang, Y. Chen, P. Yu, C. Wang, and Y. Ma, *J Power Sources*, Vol. 196, p. 5990, 2011.
204. Y. Shao, J. Wang, H. Wu, J. Liu, I.A. Aksay, and Y. Lin, *Electroanalysis*, Vol. 22, p. 1027, 2010.
205. H. Ma, D. Wu, Z. Cui, Y. Li, Y. Zhang, B. Du, and Q. Wei, *Anal Lett*, Vol. 46, p. 1, 2013.
206. M. Pumera, *Mater Today*, Vol. 14, p. 308, 2011.
207. Q. He, S. Wu, Z. Yin, and H. Zhang, *Chem Sci*, Vol. 3, p. 1764, 2012.
208. W. Zhang, M. Zhou, H. Zhu, Y. Tian, K. Wang, J. Wei, F. Ji, X. Li, Z. Li, P. Zhang, and D. Wu, *J Appl Phys*, Vol. 44, p. 205303, 2011.
209. J. Lin, L. Wang, and G. Chen, *Tribol Lett*, Vol. 41, p. 209, 2011.
210. T.S. Baby, and S. Ramaprabhu, *J Appl Phys*, Vol. 108, p. 124308, 2010.
211. T.S. Baby, and S. Ramaprabhu, *Nanoscale Res Lett*, Vol. 6, p. 289, 2011.
212. A. Ghozatloo, M.S. Nisar, and A.M. Rashidi, *Int Commun Heat Mass Transfer*, Vol. 42, p. 89, 2013.
213. W. Ai, W. Zhou, Z. Du, Y. Du, H. Zhang, X. Jia, L. Xie, M. Yi, T. Yu, and W. Huang, *J Mater Chem*, Vol. 22, p. 23439, 2012.
214. Jaidev, and S. Ramaprabhu, *J Mater Chem*, Vol. 22, p. 18775, 2012.
215. Y. Wang, Z. Shi, Y. Huang, Y. Ma, C. Wang, M. Chen, and Y. Chen, *J Phys Chem*, Vol. 113, p. 13103, 2009.
216. Z. Lin, Y. Liu, Y. Yao, O.J. Hildreth, Z. Li, K. Moon, and C.P. Wong, *J Phys Chem*, Vol. 115, p. 7120, 2011.
217. V.H. Luan, H.N. Tien, L.T. Hoa, N.T.M. Hien, E.S. Oh, J. Chung, E.J. Kim, W.M. Choi, B.S. Kong, and S.H. Hur, *J Mater Chem*, Vol. 1, p. 208, 2013.

3

Graphene and Carbon Nanotube-based Electrochemical Biosensors for Environmental Monitoring

G. Alarcon-Angeles¹, G.A. Álvarez-Romero² and A. Merkoçi³,*

¹Metropolitan Autonomous University – Xochimilco, Department of Biological Systems, Mexico City, D.F., Mexico
²Autonomous University of Hidalgo State, Academic Area of Chemistry, Hidalgo, Mexico
³ICREA and,Nanobioelectronics and Biosensors Group, Catalan Institute of Nanoscience and Nanotechnology (ICN2), Barcelona, Spain

Abstract

Interest in using electrochemical sensors for the in-field monitoring of environmental parameters has risen in recent years, due to their advantages over conventional analytical techniques. The applications of nanomaterials for electrochemical sensing are leading to the development of novel tools for environmental pollution monitoring. Carbon nanotubes (CNT) and graphene (GR) have been explored as new materials for the modification of electrodes used as biosensors due to their unique chemical, physical and mechanical properties. Problems related to the selectivity and/or specificity of biosensors have been solved using electrodes modified with various receptors. The results obtained so far have demonstrated that electrochemical sensors based on nanostructured materials are promising for monitoring environmental contaminants such as heavy metals, pesticides, phenol compounds, among others. This chapter describes the advantages of using nanomaterials like CNT and GR in the construction of electrochemical biosensors for the monitoring of chemical compounds of environmental interest.

Keywords: Graphene, carbon nanotubes, electrochemical sensors, electrochemical biosensors

Corresponding author: arben.merkoci@icn.cat

Ashutosh Tiwari and S.K. Shukla (eds.) Advanced Carbon Materials and Technology, (87–128)
2014 © Scrivener Publishing LLC

3.1   Introduction

Nowadays, the success of nanotechnology (the second industrial revolution) [1] is associated with the development and use of techniques for studying physical phenomena and structures between 1 to 100 nm in size, and the incorporation of these structures for various applications.

Nanostructured materials have properties which make them useful to enhance conductivity, strength, durability and reactivity. The physicochemical characteristics of nanoscale materials may differ substantially from the bulk material [2]. Nanomaterials behavior is different due to the high volume/surface ratio, so that a large proportion of atoms are on the surface allowing them to react easily with the adjacent atoms [3].

Nanomaterials such as carbon nanotubes (CNT) and graphene (GR) offer great opportunities in different areas of nanotechnology. Due to their unique properties, their application in analytical chemistry has been strongly considered; for example, their excellent electrochemical properties, which favor the development of new electrochemical sensors for selective and/or specific real time analysis of various analytes, which is relevant in areas related to environmental control.

3.1.1   Carbon Nanotubes (CNTs)

Since their discovery by Ijima in 1991, CNTs have become the center of attention of scientific research due to their extraordinary physical, chemical, electrical and mechanical properties, which are associated with their chemical structure [4–11].

Nanotubes belong to sp^2 carbon-bond structures and the defects in their topology make possible the formation of closed shells from graphene sheets. The geometry of CNT (Figure 3.1) shows a long cylinder made of carbon hexagons resembling a honeycomb, [4–7]. The length of the CNT is in the range of micrometers, with diameters below 100 nm. The diameter of the tube depends on the size of the semi-fullerene; for example, a nanotube based in C_{240} has a diameter of about 1.2 nm [11].

Graphene sheets can be rolled up in different ways [8] so the arrangement of the hexagonal rings is also different, and therefore, the properties of the material will change. A particular characteristic of CNTs is that they are extremely anisotropic [9], i.e., their

(a)   (b)

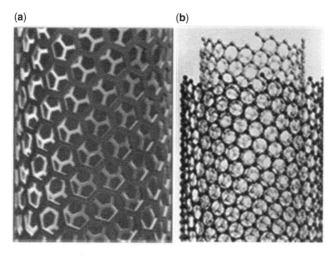

Figure 3.1 (a) Scheme of a single cylindrical layer of a carbon nanotube. b) Multiwall carbon nanotubes. The arrangement of hexagons around the circumference is helical with respect to the nanotube axis. Adapted with permission from [17].

properties such as elasticity, temperature, conductivity, speed of propagation of light, etc., vary according to the direction in which they are analyzed (axial or radial); some properties of this nanomaterial are described below.

- *Mechanical properties*: It has been reported that along the tube axis, CNTs have a Young's module of about 270–950 GPa and a tensile strength of 11–63 GPa. This feature is associated with the very small diameter of the CNTs, making them one of the strongest materials available [10], but flexible because of their great length [11].
- *Chemical properties*: CNTs are chemically stable, their large surface area and rehybridization of σ-π bonds facilitate molecular absorption, doping and high charge transfer, which explains their excellent electronic properties.
- *Thermal*: CNTs have high conductivity and thermal stability.
- *Electrical*: CNTs can behave as a semiconductor, metal and even as superconductors depending on their chirality, diameter, and torsion [12].

Compared with other materials, CNTs are more resistant than steel, their conductivity is better than that of diamonds and their electrical conductivity is greater than that of copper [13].

Another attractive property of CNTs is their tubular morphology and their relatively high strength and flexibility, which has generated great interest for use in structural applications [11]. This material is used as reinforcement in nanocomposites because it improves the mechanical properties, electrical conductivity, electrical charge, magnetic properties, crystallization properties and rheological properties (relationship between the stress and strain of the material) [14, 15].

To date, there are two kinds of CNTs that have a structure with high perfection: single-wall carbon nanotubes (SWCNTs) with diameters up to 0.4 nm, and multi-wall nanotubes (MWNTs) with diameters between 2–100 nm. The former consist of a single sheet of graphite perfectly wrapped in a cylindrical tube (Figure 3.1) [6–16], while the latter comprise a series of nanotubes concentrically nested like the rings of a tree trunk.

Generally, SWCNTs and MWCNTs are synthesized using techniques such as electric arc discharge, carbon-laser ablation, chemical-vapor deposition, template-assistant growth and solution growth [13, 16, 18,].

In general, CNT properties vary depending on the direction (angle) and the curvature [19], called chiral vector. This vector determines the nanotube geometry which can be either "zigzag," "armchair," or "chiral" (Figure 3.2).

The conductivity of the nanotube has been studied considering the CNT shape, so armchair nanotubes are metallic. If they present a zigzag and chiral shape, they can be both metallic and semiconducting [11]. Depending on the material's synthesis, if SWCNT and MWCT are perfect, they can present unique and similar properties, such as ballistic- electrons transport. These properties are why they are considered superconductors, although this depends largely on the resistance presented by the media that contain them [18].

Various applications of CNTs depend on their structural, mechanical, electronic and electrochemical properties; biomedical applications such as DNA sequencing and drug delivery; energy applications such as batteries and fuel cells, nanoreactors, etc., or; applications in sensors and biosensors for monitoring of analytes such as environmental pollutants, the last one being among the most successful applications over the last decades.

The application of CNTs in the construction of electronic devices has many advantages due to their small size and large surface area,

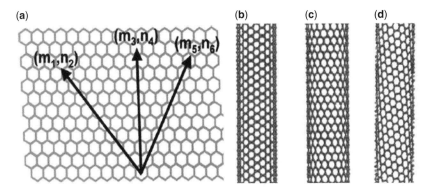

Figure 3.2 Schematic view of the honeycomb structure of a graphene sheet (A). SWCNTs can be formed by folding the sheet along the shown lattice vectors leading to armchair (B), zigzag (C), and chiral (D) tubes, respectively. Adapted with permission from [19].

excellent electron transfer and easy immobilization of the protein, while maintaining its biological activity. This nanomaterial plays an important role in the design of new electronic devices used to construct sensors, biosensors, immunosensors and genosensors, due to its properties like small size, large surface area, excellent electron transfer and easy protein immobilization without losing its biological activity.

3.1.2 Graphene (GR)

Graphene is the name given to the carbon structure with sp^2 hybridization in two dimensions, usually called carbon sheets (Figure 3.3) and is the main component of other important allotropes. These carbon sheets can be stacked to form 3D graphite or rolled to form nanotubes or fullerenes. The synthesis of a single layer of GR was reported by Novoselov in 2004 [20]; this material exhibits high crystallographic quality and is stable at room conditions. Its excellent thermal, electrical and mechanical properties are due to the π conjugation [21]. GR has an analogous structure to benzene and polycyclic aromatic hydrocarbons, so the chemistry of these compounds is similar. A determining factor in the chemistry of this nanomaterial is the formation and/or breaking of conjugated C-C bonds in the basal plane and the C-H bonds on the edges.

The different synthesis methods for GR that have been reported are mechanical exfoliation of graphite [20–25], recently used chemical vapor deposition [26, 27], liquid phase exfoliation from

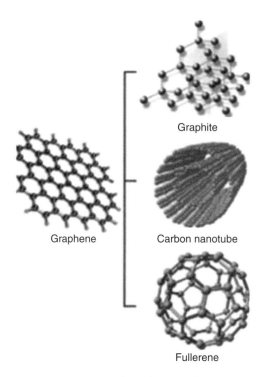

Figure 3.3 Schematic representation of graphene and its structures: graphite, carbon nanotube and fullerene. Adapted with permission from [22].

graphite [28], reduction of graphene oxide, surface segregation [29], and molecular beam epitaxy (MBE) [30].

However, the controlled mass production of GR is one of the possible drawbacks for their application, since its number of layers and its defects significantly influence the properties of the material.

For the characterization and study of some properties of GR, some of the techniques used are: scanning probe microscopy (used to study the thickness of the crystal), atomic force microscopy AFM (used to measure the mechanical properties of the material), and scanning tunneling microscopy (employed for morphological studies). Raman spectroscopy recently has been used to determine the thickness of the sheets obtained by mechanical exfoliation [30].

The main studied properties of graphene are:

- *Mechanical properties*: Surface area (2630 m^2g^{-1}); intrinsic mobility (200000 $cm^2\ v^{-1}\ s^{-1}$); high Young's modulus (1.0 TPa) [31].

- Thermal conductivity: $5000 Wm^{-1}K^{-1}$.
- Optical properties: GR shows a transmittance of 97.7%.
- Electrical properties: GR has high conductivity ($\sim 10^4$ Ω^{-1} cm^{-1}) [32]. GR can support a current density about six times higher than that of copper, the electronic characteristics of this material are mainly due to its topology; because of the size of the films at atomic levels, the electron transport can be ballistic at submicrometer distances [20, 33, 34].

Due to the above, in recent years GR has been the center of attention of many scientists because of its electrical properties. Compared with CNT, GR has some advantages such as its high thermal and electrical conductivity (due to its small thickness) and the large surface area.

This material can be used in areas such as optical [35] or electronic high-speed devices [36], storage and energy generation [36–38] and also hybrid materials [39, 40]. Because of its high electrical conductivity combined with its strength, flexibility and transparency, GR is an ideal material for optoelectronic devices, it has been demonstrated that large-scale films can be synthesized and used for touch-screen displays [41]. This material has also been used for the manufacture of high frequency transistors [36, 42]; some promising research has focused on the development of photodetectors [43], electrochemical sensors [44] and DNA sequencing [45–48].

The aim of this chapter is to summarize the research that is focused on the development of electrochemical biosensors based on either CNT or graphene for the detection and quantification of contaminants such as pesticides, phenols, heavy metals, among others, and highlighting the advantages of these biosensors.

3.1.3   Electrochemical Sensors

Since the focus of this chapter is the use of nanomaterials for the detection of different pollutants through electrochemical sensors, a brief description of the operation of biosensors is presented in this section.

In general, sensors are devices capable of detecting chemical substances, from gas molecules to biological cell molecules. These devices are constructed with an active sensing element (recognition) and a signal transducer. The output signal can be optical,

Electrochemical reaction

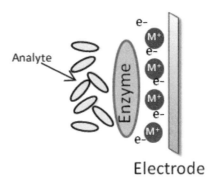

Figure 3.4 Schematic configuration of an electrochemical biosensor.

thermal, magnetic or electric (Figure 3.4). The recognition element is responsible for the selective detection of the interest analyte, and the transducer converts a chemical signal to an appropriate signal that can be used to easily determine the analyte. Biosensors have biological recognition agents such as proteins (cells, enzymes or antibodies), oligonucleotides, microorganisms and even biological tissue. Usually, biosensors are used to monitor biological processes or recognize biological molecules.

In the particular case of electrochemical biosensors, the chemical process is related to redox processes of the analyte, and the transducer generates secondary electrical signals, like current or potential.

3.1.4   Sensors and Biosensors Based on CNT and GR

Due to their electrical and electrochemical properties, CNT and GR are ideal to work as electrodes and transducers in biosensors, thus CNT and GR have been implemented considering different architectures [4, 49], for example:

- *CNT paste electrodes (CNTPE)* constructed with a mixture of mineral oil and CNT [50]. In the case of biosensors the biological material (typically enzymes, antibodies, etc.) is added to the mixture [51, 52]. These composite electrodes combine the ability of CNT to promote electron-transfer reactions combined with the attractive advantages of paste electrode materials.

- *CNT-modified electrodes* imply the modification of conventional carbon or metal electrodes (Au, Pt or Cu), being the most effective modification method for CNT growth over a metal substrate. They allow arrays of CNT that substantially improve electrochemical properties. Another method that favors the catalytic activity of the CNT is the use of metal nanoparticles during the electrode's modification.

The construction of biosensors based on CNT and GR using enzymes as recognition agents has been performed in different ways, since immobilizing the enzyme is a crucial step for the appropriate response of biosensors, presented below are the most important immobilization techniques:

- *Adsorption technique*: Nanotubes are dispersed over conventional electrodes, usually glassy carbon electrodes (GCE), and an enzyme solution is placed in the surface and then dried. However, in spite of being a simple and fast technique, problems have been reported such as poor adsorption of enzyme, leaching and homogeneity of the enzyme on the electrode's surface and poor reproducibility. Nanoparticles have represented a solution for some of these disadvantages since they favor the adsorption of enzymes. The adsorption technique has been employed for the detection of different analytes, including phenol compounds [53] and organophosphates [54].
- *Covalent binding*: This is a technique that allows direct anchoring of the enzyme on the electrode via covalent bonds, which allows direct transfer of electrons to the enzyme active site [55].
- *Electropolymerization*: This is an attractive and well-controlled method for immobilization of enzymes over the electrode's surface. This technique involves mixing the enzyme with the monomer followed by electropolymerization (Figure 3.5). The incorporation of the enzyme in the polymer matrix is often achieved through electrostatic interactions. The advantages of this technique lie in the control when producing thin layers. Usually, polypyrrole (PPy) is used for these purposes [56–58].

(a)   (b)

(c)   (d)

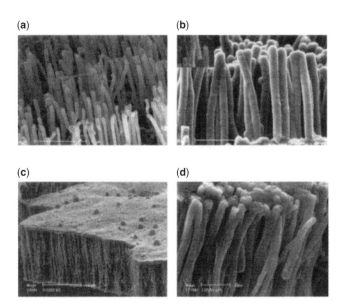

Figure 3.5 SEM images of (a) pure CNT array before PPy deposition; (b) aligned PPy-CNT coaxial nanowires, inset shows clear image of single tube coated with PPy; (c) PPy only deposited on the top of CNT surface due to the high density of the tube array; (d) polymer formed on both outside of walls and the top of the surface of the CNT array. From reference [59] with permission.

- *Encapsulation:* This technique allows the immobilization of the enzyme through hydrogel or sol-gel materials, and can be performed from a CNT-modified electrode followed by deposition of the hydrogel containing the enzyme. Alternatively, CNT can be incubated with the enzymes and then incorporated into the hydrogel or sol-gel matrix and finally deposited on the electrode's surface [60].

Protocols for the detection of proteins, antibodies and DNA which involve processes with CNT have been recently reported [61, 62]. In the amplification process the CNTs play a double role as transducers and recognition agent. In environmental areas, such sensors are essential when trying to determine the presence of microorganisms.

Due to the properties of GR such as chemical stability, low cost, wide potential window, electrochemical inertness and their electrocatalytic activity for several redox reactions, research has been focused on the design of sensors using this material.

A graphene-based sensor in combination with surfactants has recently been reported that is capable of changing the electrical properties of the electrode's interface, and so their electrochemical processes. This is caused by adsorption on the surface or the aggregation into supramolecular structures. On the other hand, polymers like chitosan (CS) have been used in combination with GR due to their high permeability and adhesive strength.

The integration of GR with metal nanoparticles is usually done in order to improve the properties of sensors. Modification of the electrode's surface with GR and biorecognition elements is usually done through composites. It has been reported that the use of GR in sensor construction is inexpensive and can be produced on a large scale in comparison with other carbon nanomaterials.

Due to their size and excellent electrochemical properties, CNT and GR continue to attract interest as components in biosensors. It has been established that these biosensors have electrochemical properties that are equal or even superior compared with other electrodes, therefore, the use of them for the determination of contaminants like heavy metals, phenolic compounds, pesticides and microorganisms, has been of great relevance. In subsequent sections, advances in the use of these biosensors for detecting these pollutants are presented.

3.2   Applications of Electrochemical Biosensors

3.2.1   Heavy Metals

Worldwide, environmental pollution caused by heavy metals is associated with indiscriminate storage of heavy metals that can be found in both the water and the air, and even in food. This is a concern since metals are not biodegradable, so many of them have detrimental effects on human health such as loss of appetite, anemia, kidney failure and nerve damage, among others.

An example of this is lead (Pb^{2+}), which can cause diseases such as nausea, convulsions, coma, kidney failure, negative effects on metabolism and cancer [63–65]. Copper (Cu^{2+}) can cause hemochromatosis, gastrointestinal catarrh, cramps in the calves, skin dermatitis brass chills [66] and Wilson's disease [67]. It is reported that heavy metal species such as hexavalent chromium Cr (VI) are commonly identified as hazardous contaminants because of their carcinogenic characteristics [68], and the U.S. Environmental Protection

Agency (EPA) has set a maximum concentration of 100 mg/L for this metal in drinking water. Also, mercury (Hg^{2+}) is neurotoxic, and is considered one of the most toxic heavy metal ions [69–72]. Even traces of mercury can induce serious diseases such as kidney and respiratory failure and damage to the gastrointestinal and nervous systems [73]. Great concern has arisen since it is estimated that, annually, about 10,000 tons of this metal generated by human activity are released into the environment.

Under this scenario, monitoring of heavy metals is a major issue in modern society. This obviously urges the development of fast quantitative analytical methods friendly to the environment. The results obtained with these methods are of considerable importance in order to take action aimed to ensure and protect human health.

The most frequent analytical methods for heavy metal analysis are: liquid chromatography (HPLC) [74, 75], gas chromatography with fluorescence detection, atomic spectrophotometry [76, 77], atomic absorption spectrophotometry [78, 79] and inductively coupled plasma-optical emission spectroscopy (ICP-OES) [80, 81]. Most of these analytical methods consume a lot of time for analysis and are generally complex instruments with a high cost. Therefore, it is necessary to develop a detection system with high sensitivity, low cost, and portability for the quantification of heavy metal.

Nowadays, electrochemical methods are an attractive alternative for heavy metal analysis due to their high sensitivity, short response time, simplicity, low cost and portability, and thus the possibility of *in situ* analysis in real time. Various electrochemical techniques can be applied to the determination of trace metals; the most common are potentiometry and voltammetry, square-wave anodic-stripping voltammetry being the most used technique. Results obtained with this last technique demonstrate that it is efficient and reliable; the effectiveness of the technique depends greatly on the preconcentration of the metal on the substrate (electrode's surface) which allows detection of very low concentrations of metal ions.

Traditionally, mercury-drop electrodes and mercury films have been widely used for such purposes; however, due to its high toxicity, new materials have been proposed to allow the determination of heavy metals. To date, gold, platinum and carbon electrodes are chemically modified to achieve this purpose, where nanomaterials have taken the challenge to solve problems related to heavy metals determination. Modification of electrodes with these materials has demonstrated great advantages in the construction of electrochemical sensors used for monitoring heavy metals [82–86]. The success of the sensors based on nanomaterials is related to their physicochemical properties and their adsorption capacity.

In particular, nanomaterials based on carbon structures play an important role as effective adsorbents of heavy metals, so they have been widely used to remove contaminants in industrial wastewater, surface water and drinking water [87–89].

Within these nanomaterials, CNT and GR stand out because of their great ability to adsorb organic and inorganic pollutants [90, 91], due to their high surface area and low cost, which makes them good candidates as contaminant adsorbents. However, GR is hydrophobic and tends to form agglomerates in water due to van der Waals interactions between neighboring sheets, reducing significantly the surface area and therefore disfavoring the adsorption. Fortunately studies to avoid restacking (agglomerate) of GR nanosheets have shown that this problem can be solved with the intercalation of inorganic nanoparticles (NPs) between the GR layers (Figure 3.6). So, the design of materials based on GR to enhance the adsorption of metal ions is in constant development.

Considering their characteristics and the ability of CNT and GR to adsorb contaminants such as heavy metals, the tendency to use these kinds of nanomaterials for electrode modification lies in their capability to be used to design nanocomposites and produce ultrasensitive sensors, which are favored with the properties of

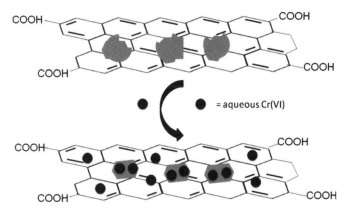

Figure 3.6 Schematic representation of synergetic adsorption of Cr (VI) on Graphene/MgAl-layered double hydroxides (G–MgAl-LDH) nanocomposite (GR-MgAl-LDH). Adapted with permission from [68].

nanomaterials such as surface area, mass transport kinetics and fast electron transfer, compared with conventional materials (bulk) [92].

3.2.1.1 CNT-Based Biosensors

In recent years, the use of CNT for the detection of contaminants has increased considerably. One of the advantages of CNT compared to other materials (such as graphite) is the surface defects that make them more reactive. Some functional groups that promote this behavior are carboxyl and hydroxyl groups; the functionalization with these groups benefits solid-phase extraction of heavy metals [93], such that CNT acts as an adsorbent and thereby as a preconcentration material for metals.

Considering the above, CNT can be used for two environmental purposes: remediation and heavy metal detection (Table 3.1), as described by Zhang *et al.* [94]. This nanomaterial serves as a capture agent and a signal amplifier in the detection of contaminants [95, 96]; it is noteworthy that CNTs have high electrocatalytic activity which promotes the detection of heavy metals in electrochemical sensors, and this reduces contamination (fouling) on the sensor's surface [97].

The key to the development of a CNT-based sensor lies in the CNT functionalization and the immobilization on the electrode's surface. To date, CNTs are functionalized incorporating amino groups and oxygen, which can coordinate with transition metals through electrostatic interactions. Considering this, MWCNTs have been functionalized with ethylenediamine [98] to quantify $Cd\,(II)$ in water. Also, adsorption in aqueous media on SWCNTs of $Pb\,(II)$, $Cd\,(II)$, and $Cu\,(II)$ improved significantly with the functionalization of SWCNT-COOH [99] (Figure 3.7). Metal absorption depends on the solution's pH due to changes in the surface charge of CNT, and the ionization degree of the metal species in aqueous solution.

Detecting heavy metals involves CNTs as transducers in conjunction with metallic or polymeric chelating agents, surfactants and/or ligands; chemical recognition can be achieved by chemical affinity or by ion exchange.

The use of polymers has some advantages due to their specific functional groups (thiols and sulfonyl groups). One of the most commonly used polymers is Nafion [100], which disperses CNT and has properties as a cation exchange material; it also has a high stability.

Table 3.1 Summary of the data for different configurations of CNT-based sensors for the determination of heavy metals in water samples.

Metal	Electrochemical platform	L.O.D	Detection technique	Sample matrix	Ref
In^{3+}	MWCNT/ Nafion/GCE	1×10^{-11} M	DPASV	—	[100]
Cd Pb	MWCNT/ Nafion	40 ng/L 25 ng/L	AdSV	tap water	[101]
Eu^{3+}	MWCNT/ Nafion	10 nM.	DPSV	vehicle exhaust particulates	[102]
Pb(II) Cd(II)	CNT polystyrene	0.04 ppb 0.02 ppb	SWV	water sample from lake	[103]
Hg^{2+}	PDAN/CNT	—	SWV	—	[108]
Cd	CNTs bismuth films	0.04ng/L (40ppt)	SWV	—	[104]

DPASV: Differential Pulse Anodic Stripping Voltammetry; AdSV: Anodic Stripping Voltammetry;

SWV: Square-Wave Voltammetry

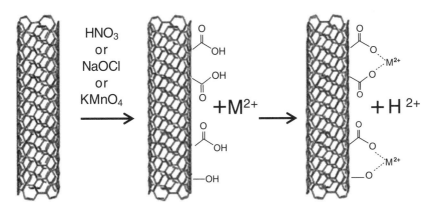

Figure 3.7 Adsorption mechanism of divalent metals ion onto SWCNT-COOH. Adapted with permission from [99].

CNTs in Nafion [100–102] have been used for the determination of heavy metals like lead, cadmium, europium, and indium. The detection is usually done with techniques such as anodic stripping voltammetry (ASV).

Besides Nafion, other polymers have been studied; a sensor based on polystyrene for the determination of heavy metals consists on a bismuth-modified polystyrene sulfonate CNT composite electrode made for simultaneous detection of lead and cadmium by anodic stripping voltammetry [103]. Wang *et al.* [104] reported a conscientious analysis of sensors based on bismuth films compared to other platforms for the study of heavy metals. However, it is important to note that with an array of nanoelectrodes based on bismuth films and CNT, very low detection limits can be achieved according to a report by Liu *et al.* [105].

Other sensors are based on CNT and metal particles, which have worked successfully in determining traces of As^{3+} with detection limits ranging up to ppt concentrations [106, 107].

The latest advances in sensor technologies for the quantification of heavy metals include CNT arrays, which are integrated through microfabrication and electropolymerization. These sensors can simultaneously detect more than one metal ion, a particular case is reported by Nguyen who describes the microfabrication of these sensors by electropolymerizing structured poly(1,8-diaminonaphthalene)/functionalized multi-walled carbon nanotubes (PDAN/CNT) thin film onto a silicon chip for square-wave voltammetry (SWV) multi-element, heavy metal ion detection [108].

An interesting characteristic in designing these sensors is the portability; this is achieved with the electrode's miniaturization, in particular with the generation of screen printed devices, which consist in a conventional three-electrode system but are all integrated in a single device. The combination of these with MWCNT has allowed the detection of metals such as Cd^{2+}, Pb^{2+} y Cu^{2+} [109].

3.2.1.2   GR-based Biosensors

Usually, the modification of the electrode's surface with GR can be achieved by covalent or non-covalent interactions; however, bonds of functional groups can destroy GR's conjugation and affect the material properties, so modification with graphene is preferred by non-covalent interactions. Even when the interactions are weak, the electrochemical processes are reversible and easy to achieve.

Recently, Kong *et al.* [110] published the modification of a gold electrode based on the functionalization with an aryl diazonium salt; the immobilization of the GR nanosheets is achieved by non-covalent stacking interaction. This electrode was used for square-wave voltammetry analysis of Cu^{2+} and Pb^{2+}, obtaining high sensitivity, good selectivity and a detection limit of 1.5 nM for Cu^{2+} and 0.4nm for Pb^{2+}.

The use of GR sheets has already been reported in the simple and economic modification of electrodes for the determination of these metals including Cd^{2+} [111], with outstanding detection limits (between 1×10^{-7} and 1×10^{-11} mol/L). These modified electrodes have demonstrated good reproducibility, sensitivity, linearity and selectivity versus other electrodes. On the same scope, the combination of polymers and nanoparticles with GR has been used for the study of Cd^{2+} and Pb^{2+}, where the detection limits are about 1×10^{-9} M. Use of GR as part of composite electrodes has been reported for the detection of $Zn^{2+} Cd^{2+}$ y Pb^{2+}, but the important drawback of this proposal is the mercury film used as an electrode [112].

It has also been demonstrated that commercial GR increases the nucleation of heavy metals but inhibits the stripping step of cadmium ions, especially when surfactants are present, which disfavors the detection of the metal [113].

From these studies, sensors based on GR films for the quantification of Hg^{2+} have been constructed, showing high selectivity and sensitivity. The platform of these sensors is based on an array of gold nanoparticles over the GR nanosheets. This composite combines the properties of these nanomaterials, favoring electro-transference processes and Hg^{2+}determination.

In general, the use of electrochemical sensors based on GR is an excellent platform for the detection of heavy metals, allowing the detection of very small concentrations (about 6 ppt) [114], even below the established concentrations by the World Health Organization (WHO). These sensors have allowed the detection of metals in rivers and drinking water, making it feasible to implement these devices for monitoring heavy metals in the environment.

3.2.2   Phenols

A large group of phenolic compounds are naturally present in many products such as food, plants, vegetables and fruits that contain antioxidant properties. These compounds may help to reduce the risk of cardiovascular diseases. Also, some phenols and dihy-droxybencene isomers like hydroquinone (HQ), catechol (CC) and resorcinol (RC) are present in nature. These are widely used in chemicals, cosmetics, pharmaceuticals, tanning processes and dyes [115], but unfortunately have high toxicity and low biodegradabil-ity, and therefore are considered major environmental pollutants [116, 117] by the U.S. Environmental Protection Agency (EPA) and the European Union (EU) [118].

Due to phenol's toxicity by contact, ingestion or inhalation it is recommended not to exceed an exposure concentration of 20 mg (average) per day for humans. Thus, it is important to develop highly efficient techniques for the detection and removal of these contaminants from wastewaters for environmental protection.

3.2.2.1 CNT-Based Biosensors

Phenolic compounds are, in general, easily oxidizable, so they are commonly detected with amperometry [119]. The disadvantage of using conventional electrodes for this purpose is the contamination due to the oxidation products of either dimmers or polymers. This problem has been solved with the use of CNT. A particular case is reported by Wang, using a MWCNT/Nafion onto a glassy carbon electrode (GCE). The sensor presented a stable response in pheno-lic solutions and maintained 85% of the initial activity after 30 min, compared with an unmodified electrode. A life time of about 6 min is reported. The contribution of this electrode based on MWCNT was the improvement of stability and a wide quantification range [120].

Another contribution of CNT for quantization of phenolic com-pounds is the development of biosensors, where enzyme tyrosinase (Tyr) is used. Tyrosinase is adsorbed on the surface of a GCE modi-fied with CNT. In the presence of oxygen, Tyr catalyzes the oxida-tion of phenolic compounds to quinones. In agreement with the reaction mechanism described in Equations 3.1–3.3 [121], phenol can be detected electrochemically. With the use of Tyr and CNT the sensitivity was improved depending on the electrode configuration.

$$\text{phenol} + O_2 \xrightarrow{\text{tyrosinase}} \text{catechol} \tag{3.1}$$

$$\text{catechol} + O_2 \xrightarrow{\text{tyrosinase}} \text{o-quione} + H_2O \tag{3.2}$$

$$o\text{-quione} + H^+ + 2e \xrightarrow{\text{tyrosinase}} \text{catechol (at electrode)} \tag{3.3}$$

Another sensor based on CNT for the detection of phenol is a paste electrode modified with polyphenol oxidase (PPOX) [122]. With this sensor it is possible to detect different phenolic compounds including phenol and catechol in real samples.

Table 3.2 shows some of the configurations based on CNT and a comparison with other matrices as reported by Alarcon *et al.* [123], where low detection limits can be observed when using CNT to detect phenol in water, with concentrations even below those established by the EPA in accordance with the authors.

One of the advantages of the electrode based on a MWCNT/ dimethylditetradecylammonium bromide (DTDAB)/Tyr film is the capability of detecting different compounds such as phenol, catechol, *p*-cresol, and *p*-chlorophenol [129].

Another promising electrode was proposed by Yang, where there are synthetized gold nanoparticle–graphene nanohybrids (Au–GR) and 3-amino-5-mercapto-1,2,4-triazole-functionalized MWCNT (MWCNT–SH). Due to the synergistic effects between MWCNT and Au–GR an excellent film was obtained. This electrode was used for the simultaneous determination of hydroquinone, catechol and resorcinol using differential pulse voltammetry (DPV). Simultaneous determination in real samples was made in lake water, tap water and river water using the standard addition method.

Table 3.2 Electrochemical strategies for phenol detection using CNT.

Phenol compound	Electrochemical platform/label	L.O.D	Sample matrix	Ref
Phenol	PPOx/MWCNT/ PE	—	water	[122]
Phenol	Tyr/MWCNTs/ SPE	0.140	sea water	[123]
Bisphenol	Tyr/SWCNTs/ CPE	0.020	—	[124]
Phenol	MNPs / CNTs /polyelectrolyte PDDA/ Tyr	0.005	—	[125]
Phenol	Tyr/SPE/PE	0.410	wastewater	[126]
Phenol	Tyr/ SWCNT/ Nafion/ GCE	0.020	—	[127]
Phenol	Tyr/MWCNT/GCE	—	water	[128]

3.2.2.2   GR-Based Biosensors

Recent research for improving the detection of phenolic compounds has arisen with the arrival of GR. There have been reports in literature where this nanomaterial is used for the development of new electrochemical sensors allowing the measurement of these pollutants in soil and water samples.

Therefore, a GR-based sensor (already discussed in Section 3.1.3) was reported in combination with surfactants, whose surfactants are capable of changing the electrical properties of the electrode's interface and so its electrochemical processes, due to its adsorption on the electrode's surface or aggregation into supramolecular structures. This sensor was used to monitor the phenol biodegradation by microorganisms; a glassy carbon electrode modified with GR in the presence of cetyltrimethylammonium bromide (CTAB) (GR-/GCE/) was used for this purpose with satisfactory results when tested in water samples contaminated with phenol [130].

Sensors used for the quantification of phenolic compounds include GR and polymers like chitosan (CS) [131] due to its high permeability and adhesive strength. Results obtained with a GCE modified with a GR-CS composite exhibit excellent electrocatalytic activity for the oxidation of catechol (CT), resorcinol (RS) and hydroquinone (HQ), so that the oxidation potential is significantly decreased and the oxidation currents increased, although the detection limit is lower compared to sensors based on CNT [132–137]. These pollutants have been studied simultaneously by DPV for tap water samples, river water, lake water and sanitary wastewater, using this electrode.

In order to improve the sensitivity and detection limits of electrochemical sensors, REDOX mediators have been used; the best known and used is ferrocene (Fc) [138, 139] due to its REDOX properties, so it is not surprising that this has already been used along with GR for the detection of phenolic compounds. For this purpose, an electrode modified with GR-(4-ferrocenylethyne) phenylamine-carbon-GR nanoparticles (FEPA-CNP-GR) using Nafion as immobilizer (Figure 3.8) is designed. The authors report that the composite construction is fast and easy, with a considerable improvement of specificity, sensitivity, reproducibility and electroconductivity. Moreover, the Nafion membrane is highly permeable to cations but impermeable to anions, which enhances the selectivity of the sensor. The developed method was successfully applied

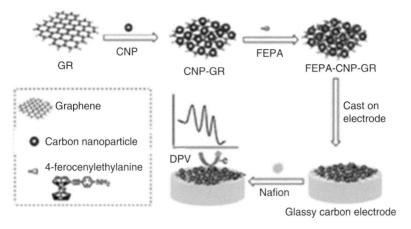

Figure 3.8 Electrochemical sensor bases on Nafion-CarboNanoparticle-Graphene modified GCE for simultaneous detection of dihydroxybenzene isomers. Adapted with permission from [139].

to the simultaneous determination of dihydroxybenzene isomers in synthetic and real samples, with satisfactory results [140–141]

An alternative to the sensors based on GR nanosheets is the use of conductive polymers such as poly(4-vinylpyridine) (P4VP), which improves the electrocatalytic activity and stability, in addition to the dispersion and orientation of the nanomaterial. This polymer also acts as mediator. The sensor's configuration based on GR/P4VP showed a higher sensitivity to hydroquinone and catechol, with detection limits ranging up to nanomolar concentrations. The proposed electrode has displayed a synergistic effect of P4VP and GR on the electrocatalytic oxidation of CC and HQ in a sodium sulphate buffer solution (pH 2.5). This sensor works even in the presence of common interfering species such as nitrophenol, aminophenols, bisphenol A (BPA) and chlorophenols. The GR-P4VP/GCE was successfully applied for simultaneous detection of spiked HQ and CC in tap water and lake water with encouraging results [142].

The combination of nanomaterials with GR is commonly known as *nanohybrids* (e.g., Au-Gr-MWCNT). These can be synthetized from MWCNT functionalization with 3-amino-5-mercapto-1,2,4-triazole, which coordinates covalently with five sites to form MWCNT–SH bridged Au–GR through the interaction between the SH groups of MWCNT–SH and Au nanoparticules of Au–GR. The new material MWCNT–SH@Au–GR (Figure 3.9) was obtained and used for the determination of hydroquinone, catechol and

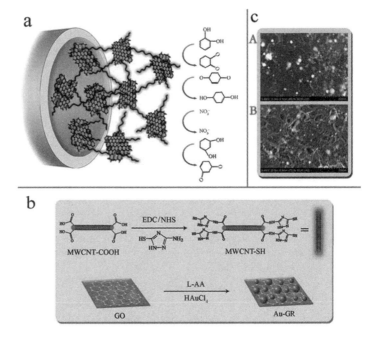

Figure 3.9 Schematic representation of (a) MWCNT–SH@Au–GR modified electrode: (b) The preparation process for MWCNT–SH and Au–GR; (c) SEM images of (A) Au–GR nanohybrid and (B) MWCNT–SH@Au–GR hybrid material. Adapted from [143].

resorcinol by DPV, and tested in lake water, tap water and river water [143].

Moreover, the combination of nanomaterials like gold NPs and GR is a favorable strategy in the construction of GR-gold NPs films. This strategy consists in the electrodeposition of GR and gold NPs alternately over the GCE surface. The thickness and amount of GR and gold NPs can both be controlled with this technique, obtaining high reproducibility, and low detection limits compared to other configurations based on GR; the sensor was used to determine pollutants in tap water, ground water and lake water [144].

Among the phenolic compounds of great interest is BPA which can damage (disrupt) the endocrine system and cause cancer. Usually this compound is used as raw material in plastics industries, specifically for coatings in cans used for food storage. BPA is an electroactive compound, but the electrochemical response is very poor, so the development of a GR-based sensor to increase the electrocatalytic activity has been of great importance. It has been reported that N-doped GR sheets (NGS), and NGS-CS substantially improve the response towards BPA. The sensor was used in the evaluation of BPA in river water [145].

Phenolic compounds have also been determined using Tyr-based biosensors. A new biosensor was constructed using a GR-nanosheets matrix, and the enzyme was immobilized with glutaraldehyde. Lower detection limits where obtained compared to those reported with other methods based on GR sensors. Phenol was measured in plastic drinking bottles and the amount of phenol found was 6.65 μM [146].

Table 3.3 presents some of the configurations based on GR and a comparison with other matrices reported, where low detection limits can be observed when using GR to detect phenol.

The challenge in the construction of new sensors for the detection of phenol is the balance between stability and the enzyme activity, and also the simultaneous determination of different compounds in real samples. However, the successful use of nanomaterials for the analysis of this kind of compound lies in the enzyme availability due to nanometer structures, which contribute to the catalytic effect. Also, the construction of smaller devices is favored, allowing real-time analysis.

3.2.3   Pesticides

The term *pesticide* generally refers to chemicals used to control and/or eliminate pests and diseases of plants or animals; these are generally used in agriculture. Large amounts of these organic chemicals are very persistent in the environment; constant and prolonged use of these leads to risks like their retention in soil which can cause contamination of surface and groundwater due to washing and leaching processes [147].

For this reason, interest has grown in the knowledge of health risks associated with exposure to pesticides that occur in the environment. According to the EPA, the risk to human health from pesticide exposure depends on both toxicity of the pesticide and the likelihood of people coming into contact with it.

Among the compounds considered unsafe are organophosphates (OPs), N-methyl carbamates, triazines, chloroacetanilides and pyrethrins/pyrethroids. To determine their risk levels, it is important to establish adequate methods for detection and quantification; usually HPLC is considered the best technique. However,

Table 3.3 Electrochemical strategies for phenol detection using GR-based biosensors.

Analyte	Electrochemical platform	L.O.D	Detection technique	Sample matrix	Ref
Phenol biodegradation	GR/CTAB/GCE	0.02 mg L^{-1}	DPV	Wastewater	[130]
HQ CC RC	GR-CS/GCE	0.75 μM	DPV	tap water, river water, lake water and sanitary wastewate	[131]
HQ CC RC	FEPA-CNP-GR	0.1 μM 0.2 μM 0.7 μM	DPV	tap water	[139]
HQ CC	GR-P4VP/GCE	8.1 nM 26 nM	DPV	tap water, lake water	[142]
HQ CC RC	Au-GR	4.17 μM 1.00 μM 7.80 μM	DPV	lake water tap water river water	[143]
HQ RC	graphene-gold nanocomposite film	5.2 nM 2.2 nM	DPV	tap water ground water lake water	[144]
BPA	CS/N-GS/GCE	5.0 nM	amperometry	river water	[145]
CC Phenol BPA	Gr–SP–Tyr/GCE	0.23 nM 0.35 nM 0.72 nM	amperometry	commercial plastic drinking	[146]

HQ: hydroquinone; CC: catechol; RC: resorcinol; GR: Graphene; CTAB cetyltrimethylammonium bromide; GCE: glassy carbon electrode; CS: Chitosan; P4VP: poly (4-vinylpyridine); SH :3-amino-5-mercapto-1,2,4-triazole; FEPA-CNP-GR: -4-ferrocenylethyne phenylamine-carbon-graphene nanoparticles; N: Nafion; Tyr: tyrosinase.

new methodologies based on nanotechnology allow analysis of different pesticides by means of electrochemical techniques such as amperometry, chronoamperometry, voltammetry cyclic voltammetric stripping (SV) and square-wave voltammetry (SWV).

Since the detection of OPs is essential to protect water resources and food, as well as for monitoring detoxification processes, the trends in pesticide analysis focus on the development of biosensors based on composite nanomaterials and the use of enzymes as OP recognition agents. The challenge in constructing biosensors is the immobilization of the enzyme on the electrode's surface; however, it has been proven that the use of nanomaterials promotes enzyme activity. The most commonly used enzymes for this purpose are choline oxidase (ChO), horseradich peroxidase (HRP), tyrosinase (Tyr), organophosphorus hydrolase (OPH) and acetylcholinesterase (AChE).

Among the most frequently used enzymes for detecting pesticides stands organophosphorus hydrolase (OPH), which converts OP compounds to p-nitrophenol; the oxidation is carried out at the electrode's surface, where the enzyme catalyzes the chemical reaction enabling the electrochemical detection.

3.2.3.1 CNT-Based Biosensors

Generally, methods based on electrochemical biosensors do not depend only on selecting a suitable enzyme to ensure pesticide detection, but also on the study of different parameters such as the optimum amount of nanomaterial and enzyme, and the dispersant agent which prevents the formation of agglomerates. Other parameters important in pesticide detection are the optimal pH related to the enzyme activity and the instrumental parameters which depend on the electrochemical technique used (applied potential E_{ap} duration of the E_{ap} etc.).

Wang [4], in addition to the above parameters, studied the influence of the synthesis of MWCNT in the response to p-nitrophenol. It was demonstrated that the sensor constructed with MWCNT obtained from vapor deposition, exhibited better response and stability with respect to those obtained by ARC discharge. The biosensor designed for pesticide detection was constructed from the modification of a GCE with MWCNT and the OPH enzyme. The modification consisted of two layers: the first with MWCN dispersed in Nafion and the next with enzyme dissolved in Nafion.

The response was amperometric and the biosensor was used to determine paraoxon and methyl parathion. The author notes that both the detection limit and sensitivity were better for the first pesticide due to the selectivity of the enzyme.

The OPH has been employed in different biosensor configurations [148]. However, a very particular one was proposed by Choi et al. [149] since they were able to apply the so called bucky gels (BGs), which are composed of CNT and ionic liquids (IL). Both materials offer a combination of characteristics such as good biocompatibility, high chemical stability, high dispersion of CNT, and easy preparation. It has been demonstrated that electron transference improves using these materials, and when modifying the electrode with CNT/IL and the enzyme, the biosensor's sensitivity improves. The authors attribute this improvement to the presence of OPs, and the low detection limit obtained (Table 3.4) is due to the generated microenvironment and conformational changes of the enzyme OPH, favoring biocatalysis of paraoxon.

Besides the adequate selection of the enzyme for determination of pesticides, the selection of the electrochemical technique is also important. Square-wave voltammetry is widely used because it is highly sensitive. Garcia et al. [150] proposed a method involving SWV and a biosensor based on a GCE modified with MWCNT and the immobilized enzyme dihexadecylhydrogenphosphate (DHP). The authors reported an increment in the electrocatalytic activity related to the detection of paraquat (organophosphate pesticide) and the detection limit registered with the biosensor was 1×10^{-8} mol L^{-1}. The method was used for the determination of paraquad in natural water. It is important to note that the results obtained agreed with those obtained using HPLC with a confidence level of 95%.

Development of biosensors for pesticide analysis has taken different investigative routes depending on the nanomaterial, immobilization technique, and the enzyme used as recognition agent; the last will determine the biosensor's sensitivity and affinity to a specific pesticide. Among the enzymes used for detecting pesticides stands acetylcholinesterase (AChE), which is one of the most used. Next, we cite the most relevant investigations related to the immobilization of the enzyme AChE.

In the work reported by Du et al. [151], β-cyclodextrin and MWCNT are used since the combination of these materials generate porous structures, which provide a favorable environment for maintaining the bioactivity of AChE. Cyclodextrins stand out because of their

Table 3.4 Electrochemical strategies for pesticides detection using CNT and GR.

Analyte	Electrochemical platform/label	L.O.D	Detection technique	Sample matrix	Ref
Paraoxon methyl parathion	OPH /CNT	0.15 μM 0.8 μM	amperometry	—	[4]
p-nitrophenol	OPH -ionic liquid (IL) /CNT	2–20μM	chronoamperometry	paraoxon	[148]
Paraquat	MWCNT/DHP	0.01 μM	square-wave voltammetry	natural waters	[149]
Dimethoate	AChE /MWCNT-CD	2 nM	amperometry	—	[150]
p-nitrophenol methyl parathion	MWCNTs-CS	7.5×10^{-13} M	amperometry	water soil	[151]
Dichlorvos	(MWCNTs/ALB)n	0.68±0.076ug/L	CV	—	[152]
Organophosphate pesticides	ChE -CNT Dice: ChE-CNT Debe decir: ChE/CNT	2 pM	amperometry	—	[153]

(Continued)

Table 3.4 (Cont.)

Analyte	Electrochemical platform/label	L.O.D	Detection technique	Sample matrix	Ref
Carbamate pesticides	MWCNT/PANI/AChE	1.4 μM 0.95 μM	chrono-amperometry	fruit vegetables	[156]
Organophosphate pesticides	GR/CNTs/CS	0.5 ng mL^{-1}	amperometry	—	[158]
Dichlorvos	Er-NGRO/AChE	2.0 ng mL^{-1}	amperometric	—	[171]
Paraoxon	RGRON films/OPH	1.37 ×10^{-7} M	amperometric	—	[173]
Methyl parathion chlorpyrifos Carbofuran	NiO NPs-GR-N/AChE	5×10^{-14} M 5×10^{-13} M 5×10^{-13} M	amperometric	cabbage apple, tap water and lake water	[175]

OPH: organophosphates; DHP: dihexadecylhydrogenphosphate; CD: Cyclodextrin; SPE: Screen printed electrode ALB:phosphatidylcholine liposomes; PANI: polyaniline; Er: electrochemical reduction RGRON: reduction graphene-oxidized nafion; NiO NPs: nickel nanoparticles-graphene.

very particular structure, with a cavity in which various molecules can be included depending on their size and polarity; a feature of this substance is that inside the cavity there is a polar environment while outside it is apolar. The use of cyclodextrins has proved to be an easy and nontoxic way for CNT dispersion; on the other hand, the role of MWCNT is related to an improvement of low potential electron transfer. Catalysis of electro-oxidation of thiocholine causes an increment in sensitivity. The OPs determination is based on monitoring the activity inhibition of AChE, using dimethoate as model compound obtaining a very low detection limit.

Another interesting configuration in biosensor development is the immobilization of AChE over MWCNT-chitosan. Chitosan (CS) is a hydrophilic polysaccharide, a biopolymer attractive due to its biocompatibility, biodegradability and nontoxicity; it is frequently used because it has the ability to form films and is capable of dispersing nanomaterials, presenting a good biocompatible microenvironment when constructing sensors. It has been reported that the bioactivity of the enzyme can reach up to 96% in relation to free enzyme activity.

Dong *et al.* described a modification of the Ellman method, where the nitro groups of 5,5-dithiobis 2-nitrobenzoic acid (DTNB) are reduced in the presence of compounds with thiol groups and with the inhibition of AChE. The biosensor demonstrated excellent sensitivity, conductivity and biocompatibility. The analysis was focused in methyl-parathion; its concentration is related to the inhibition of the enzyme associated with the electrochemical response in the reduction of DTNB. The detection limit obtained with this methodology was 7.5×10^{-13} M. This biosensor was successful in the determination of methyl-parathion in water and soil samples [152].

Recently a novel strategy was reported which consists of the creation of bioreactors [153], where the AChE is encapsulated in phosphatidylcholine liposomes (ALB) (Figure 3.10a–d). The microscopic characterization showed that these bioreactors have spherical shapes with a diameter of 7.3 microns (Figure 3.10e–h). The biosensor is based on the assembly of multiple layers of MWCNT-CS-ALB over a GCE. However, the efficiency of the bioreactor depends on the layers in the biosensor. The results showed that the biosensor is more efficient and sensitive with six layers (Figure 3.10i). Among the characteristics of the biosensor were found great stability and a quick response time when used to determine dichlorvos, with a low detection limit (0.68 ± 0.076 µg/L).

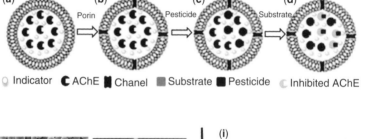

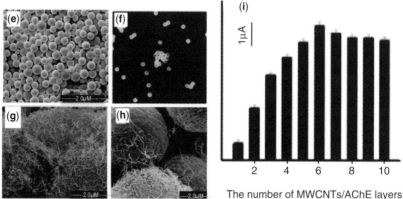

Figures 3.10 Schematic diagram of the AChE-based inhibitor liposome bioreactor. (a) The encapsulated enzyme and pyranine were entrapped in liposomes; (b) porins embedded into the lipid membrane allow for the free substrate and pesticide transport into the liposomes; (c) the pesticides measurements were performed after incubation of the sample in the AChE/liposome biosensor; (d) after incubation, the substrate was added and the inhibition reaction was monitored over time. SEM images of (e) AChE bioreactor; (f) confocal laser scanning microscopy images of AChE bioreactor; (g) SEM images of MWCNTs, and; (h) the (MWCNTs/ALB) composite. (i) Effect of the number of (MWCNTs/ALB) bilayers on the magnitude of the output current of the GCE. Adapted from [153].

Another interesting investigation is based on the development of biosensors using biological fluids, since they contain enzymes that may be useful in determining pesticides. A good example is the biosensor that uses saliva as biomarker. Biomonitoring through saliva is a non-invasive and simple method for real-time analysis. The biosensor is based on the modification of a SPE electrode with CNT and, as already mentioned, saliva as recognition agent due to the presence of AChE in it which is used for the recognition of OP. This electrode was compared with one constructed with commercial enzyme, demonstrating that it can be used to detect exposure to organophosphates [154].

Nowadays it is very common to combine nanomaterials when constructing biosensors, particularly MWCNT and GR. Their combination can produce a synergic effect which can enhance the electron transfer process, and thus, improve the detection of chemicals. However, the formation of agglomerates limits the biological applications of these biosensors [155, 156], so effective dispersants like polymers such as polyaniline (PANI) are required to solve this problem successfully.

Development of biosensors based on hybrid nanobiomaterials has allowed the detection of pesticides such as carbaryl (carbamate pesticide). Because of its high effectiveness in controlling insects, this pesticide has been widely used in agriculture. The principal problem with this pesticide is long-term accumulation in the atmosphere, as it represents a serious risk to human health due to its high toxicity to AChE; this enzyme is the key to some neural functions of the human central nervous system. A biosensor for the determination of this pesticide is based on the modification of a GCE by electropolymerization of PANI with MWCNT being AChE, the immobilized enzyme. The biosensor showed a detection limit that accomplishes the standard regulations of Brazil. Pesticide presence was evaluated in fruits and vegetables and the results were consistent with those obtained with the traditional HPLC method [157].

One of the important contributions of this GR/CNT/CS sensor is that no enzyme is used to detect OPs like methyl-parathion (MP), even though the detection limit obtained was $0.5 ngmL^{-1}$, which is much lower compared with sensors using enzymes. Reproducibility and stability also were better and the interference analysis that resulted was satisfactory. In general, the sensor presented high sensitivity and fast response time [158].

Furthermore, research in nanomaterials for the construction of biosensors sensitive to pesticides can be found in the review of Aragay *et al.*, where interesting results are reported when using CNT with Prussian blue or gold nanoparticles [159].

3.2.3.2   GR-Based Biosensors

Graphene along with its derivatives (graphene oxide, graphene nanoribbon, chemically-reduced graphene oxide and nitrogen-doped graphene), have demonstrated several advantages in electrochemical devices such as sensors and biosensors. This material is used due to its interactions with an enzyme type of π-stacking which induces non-specific bonds. These bonds are also produced by electrostatic interactions of GR-enzyme, which generate a better adsorption of the enzyme [160].

It is well known that the success of biosensors depends on the correct choice of the enzyme and its immobilization, along with the rate of electron transference, so the material used in the biosensor interface is crucial. Detection of a substance at high potentials means a poor selectivity. The emergence of nanomaterials such as GR, which has excellent electrocatalytic properties, produces an improvement in the analytical response, including the detection limits. However, the direct use of GR sheets is not suitable for sensor construction due to their low solubility, so instead GR derivatives are used, especially by reducing GR sheet (reduced graphene nanosheet [rGR]), popular because of their simple functionalization and biocompatibility [161].

Graphene can maintain the specific bioactivity of biomolecules due to the large amount of oxygen-containing functional groups, making it a good material for the immobilization of enzymes such as acetylcholinesterase (AChE) [162]. Pesticides can be monitored through the inhibition of the enzyme, so that the concentration of OPs is determined by the oxidation current of thiocholine (TCh) before and after inhibition; the corresponding equation is:

$$inhibition(\%) = \frac{i_{P,control} - i_{P,exp}}{i_{P,control}} \times 100\% \qquad (3.4)$$

where $i_{P,control}$ is the peak current of TCh registered with the biosensor, and $i_{P,exp}$ is the amperometric response of TCh with pesticide inhibition.

Moreover, it is important to note that the analytical response also depends on the interfering species considered in the pesticide's analysis, usually NO^{2-}, F^-, SO_4^{2-}, Cl^-, K^+, Ca^{2+}, Mg^{2+}, sucrose and fructose, which can coexist with OPs in fruits, vegetables and water.

As already mentioned, the key to improving the sensitivity and selectivity of biosensors is based on the design and the electrochemical detection at low potentials of the enzyme product. In order to reduce the oxidation potential of TCh, mediators have been used, for example, cobalt (II) phthalocyanine, Prussian blue, TiO_2, SnO_2 [162–165] and, recently, metal nanoparticles such as Au, [166–169].

Strategies for enzyme immobilization along with the use of GR are based on the use of polymers such as Nafion. This polymer is used because it can disperse GR sheets and due to its specific structure, a three-dimensional interpenetrating and conducting network (conductive three-dimensional interpenetrating network) is formed [170].

Particularly, in the biosensor designed for the detection of OPs like dichlorvos, [171] a mixture of Nafion and GR oxide (GRO) is used to form the nanocomposite graphene-oxidized Nafion (GRON) [172]; a GCE electrode is sequentially modified by placing a layer of the nanocomposite followed by the enzyme (AChE). The biosensor AChE-ERGRON/GCE was obtained by a reduction process using cyclic voltammetry (CV). The nanobiomolecular interface AChE-ERGRON/GCE presents a fast electron transfer and good biocompatibility. This biosensor presented high catalytic activity for the anodic oxidation of TCh. With this configuration it is possible to quantify using much lower potentials (+50 mV) compared with other biosensors (+700 mV), besides having a high sensitivity, wide linear range and good stability.

A similar configuration was used for the determination of pesticides [173]. The OPH is immobilized in Nafion and the films obtained with this method are robust and flexible because of Nafion's hardness. In these films, Nafion provides a fast charge transport and low interfacial resistance which helps to achieve a high performance for the biosensor. These films were used for the detection of paraoxon, which is a potent insecticide. The OPH enzyme was immobilized over the surface of ErGRON and the design of this biosensor presented excellent electrochemical properties, good sensitivity, low detection limit and fast response (Table 3.4).

Even when the use of GR has substantially improved detection limits for pesticide quantification, its use in combination with metal nanoparticles improves detection limits with values in the picomolar range [174]. A strategy for constructing these biosensors is based on a composite which consists of chemically reduced GR sheets (RGR) and gold nanoparticles (AuNP) with poly-(diallyldimethylammonium chloride) (PDDA) to produce the nanohybrid AuNP/RGR (Figure 3.11). The PDDA helps to prevent the formation of agglomerates and to stabilize the nanoparticles on the GR; also, this polymer is used for the immobilization of the enzyme AChE. This configuration greatly increases the enzyme activity, thereby generating ultrasensitive biosensors to detect organophosphate pesticides. In general, the use of metal nanoparticles [175, 176]

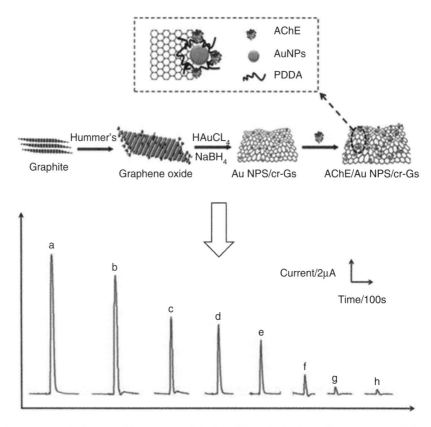

Figure 3.11 Schematic illustration of Au NP/RGO hybrid synthesis and AChE/Au NP/RGO nanoassembly generation by using PDDA. Typical amperometric responses of AChE/Au NP/RGR-based SPEs (screen printed electrode) to paraoxon by using flow injection analysis system. Adapted from [174].

in combination with GR helps reinforce the electrocatalytic activity obtaining very low detection limits.

Table 3.4 shows some of the configurations based on CNT and GR and a comparison with other matrices reported, where ultrasensitive responses and low detection limits can be observed

In summary, we can conclude that progress on the development of new biosensors for the quantitative detection of pesticides through the use of nanomaterials like GR and CNT has allowed the analysis of different kinds of pesticides. Selectivity, response time and stability of the biosensors are closely related to the selected enzyme, while the electrochemical response is largely associated with the presence of nanomaterials. The combination of both enzyme and nanomaterials for creating nanobiomaterials has allowed the obtainment of low detection limits, making possible the detection of pesticides in real samples from various sources such as water, soil, fruits and vegetables.

As expected, in the near future these substances will be detectable *in situ* and monitored in real-time, allowing rapid decisions and actions for the prevention and correction of potential health risks from exposure to these substances.

3.3   Conclusions and Future Perspectives

Electrochemical biosensors using CNT or GR in different configurations have been successfully used for the determination of different pollutants present in waters, soil and food samples (e.g., fruits and vegetables).

Both nanomaterials have been demonstrated to be very effective when used as electrocatysts enhancing the electron's transference. In addition, CNT and GR, due to their rich chemistry, have been applied as immobilization surface either by physical adsorption, covalent bonding or composite mixing/matrix phase. The developed electrochemical biosensors based on CNT or GR showed clear advantages in terms of analytical performance for the detected environmental pollutants. Higher stability, lower detection limits and high sensitivity were the improved parameters when comparing the developed CNT/GR-based designs with other conventional ones.

The challenge in the construction of new sensors for the detection of chemicals such as phenols or pesticides is to achieve a correct balance between stability and enzyme activity, as well as the simultaneous determination of different compounds in real samples. The success of nanomaterials for the analysis of these kinds of compounds lies in the enzyme availability due to the nanometer structures which contribute to the catalytic effect. Also, the construction of smaller devices is favored, allowing real-time analysis.

References

1. M.C. Roco, *Journal of Nanoparticle Research*, Vol. 7, p. 129, 2005.
2. O. Preining, *Journal of Aerosol Science*, Vol. 29, p. 48, 1998.
3. D.A. Jefferson, *Philosophical Transactions of the Royal Society of London A*, Vol. 358, p. 2683, 2000.

4. T.W. Ebbesen, *Carbon Nanotubes: Preparation and Properties*, Florida CRC Press, 1997.
5. C.H. Kiang, W.A. Goddard, R. Beyers, D.S. Bethune, *Carbon*, Vol. 33, p. 903, 1995.
6. P.M. Ajayan, T.W. Ebbesen, *Reports on Progress in Physics*, Vol. 60, p. 1025, 1997.
7. B.I. Yakabson, R.E. Smalley, *American. Science*, Vol. 85, p. 324, 1997.
8. S. Ijima, *Physics B*, Vol. 323, p.1, 2002.
9. R.B. Patel, J. Liu, J.V. Scicolone, S. Roy, S. Mitra, R.N. Dave, Z. Iqbal, *Journal of Materials Science*, Vol. 48, p. 1387, 2013.
10. R. Saito, M.S. Dresselhaus, G. Dresselhaus, *Physical Properties of Carbon Nanotubes*, New York, World Scientific Press, 1998.
11. P.M. Ajayan, *Chemical Reviews*, Vol. 99, p. 1787, 1999.
12. N. Karousis, N. Tagmatarchis, *Chemical Reviews*, Vol.110, p. 5366, 2010.
13. A. Merkoçi, M. Pumera, X. Llopis, B. Pérez, M. Del Valle, S. Alegret, *Trends in Analytical Chemistry*, Vol. 24, p. 826, 2005.
14. M.F. Yu, O. Lourie, M.J. Dryer, K. Moloni, T.F. Kelly, R.S. Ruoff, *Science*, Vol. 287, p. 637, 2000.
15. X. Guo, D. Yu, J. Wu, Ch. Min and R. Guo, *Polymer Engineering and Science*, Vol. 53, p. 370, 2013.
16. R.H. Baughman, *Science*, Vol. 297, p. 787, 2002.
17. S. Ijima, *Physica B: Condensed Matter*, Vol. 323, p. 1, 2002.
18. Y. Zhang, J. Zhu, *Nanomaterials*, Vol. 33, p. 523, 2002.
19. A. Merkoci, *Microchimica Acta*, Vol. 152, p. 157, 2006.
20. K.S. Novoselov, A.K. Geim, S.V. Morozov, D. Jiang, Y. Zhang, S.V. Dubonos, I.V. Grigorieva, A.A. Firsov, *Science*, Vol. 306, p. 666, 2004.
21. M.J. Allen, V.C. Tung, and R.B. Kaner, *Chemical Reviews*. Vol. 110, p.132, 2010.
22. T. Gan and H. Shengshui, *Microchimica Acta*, Vol. 175, p.1, 2011.
23. K.S. Novoselov, D. Jiang, F. Schedin, T.J. Booth, V.V. Khotkevich, S.V. Morozov, A.K. Geim, *Proceedings of the National Academy of Sciences, U.S.A.* Vol. 102, p. 10451, 2005.
24. S. Stankovich, D.A. Dikin, R.D. Piner, K.A. Kohlhaas, A. Kleinhammes, Y. Jia, Y. Wu, S.T. Nguyen, R.S. Ruoff, *Carbon*, Vol. 45, p. 1558, 2007.
25. I. Jung, D. Dikin, S. Park, W. Cai, S.L. Mielke, R.S. Ruoff, *The Journal of Physical Chemistry C*, Vol. 112, p. 20264, 2008.
26. A.N. Obraztsov, *Nature Nanotechnology*, Vol. 4, p. 212, 2009.
27. X.S. Li, *Science*, Vol. 324, p. 1312, 2009.
28. S. Das, A.S. Wajid, J.L. Shelburne, Y. Liao, and M.J. Green, *Applied Materials & Interfaces*, Vol. 3, p.1844, 2011.
29. D. Fujita, *Science and Technology of Advanced Materials*, Vol.12, p.611, 2011.
30. J. Park, W.C. Mitchel, L. Grazulis, H.E. Smith, K.G. Eyink, J.J. Boeckl, D.H. Tomich, S.D. Pacley, J.E. Hoelscher, *Advanced Materials*, Vol. 22, p.4140, 2010.

31. Ch. Lee, X. Wei, J.W. Kysar, J. Hone, *Science,* Vol. 321, p. 385, 2008.
32. Y. Zhu, S. Murali, W. Cai, W. Li, J. Suk, J.R. Potts, R.S. Ruoff, *Advanced Materials,* Vol. 22, p.1, 3906, 2010.
33. A.K. Geim, K.S. Novoselov, *Nature Materials,* Vol. 6, p.183, 2007.
34. K.S. Novoselov, V.I. Falko, L. Colombo, P.R. Gellert, M.G. Schwab, K. Kim, *Nature,* Vol. 490, p.192, 2012.
35. M. Liu, X. B. Yin, E. Ulin-Avila, B.S. Geng, T. Zentgraf, L. Lu, F. Wang, X. Zhang, *Nature,* Vol. 474, p.64, 2011.
36. Y.M. Lin, C. Dimitrakopoulos, K.A. Jenkins, D.B. Farmer, H.Y. Chiu, A. Grill, P. Avouris, *Science,* Vol. 327, p. 662, 2010.
37. K.S. Kim, Y. Zhao, H. Jang, S.Y. Lee, J.M. Kim, K.S. Kim, J.Y. Ahn, P. Kim, J.Y. Choi, B.H. Hong, *Nature,* Vol. 457, p. 706, 2009.
38. Y W. Zhu, S. Murali, M.D. Stoller, K.J. Ganesh, W.W. Cai, P.J. Ferreira, A. Pirkle, R.M. Wallace, K.A. Cychosz, M. Thommes, D. Su, E.A. Stach, R.S. Ruoff, *Science,* Vol. 332, p.1537, 2011.
39. M.F. El-Kady, V. Strong, S. Dubin, R.B. Kaner, *Science,* Vol. 335, p. 6074, 2012.
40. X. Yang, M.S. Xu, W.M. Qiu, X.Q. Chen, M. Deng, J.L. Zhang, H. Iwai, E. Watanabe, H.Z. Chen, *Journal of Materials Chemistry,* Vol. 21, p. 8096, 2011.
41. S. Bae, *Nature Nanotechnology,* Vol. 5, p.574, 2010.
42. L. Liao, *Nature,* Vol. 467, p. 305, 2010.
43. F. Bonaccorso, Z. Sun, T. Hasan, A.C. Ferrari, *Nature Photon,* Vol. 4, 611, 2010.
44. M. Deng, X. Yang, M. Silke, W.M. Qiu, M.S. Xu, G. Borghs, H.Z. Chen, *Sensors and Actuators B: Chemical,* Vol. 158, p. 176, 2011.
45. M.S. Xu, D. Fujita, N. Hanagata, *Small,* Vol. 5, p. 2638, 2009.
46. S. Garaj, W. Hubbard, A. Reina, J. Kong, D. Branton, J.A. Golovchenko, *Nature,* Vol. 467, p. 190, 2010.
47. M.S. Xu, Y. Gao, X. Yang, H.Z. Chen, *Chinese Science Bulletin,* Vol. 57, p. 3000, 2012.
48. B. Pérez-López and A. Merkoçi, *Microchim Acta*. Vol. 175, p.1, 2011.
49. J. Wang, *Electroanalysis,* Vol. 17, p. 7, 2005.
50. M.D. Rubianes, G.A. Rivas, *Electroanalysis,* Vol. 17, p.73, 2005.
51. F. Valentini, A. Amine, S. Orlanducci, M. Terranova, and G. Palleschi, *Analytical Chemistry,* Vol. 75, p. 5413, 2003.
52. M.D. Rubianes, G.A Rivas, *Electrochemistry Communications,* Vol. 5, p. 689, 2003.
53. Q. Zhao, L. Guan, Z. Gu, Q. Zhuang, *Electroanalysis,* Vol. 7, p. 85–88, 2005.
54. R.P. Deo, J. Wang, I. Block, A. Mulchandani, K.A. Joshi, M. Trojanowicz, F. Scholz, W. Chen, Y. Lin, *Analytica Chimica Acta,* Vol. 530, p. 185, 2005.
55. J.Liu, A. Chou, W. Rahmat, M.N. Paddon-Row, J.J. Gooding, *Electroanalysis,* Vol. 17, p. 38, 2005.

56. P.N. Bartlett, J.M. Cooper, *Journal of Electroanalytical Chemistry*, Vol. 362, p. 1, 1993.
57. J.Wang, R.P. Deo, M. Musameh, *Electroanalysis*, Vol. 15, p. 1830, 2003.
58. J. Wang, M. Musameh, *Analytica Chimica Acta*, Vol. 539, p. 209, 2005.
59. M. Gao, L. Dai, G.G. Wallace, *Electroanalysis* Vol.15, p. 1089, 2003.
60. P.P. Joshi, S.A. Merchant, Y. Wang, D.W. Schmidtke, *Analytical Chemistry*, Vol. 77, p. 3183, 2005.
61. J.S. Ye, Y. Wen, W.D. Zhang, H.F. Cui, G.Q. Xu, F.S. Sheu, *Electroanalysis*, Vol. 17, p. 89, 2005.
62. J. Wang, G. Liu, M. Jan, *Journal of the American Chemical Society*, Vol. 126, p. 3010, 2004.
63. S. Ayata, S. Seyhan Bozkurt, and K. Ocakoglu, *Talanta*, Vol. 84, p. 212, 2011.
64. J. Knapek, J. Komarek, and K. Novotny, *Microchimica Acta*, Vol. 171, p. 145, 2010.
65. S. Chen, M. Xiao, D. Lu and Z. Wang, *Spectrochimica Acta, Part B*, Vol. 62, p. 1216, 2007.
66. F. Zhao, Z. Chen, F. Zhang, R. Li, and J. Zhou, *Analytical Methods*, Vol. 2, p. 408, 2010.
67. C. Fontas, E. Margui, M. Hidalgo, I. Queralt, *X-Ray Spectrom*, Vol. 38, p.9, 2009.
68. X. Yuan, Y. Wang, J. Wang, Ch. Zhou, Q. Tang, X. Rao, *Chemical Engineering Journal*, Vol. 221, p. 204, 2013.
69. J.R. Miller, J. Rowland, P.J. Lechler, M. Desilets, L.C. Hsu, *Water, Air, Soil Pollution*, Vol. 86, p. 373, 1996.
70. R. Eisler, *Environmental Geochemistry and Health*, Vol. 25, p. 325, 2003.
71. Q.R. Wang, D. Kim, D.D. Dionysiou, G.A. Sorial, D. Timberlake, *Environmental Pollution*, Vol. 131 p. 323, 2004.
72. P.B. Tchounwou, W.K. Ayensu, N. Ninashvili, D. Sutton, *Environmental Toxicology*, Vol. 18, p. 149, 2003.
73. F. Zahir, S.J. Rizwi, S.K. Haq, R.H. Khan, *Environmental Toxicology and Pharmacology*, Vol. 20, p. 351, 2005.
74. A. Montero Álvarez, J.R. Estevez Álvarez, and R. Padilla Alvarez, *Journal of Radioanalytical and Nuclear Chemistry*, Vol. 245, p. 485, 2000.
75. M. Pesavento, G. Alberti, and R. Biesuz, *Analytica Chimica Acta*, Vol. 631, p. 129, 2009.
76. H. Xu, L.P. Zeng, S.J. Xing, G.Y. Shi, J.S. Chen, Y.Z. Xian, L.T. Jin, *Electrochemistry Communications*, Vol. 10 p. 1893, 2008.
77. X.P. Yan, X.B. Yin, D.Q. Jiang, X.W. He, *Analytical Chemistry*, Vol. 75, p. 1726, 2003.
78. S.J. Christopher, S.E. Long, M.S. Rearick, J.D. Fassett, *Analytical Chemistry*, Vol.73, p. 2190, 2001.
79. R.M. Kazem, S. Alireza, A. Asieh, *Journal of Hazardous Materials*, Vol. 192, p. 1358, 2011.

80. S. Gil, I. Lavilla, C. Bendicho, *Talanta*, Vol. 76, p. 1256, 2008.
81. R. Clough, S.T. Belt, E.H. Evans, B. Fairman, T. Catterick, *Analytica Chimica Acta*, Vol. 500, p. 155–170, 2003.
82. Y. Wei, R. Yang, X.Y. Yu, L. Wang, J.H. Liu, X.J. Huang, *Analyst*, Vol. 137, p. 2183, 2012.
83. Y. Wei, Z.G. Liu, X.Y. Yu, L. Wang, J.H. Liu, X.J. Huang, *Electrochemistry Communications*, Vol. 13, p. 1506, 2011.
84. Y. Wei, R. Yang, Y.X. Zhang, L. Wang, J.H. Liu, X.J. Huang, *Chemical Communications*, Vol. 47, p. 11062, 2011.
85. Z.Q. Zhao, X. Chen, Q. Yang, J.H. Liu, X.J. Huang, *Chemical Communications*, Vol. 48, p. 2180, 2012.
86. M. Lehmann, A.I. Zouboulis, K.A. Matis, *Chemosphere*, Vol. 39, p. 881, 1999.
87. N. Savagel, M.S. Diallo, *Journal of Nanoparticle Research*, Vol. 7, p. 331, 2005.
88. X. Zhao, L. Lv, B. Pan, W. Zhang, S. Zhang, Q. Zhang, *Chemical Engineering Journal*, Vol. 170, p. 381, 2011.
89. J. Liu, S.Z. Qiao, Q.H. Hu, G. Qing, *Small*, Vol.7, p. 425, 2011.
90. X. Ren, C. Chen, M. Nagatsu, X. Wang, *Chemical Engineering Journal*, Vol. 170, p. 395, 2011.
91. M. Zhu, G. Diao, *Nanoscale*, Vol. 3, p. 2748, 2011.
92. G. Aragay, A. Merkoci, *Electrochimica Acta*, Vol. 84, p. 49, 2012.
93. M.M. Liang, L.H. Guo, *Journal of Nanoscience and Nanotechnology*, Vol. 9, p. 2283, 2009.
94. L.D. Zhang, M. Fang, *Nano Today*, Vol. 5, p. 128, 2010.
95. X. Chen, Y. Dong, L. Fan, D. Yang, *Analytica Chimica Acta*, Vol. 582, p. 281, 2007.
96. W. Xu, J. Zhang, L. Zhang, X. Hu, X. Cao, *Journal of Nanoscience and Nanotechnology*, Vol.9, p. 4812, 2009.
97. Q. Zhao, Z.H. Gan, Q.K. Zhuang, *Electroanalysis*, Vol. 14, p. 1609, 2002.
98. G.D. Vukovic, A.D. Marinkovic, M. Colic, M.D. Ristic, R. Aleksic, A.A. Peric-Grujic, P.S. Uskokovic, *Chemical Engineering Journal*, Vol. 157, p. 238, 2010.
99. O. Moradi, K. Zare, M. Monajjemi, M. Yari, H. Aghaie, *Fullerenes Nanotubes and Carbon Nanostructures*, Vol. 18, p. 285, 2010.
100. J.H. Li, F.X. Zhang, J.Q. Wang, Z.F. Xu, R.Y. Zeng, *Analytical Sciences*, Vol. 25, p. 653, 2009.
101. H. Xu, L.P. Zeng, S.J. Xing, Y.Z. Xian, G.Y. Shi, *Electroanalysis* Vol. 20, p. 2655, 2008.
102. S.A. Yuan, Q.O. He, S.J. Yao, S.S. Hu, *Analytical Letters*, Vol. 39, p. 373, 2006.
103. X.F. Jia, J. Li, E.K. Wang, *Electroanalysis*, Vol. 22, p. 1682, 2010.
104. N. Wang, X. Dong, *Analytical Letters*, Vol. 41, p. 1267, 2008.
105. G. Liu, Y. Lin, Y. Tub, Z. Ren, *Analyst*, Vol. 130, p. 1098, 2005.

106. H. Xu, L. Zeng, S. Xing, Y. Xian, L. Jin, *Electrochemistry Communications*, Vol. 10, p. 551, 2008.

107. H. Xu, L. Zeng, S. Xing, G. Shi, J. Chen, Y. Xian, L. Jin, *Electrochemistry Communications*, Vol. 10, p. 1893, 2008.

108. D.T. Nguyen, L.D. Tran, H.L. Nguyen, B.H. Nguyen, N.V. Hieu, *Talanta*, Vol. 85, p. 2445, 2011.

109. W. Song, L. Zhang, L. Shi, D. Li, Y. Li, *Microchimica Acta*, Vol. 169, p. 321, 2010.

110. N. Kong, J. Liu, Q. Kong, R. Wang, C.J. Barrow, W. Yang *Sensors and Actuators B*, Vol. 178, p. 426, 2013.

111. B. Wang, Y. Chang, L. Zhi, *New Carbon Materials*, Vol. 26, p. 31, 2011.

112. Z. Wang, L. Li, E. Liu, *Thin Solid Films*, In Press, Corrected Proof. Available online 5 March, 2013.

113. D.A.C. Brownson, *Electrochemistry Communications*, Vol. 13, p. 111, 2013.

114. J. Gong, T. Zhou, D. Song, L. Zhang, *Sensors and Actuators B*, Vol. 150, p. 491, 2010.

115. Y.P. Ding, W.L. Liu, Q.S. Wu and X.G. Wang, *Journal of Electroanalytical Chemistry*, Vol. 575, p. 275, 2005.

116. M.A. Ghanem, *Electrochemistry Communications*, Vol. 9, p. 2501, 2007.

117. S. Mu, *Biosensors and Bioelectronics*, Vol. 211, p. 237, 2006.

118. T. Xie, Q. Liu, Y. Shi, *Journal of Chromatography A*, Vol. 1109, p. 317, 2006.

119. L.R. Sharma, G. Singh, A. Sharma, *Indian Journal of Chemistry*, Vol.25A, p. 345, 1986.

120. J. Wang, R.P. Deo, M. Musameh, *Electroanalysis*, Vol. 15, p. 1830, 2003.

121. Y. Cheng, Y. Liu, J. Huang, Y.X.Z. Zhang, L. Jin, *Electroanalysis* Vol. 20, p. 1463, 2008.

122. M.D. Rubianes, G.A. Rivas, *Electroanalysis*, Vol. 17, p. 73, 2005.

123. G. Alarcón, M. Guix, A. Ambrosi, M.T. Ramirez Silva, M.E. Palomar Pardave, A. Merkoci, *Nanotechnology*, Vol. 21, p. 245502, 2010.

124. D.G. Mita, A. Attanasio, F. Arduini, N. Diano, V. Grano, U. Bencivenga, S. Rossi, A. Amine, and D. Moscone, *Biosensors and Bioelectronics*, Vol. 23, p. 60, 2007.

125. Y.X. Cheng, Y.J. Liu, J.J. Huang, Y.Z. Xian, Z.G. Zhang, L.T. Jin, *Electroanalysis*, Vol. 20, p. 1463, 2008.

126. R. Solna, E. Dock, A. Christenson, M. Winther-Nielsen, C. Carlsson, J. Emneus, T. Ruzgas, and I. Sklada, *Analytica Chimica Acta*, Vol. 528 P, p. 9, 2005.

127. J. Wang, *Electroanalysis*, Vol. 17, p. 7, 2005.

128. Q. Zhao, L. Guan, Z. Gu, Q. Zhuang, *Electroanalysis*, Vol.17, p. 85, 2005.

129. S. Hashemnia, S. Khayatzadeh, M. Hashemnia, *Journal of Solid State Electrochemistry*, Vol. 16, p. 473, 2012.

130. Y. Zhang, D. Lu, T. Ju, L. Wang, S. Lin, Y. Zhao, C. Wang, Y. Du, *International Journal of Electrochemical Science*, Vol. 8, p. 504, 2013.

131. H. Yin, *Electrochimica Acta*, Vol. 56, p. 2748, 2011.

132. Y. Ding, W. Liu, Q. Wu, X. Wang, *Journal of Electroanalytical Chemistry*, Vol. 575, p. 275, 2005.

133. H. Yin, Q. Ma, Y. Zhou, S. Ai, L. Zhu, *Electrochimica Acta*, Vol. 55, p. 7102, 2010.

134. H. Qi, C. Zhang, *Electroanalysis*, Vol. 17, p. 832, 2005.

135. Y. Zhang, J. Zhang, H. Wu, S. Guo, J. Zhang, *Journal of Electroanalytical Chemistry*, Vol. 681, p. 49, 2012.

136. D. Zhang, Y. Peng, H. Qi, Q. Gao, C. Zhang, *Sensors and Actuators B*, Vol. 136 p. 113, 2009.

137. Z. Wang, S. Li, Q. Lv, *Sensors and Actuators B*, Vol. 127, p. 420, 2007.

138. P.T. Radford, S.E. Creager, *Analytica Chimica Acta*, Vol. 449, p.199, 2001.

139. L. Wang, *Electrochimica Acta*, Vol. 92, p. 216, 2013.

140. J. Razumiene, A. Vilkanauskyte, V. Gureviciene, *Journal of Organometallic Chemistry*, Vol. 668, p.83, 2003.

141. M. Xu, J. Zhu, H. Su, J. Dong, Sh. Ai, R. Li, *Journal of Applied Electrochemistry*, Vol. 42, p 509, 2012.

142. R.M.A. Tehrani, H. Ghadimi, S. Ghani, *Sensors and Actuators B: Chemical*, Vol. 177, p. 612, 2013.

143. C. Yang, Y. Chai, R. Yuan, W. Xu, S. Chen, *Analytical Methods*, Vol. 5, p. 666, 2013.

144. L. Zaijun, S. Xiulan, X. Qianfang, L. Ruiyi, F. Yinjun, Y. Shuping, L. Junkang, *Electrochimica Acta*, Vol. 85, p. 42, 2012.

145. H. Fan, Y. Li, D. Wu, H. Ma, K. Mao, D. Fan, B. Du, Q. Wei, *Analytica Chimica Acta*, Vol. 711, p 24, 2012.

146. Y. Qu, M. Ma, Z. Wang, G. Zhan, B. Li, X. Wang, H. Fang, H. Zhang, L. Chuny, *Biosensors and Bioelectronics*, Vol. 44, p. 85, 2013.

147. A. Aminea, H. Mohammadia, I. Bourais, G. Palleschi, *Biosensors and Bioelectronics* Vol. 21, p. 1405, 2006.

148. B.V. Chai, Y. Niu, X. Chen, C. Zhao, H.M. Lan, *Analytical Letters*, Vol. 46, p. 803, 2013.

149. B.G.l Choi, H.S. Park, T.J. Park, D.H. Kim, S.Y. Lee, W.H. Hong, *Electrochemistry Communications*, Vol. 11, p. 672, 2009.

150. L.L.C. Garcia, L.C.S. Figueiredo-Filho, G.G. Oliveira, C.E. Banks, *Sensors and Actuators B: Chemical*, Vol. 181, p. 306, 2013.

151. D. Du, M. Wang, J. Cai, A. Zhang, *Sensors and Actuators B*, Vol. 146, p. 337, 2010.

152. J. Dong, X. Fan, F. Qiao, S. Ai, H. Xin, *Analytica Chimica Acta*, Vol. 761, p. 78, 2013.

153. J. Yan, H. Guan, J. Yu, D. Chi, *Pesticide Biochemistry and Physiology*, Vol. 105, p. 197, 2013.

154. J. Timchalk, C. Lin, Y. Wang, *Environm Science and Technology*, Vol. 42, p. 2688, 2008.
155. J. Manso, M.L. Mena, P. Yañez-Sedeño, J.M. Pingarrón, *Electrochimica Acta*, Vol. 53, p. 4007, 2008.
156. Y. Chai, X. Niu, C. Chen, H. Zhao, M.. Lan, *Analytical Letters*, Vol. 46, p. 803, 2013.
157. I. Cesarino, F. Morales, M.R.V. Lanza, S.A.S Machado, *Food Chemistry*, Vol. 135, p. 873, 2012.
158. Y. Liu, S. Yang, W. Niu, *Colloids and Surfaces B: Biointerfaces*, Vol. 108, p. 266, 2013.
159. G. Aragay, F. Pino, A. Merkoçi, *Chemical Reviews*, Vol. 112, p. 5317, 2012.
160. X. Hu, Q. Zhou, *Chemical Reviews*, Vol. 113, p. 3815, 2013.
161. C. Nethravathi, J.T. Rajamathi, N. Ravishankar, C. Shivakumara and M. Rajamathi, *Langmuir*, Vol. 24, p. 8240, 2008.
162. S. Andreescu, J.L. Marty, *Biomolecular Engineering*, Vol. 23, p. 1, 2006.
163. J.P. Hart and I.C. Hartley, *Analyst*, Vol. 119, p. 259, 1994.
164. W.A. Collier, M. Clear, and A.L. Hart, *Biosensors and Bioelectronics*, Vol. 17, p. 815, 2002.
165. F. Ricci, F. Arduini, A. Amine, D. Moscone, and G. Palleschi, *Journal of Electroanalytical Chemistry*, Vol. 563, p. 229, 2004.
166. Y.X. Fang, S.J. Guo, C.Z. Zhu, Y.M. Zhai, and E.K. Wang, *Langmuir*, Vol. 26, p. 11277, 2010.
167. S. Zhang, Y.Y. Shao, H.G. Liao, J. Liu, I.A. Aksay, G.P. Yin, and Y.H. Lin, *Chemistry of Materials*, Vol. 23, p. 1079, 2011.
168. Y.C. Qiu, K.Y. Yan, S.H. Yang, L.M. Jin, H. Deng, and W.S. Li, *ACS Nano*, Vol. 4, p.6515, 2010.
169. O. Akhavan, M. Abdolahad, A. Esfandiar, and M. Mohatashamifar, *The Journal of Physical Chemistry C*, Vol. 114, p. 12955, 2010.
170. B.G. Choi, H.S. Park, T.J. Park, M.H. Yang, J.S. Kim, S.Y. Jang, *ACS Nano*, Vol. 4, p. 2910, 2010.
171. S. Wu, F. Huang, X. Lan, X. Wang, J. Wang, C.E. Meng, *Sensors and Actuators B: Chemical*,Vol. 177, p. 724, 2013.
172. W.S. Hummers, R.E. Offeman, *Journal of the American Chemical Society*, Vol. 80, p. 1339, 1958.
173. B.G. Choi, H.S. Park, T.J. Park, M.H. Yang, J.S. Kim, S.Y. Jang, *ACS Nano*, Vol. 4, p. 2910, 2010.
174. Y. Wang, S. Zhang, D. Du, Y. Shao, Z. Li, J. Wang, M.H. Engelhard, J. Li, Y. Lin, *J. Materials Chemistry and Physics*, Vol. 21, p. 5319, 2011.
175. Y. Long, W. Guangcan, L. Yongjun, W. Min, *Talanta*, Vol. 113, p. 135, 2013.
176. L. Zhang, A. Zhang, D. Du, Y. Lin, *Nanoscale*, Vol. 4, p. 4674, 2012.

4

Catalytic Application of Carbon-based Nanostructured Materials on Hydrogen Sorption Behavior of Light Metal Hydrides

Rohit R Shahi* and O.N. Srivastava

Nanoscience and Nanotechnology Unit, Hydrogen Energy Center, Department of Physics, Banaras Hindu University, Varanasi, India

Abstract

The development of hydrogen storage materials with favorable thermodynamics (e.g., kinetics, desorption/adsorption temperature) has attracted considerable attention in recent years. Alanates, amide-hydride mixtures and magnesium hydride are the candidates with the most potential storage material due to their high hydrogen storage capacity and good reversibility, but each has its own limitations (e.g., high desorption temperature and sluggish kinetics). Carbon has many allotropes such as graphite, activated carbon, fullerenes, carbon nanotubes, and the most recent, graphene, etc. These have novel properties which are useful in many new innovative applications. Several recent investigations have also demonstrated the beneficial effect of carbon materials as catalyst for enhancing sorption behavior of different light hydrogen storage materials. Carbon with a small curvature radius exhibits prominent "catalytic" effect for light metal and complex hydrides. The reduction in curvature radius of carbon nanostructures enhances the electron affinity and interaction of carbon with hydrogen because the hydrogen release/combination energy has been changed, and consequently, the de-/rehydrogenation kinetics of the material is improved. In this chapter, we will highlight the current advances (including our recent works) in the hydrogen sorption enhancement of metal and complex hydrides by incorporating carbon nanomaterials as a catalyst. There will be a particular emphasis on carbon nanotubes,

Corresponding author: rohitrshahi@gmail.com

Ashutosh Tiwari and S.K. Shukla (eds.) Advanced Carbon Materials and Technology, (129–172)
2014 © Scrivener Publishing LLC

carbon nanofibers and graphene employed as a catalyst for the aforesaid hydrogen storage materials.

Keywords: Hydrogen storage, MgH_2, Li-Mg-N-H, CNTs, CNFs, graphene, catalyst

4.1   Introduction

Increasing energy needs and global warming urgently demand a shift from fossil fuel to renewable energy resources. In terms of mobile applications, hydrogen is seen as the main contender. Unlike petroleum fuels, hydrogen is clean (pollution free), renewable and environmental friendly. Cold combustion of hydrogen in fuel cell leads to the creation of electrical power, and hot combustion in internal combustion (IC) engines of motor vehicles provides power in the same way as fossil fuels do [1–4]. Both the cold and hot combustion processes lead to the emission of water. The main advantage of hydrogen as a fuel is the absence of CO_2 emissions. Thus hydrogen is considered as a renewable and sustainable solution for reducing global fossil fuel consumption and combating global warming. Energy from sunlight is converted into electricity, for example, by means of photovoltaic cells. This electricity (from a renewable energy source) is used for the electrolysis of water (Figure 4.1a). The oxygen is released into the atmosphere and hydrogen is stored, transported and distributed (Figure 4.1b). Finally, hydrogen, together with the oxygen, is combusted and the energy is released as heat, and work leaving water or steam which can be further used for the production of hydrogen (Figure 4.1c). Therewith the hydrogen cycle is closed (it is produced from water and burns back to water) as shown Figure 4.1.

Hydrogen is the most abundant element in the universe, and molecular hydrogen (H_2) the most abundant molecule. It is the ideal fuel for the future because it significantly reduces greenhouse gas emissions, global dependence on fossil fuels, and increases the efficiency of the energy conversion process for both internal combustion engines and proton exchange membrane fuel cells [5]. The energy per mass of hydrogen (120 MJkg^{-1}) is three times higher than that of other chemical fuels, i.e., gasoline (44.4 MJkg^{-1}). Storage is a key issue for hydrogen economy. It cuts across the production, distribution, safety and applications aspects. One gram of hydrogen gas occupies ~11 liters (2.9 gallons) of space

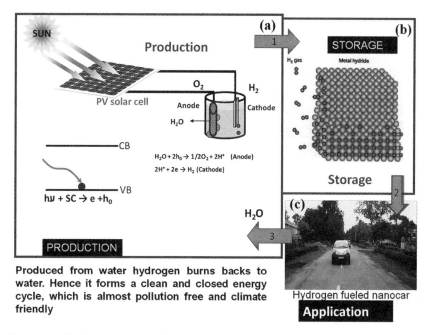

Figure 4.1 Hydrogen cycle: (a) production from electrolysis of water; (b) storage as form of hydride (from Google Images); (c) used as fuel in vehicles.

at atmospheric pressure, so storage implies a need to reduce the enormous volume occupied by hydrogen [1]. For on-board energy storage, vehicles need compact, light, safe and affordable containment. A modern, commercially available car optimized for mobility and not prestige with a range of 400 km burns about 24 kg of petrol in a combustion engine to cover the same range, 8 kg hydrogen are needed for the combustion engine version or 4 kg hydrogen for an electric car with a fuel cell [1]. At room temperature and atmospheric pressure, 4 kg of hydrogen occupies a volume of 45 m^3. This corresponds to a balloon of 5 m diameter, hardly a practical solution for a vehicle. Thus the storage of hydrogen is one of the crucial factors for successful realization of hydrogen-fueled IC engines/fuel cells. Thus developing safe, reliable, compact, and cost-effective hydrogen storage technologies is one of the most technically challenging barriers to the widespread use of hydrogen as a form of energy.

A further distinctive feature of hydrogen is its very low boiling point, which results in added cost (and technical complications) for hydrogen storage and handling in the liquid state. Similarly, hydrogen transported as a compressed gas in high pressure cylinders also meets with the limitations of safety and added cost [6]. Hence safe and cost-effective storage are needed for widespread usage of hydrogen for not only in portable, but also in stationary applications. All practical storage options have disadvantages, but still, the most promising appears to be solid state storage. In general terms, solid materials are most actively investigated and divided into two distinct groups: (a) physisorption reversibly adsorb molecular hydrogen, and (b) chemisorptions of hydrogen followed by compound formation.

The first group is typified by highly porous solids and comprises mainly active carbons (including carbon nanostructures) and solids formed by open metal–organic frameworks (MOFs). However, other porous solids such as zeolites and (organic) polymers having intrinsic microporosity (PIMs) are also under active research [7–11]. From the applications point of view, the most remarkable difference between metal hydrides and microporous adsorbents is their operational temperature. Because of their relatively high stability, metal hydrides usually have to be heated at a temperature higher than about 500 K for thermal decomposition and consequent hydrogen release. By contrast, a problem with most (physical) adsorbents is that they usually need a low temperature (about 77 K) for storing hydrogen in a reasonable amount. Since operation at (or near) ambient temperature is highly desirable, main strategic aim is for either the lowering of operational temperature for hydrides or increasing it for porous adsorbents.

Apart from physisorptive materials, the most extensively investigated light weight materials are referred to as complex hydrides (e.g., $NaAlH_4$, $LiNH_2$, $LiBH_4$, NH_3BH_3 etc). In the periodic table, elements which have $Z \leq 13$ are known as light elements such as Li, B, C, N, Na, Mg, and Al. These light weight hydrogen storage materials have high gravimetrical capacity and can meet the DOE target [12]. After decades of extensive studies, it has been found that among different complex hydrides, $NaAlH_4$ and Li-Mg-N-H systems show promising storage behavior. The MgH_2 is also known as a potential hydrogen storage material because of its high gravimetric hydrogen storage capacity (7.6 wt% and 0.11kg/L) [13, 14], high abundance (abundance in both earth crust and in sea water), acceptable cost, nontoxicity and high safety [13-17]. The sluggish sorption kinetics and high desorption temperature are the major limitations for these promising materials.

The ability of hydrogen storage in solid materials can be altered by changing physiochemical properties such as composition, structure, size, phase transformation, grain boundaries, defects, vacancies, catalytic effect, and so on [18]. Thus the improvement in sorption behavior and decrement in desorption temperature can be achieved by the material tailoring and deployment of various suitable catalysts [19–40].

Nowadays for hydrogen storage materials, apart from the traditional catalyst (having high density due to which dead weight is created for storage application) the focus is now shifted towards the light weight, high surface area and extremely high active site carbon materials (such as graphite, activated carbon, fullerenes, carbon nanotube, carbon nanofiber and, recently, graphene) [41, 42]. Additionally, due to their light weight and variable hybridization states (sp, sp^2 and sp^3), carbon-based nanostructured materials (CNSs) serve as an effective catalyst to destabilize the bonds in the hydrogen storage materials. The focus of the present chapter is to review the effect of doping or addition of carbon on enhancing the hydrogen storage performance of the most promising hydrogen storage materials (MgH_2, $NaAlH_4$ and Li-Mg-N-H system), and our attempt to gain a deep insight into the mechanisms by reviewing these state-of-the-art materials.

4.2   Different Carbon Allotropes

Carbons have novel properties that make them potentially useful in many new innovative applications. Carbon atoms can bond to one another in chains, rings, and branching networks to form a variety of structures, including minerals, solid carbon materials, fossil fuels and organic compounds. In solid carbon materials, carbon atoms can form C-C single, double and triple bonds [43]. Diamond and graphite are three-dimensional carbon allotropes that have been well known for a long time. The other ordered allotropes which constitute the backbone of new carbon chemistry and nanotechnology (shown in Figure 4.2) are:

- Fullerenes with perturbed sp^2 bonding.
- Carbon nanotubes (CNTs) with sp^2 bonding.
- Most recent graphene, exfoliated single sheets of atoms with sp^2 bonds.

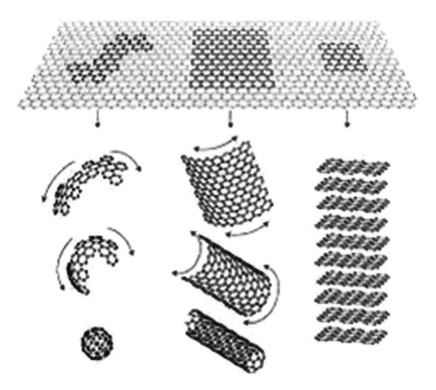

Figure 4.2 Carbon allotropes (from Google images).

Carbon nanotubes (CNTs) are allotropes of carbon with a nano-structure that can have a length-to-diameter ratio greater than 1,000,000. These cylindrical carbon molecules have novel properties that make them potentially useful in many applications in nanotechnology [44]. They are nanometers in diameter and several millimeters in length and have a very broad range of electronic, thermal, and structural properties. These properties vary with the kind of nanotube defined by its diameter, length, chirality or twist and wall nature. Several techniques have been developed to produce nanotubes in sizeable quantities, including arc discharge, laser ablation, chemical vapor deposition, silane solution method and flame synthesis method [45–49]. The bonding in carbon nano-tubes is sp^2, with each atom joined to three neighbors, as in graph-ite. The tubes can therefore be considered as rolled-up graphene sheets (graphene is an individual graphite layer) [50]. This bond-ing structure, which is stronger than the sp^3 bonds found in dia-mond, provides the molecules with their unique strength [51].

Graphene is known as the parent of all graphitic forms of carbon nanomaterials (Figure 4.2). It has now become the most exciting topic of research in the last four years. The two-dimensional single planar sheet of sp^2-bonded carbon atoms corresponds to one hexagonal basal plane of graphite and is termed graphene. It is distinctly different from carbon nanotubes (CNTs) and fullerenes and has unique innovative properties. Recently, a successful attempt to isolate such a graphene layer has been reported [51], and apparently, it has become possible to produce a new graphene phase by exfoliation or mechanical peeling of layers from bulk graphite. If separated completely, graphene layer should provide the specific surface area (SSA) of 2,622 g/m^2 for two sides of the graphene [52]. Three different types of graphenes can be defined: single-layer graphene (SG), bilayer graphene (BG), and few-layer graphene (FG, number of layers ≤10) [50].

4.3   Carbon Nanomaterials as Catalyst for Different Storage Materials

Recently, carbon nanomaterials (CNSs) have been explored as catalyst for improving the de/rehydrogenation of different hydrogen storage materials due to the presence of delocalized Π bonds and the possible interaction with hydrogen atoms [41, 42]. Carbon nanomaterials have certain properties that make them a suitable additive for enhancing sorption behavior of different hydrogen storage materials. Since the light metal hydrides are chemically reactive, the specific chemical inertness is the most important property, which makes them an effective catalyst. Some important reactions with carbon matrix exist because of the enhancement in sorption behavior, which has been observed. The details of these properties are described by Adelhelm *et al.* [41]. These are (a) intercalation, (b) carbide formation, and (c) interaction with terminating group, defects and contaminations; there are also some important physical properties such as (d) H_2 diffusion promoter, (e) thermal conductivity, and (f) dispersion of catalysts, which are responsible for the beneficial effect of carbon materials [41].

  a. **Intercalation**
     A well known property of graphite is the formation of lamellar intercalation compound with different species [53-55]. This process is highly reversible and

allows thermodynamical tailoring of a metal hydride. This process is exothermic, the metal phase is stabilized towards the hydride phase and ΔH for de-/rehydrogenation has been reduced [56].

b. Carbide formation

In some cases, at high temperature and pressure the carbide phase is formed. It has been reported that for the case of graphite/LiH composite [57, 58] and carbon/LiBH$_4$ composite, formation of Li$_2$C$_2$ has occurred [59]. However, carbide formation is undesired, because the carbon structure is destroyed and formation of hydride is hindered.

c. Terminating group defect and contamination

The terminating species also provide a possible site for chemical reaction with metal hydride. Ideally, CNTs and fullerenes exhibit no terminating groups and thus represent the purest form of CNS. Also, structural defects within the graphene might be catalytically active [60, 61]. It has been suggested that the defect in the form of non-hexagonal ring can ease the dissociation. If the same holds for dissociation of H$_2$ structure defect in carbon materials, it would be a powerful tool to improve the hydrogen sorption of carbon-supported metal hydride.

d. Hydrogen diffusion promoter

The rate limiting step in the hydrogenation reaction is the slowest step; in most of the hydride the hydrogen diffusion through bulk metal is the slowest. In both cases either metal hydride dispersed over a porous carbon structure or carbon materials are dispersed in the metal hydride phase; the fast hydrogen diffusion through or along the carbon phase would be responsible for improvement in kinetics of hydrogen release and uptake.

e. Thermal conductivity

The thermal conductivity of hydrides is poor, thus the ability of carbon materials to effectively conduct heat is an important aspect. The thermal conductivity of graphite is highly anisotropic and typically found to be ~12-175W m^{-1}K^{-1} It is found that even a small amount of graphite significantly improves the heat conduction of metal hydride. The thermal conductivity of metal hydride is higher in the presence of carbon [62].

f. **Dispersion of catalysts**

Carbon materials can also be incorporated by the addition of metal catalyst to improve the sorption kinetics. The Ni and Co are known catalysts for dissociation of hydrogen molecules and can easily be dispersed in carbon support [63, 64]. Also, CNTs and CNFs usually contain Ni or Fe nanoparticles that get distributed in the hydride phase during ball-milling [65].

4.4 Key Results with MgH$_2$, NaAlH$_4$ and Li-Mg-N-H Systems

Currently MgH$_2$ and complex hydrides such as NaAlH$_4$ and Li-Mg-N-H have attracted much attention because of their high capacity and reversibility. The application of these complex hydrides are plagued by (a) high enthalpy for desorption because hydrogen is covalently bonded with anions, and (b) high kinetic barriers rehydrogenation due to chemical inertness and/or phase segregation [66]. These drawbacks resulted in high temperature and pressure requirement for hydrogen release and uptake. These problems are alleviated by material tailoring and deployment of catalyst. Surprisingly, CNSs act as catalyst for enhancing the sorption behavior of aforesaid hydrogen storage materials. We reviewed the beneficial effect of carbon materials on sorption behavior of aforesaid storage materials.

4.4.1 Magnesium Hydride

Magnesium hydride (MgH$_2$) is a potential hydrogen storage material. It satisfies most of the requisites for a viable hydrogen storage system for mobile as well as stationary hydrogen devices. The slow hydriding/dehydriding are the major drawbacks of the magnesium hydride as hydrogen storage material. However, magnesium hydride still holds interest for many researchers due to its low atomic weight, high hydrogen storage capacity (7.2 wt%), low cost and simplicity of structure despite its slow kinetics and high operating temperature of 600 K [67–69]. Sakintuna *et al.* [70] have tabulated a list of modified magnesium-based systems; among these most involved ball-milling for material preparation. This leads to increase of surface area, creation of defects and nucleation sites, and also reduces the hydrogen diffusion path.

Recently, some reports have appeared to describe the various MgH_2 admixed (catalyzed) carbon nanostructures [71–77]. These studies bring out the beneficial effect of carbon materials on both the hydriding and dehydriding behaviors. Here we will discuss and describe the important studies (including our recent study) on the catalytic effect of carbon nanostructures for Mg/MgH_2.

4.4.1.1   Effect of Graphite Addition on Sorption Behavior of Mg/MgH₂

Imamura *et al.* [78–82] reported the beneficial effect of graphite on Mg/MgH_2. Imamura *et al.* reported that the composite of Mg with graphite carbon synthesized by ball-milling in the presence of organic solvent improved the sorption behavior of Mg [82]. They reported that the presence of organic additive is important for the cleavage of graphite along its layer through which there is the occurrence of synergetic interaction with Mg, which is responsible for beneficial effects. Without the presence of organic additive, graphite was broken in an irregular and disorderly manner. Furthermore, metallic doping has also been performed through the ball-milling with organo-metallic solutions ($Al(C_2H_5)$), $Ti(oC_3H_7)_4$, $Fe(C_5H_5)_2$ and $Zn(C_2H_5)_2$) in benzene. The best results were found by the doping with Ti. The Ti-doped Mg/C system have about ten-fold sorption improvements as compared to metal-free composite.

In 2004, Shang *et al.* [83] also investigated the effect of graphite on sorption behavior of MgH_2. Different concentrations of graphite powder were mechanically milled with MgH_2. The results demonstrated that graphite showed negligible improvement in desorption properties of MgH_2. But it significantly improved the absorption behavior of MgH_2, 5 wt% of hydrogen reabsorbed within 30 min at 250°C by the 10 h ball-milled MgH_2+10wt% graphite. The small reduction in desorption temperature was explained mainly by the milling effect as a result of reduced particle size. The beneficial effect for absorption kinetics may be attributed to the interaction between crystalline graphite and dissociation of hydrogen molecule at the MgH_2 or Mg surface. Also, graphite coating on the powder particles inhibited the formation of a new oxide layer on Mg powder.

4.4.1.2   Effect of Carbon Nanotubes on Sorption Behavior of Mg/MgH_2

Chen *et al.* [84] investigated the effect of multi-walled carbon nanotubes on de/rehydrogenation behavior of MgH_2. They reported that at certain temperature and pressure the concentration of MWCNTs should be in a proper ratio. The improvement for hydrogen sorption kinetics of MgH_2-5 wt% MWCNTs may arise due to two possible reasons, one is the catalyst reaction with parent material in which ball-milling under the hydrogen atmosphere of the composite is fully mixed and directly hydrogenated during the preparation. The other is the carbon content present in the composite. This consists of the broken micro-MWCNTs. With the absorption of carbon component on freshly exposed Mg surfaces, the impermeable surface oxide/hydroxide layer is not reformed and composite is readily accessible to hydrogen. Therefore the carbon component in the composite may play an important role in hydrogen uptake and release.

Wu *et al.* [71] compared the catalytic effect of different carbon/non-carbon additives on dehydrogenation characteristics of Mg. They reported that all carbon additives show a beneficial effect as compared to non-carbon additives. Among different studied carbon additives, the SWCNTs showed the most promising results with respect to lowering of desorption temperature and enhancing the sorption kinetics. The initial hydrogen desorption temperature of MgH_2 is lowered by 60K with the addition of SWCNTs. The hydrogen storage capacity of all Mg/C composite is over 6 wt% within 5 min, in particular the H storage capacity for 5 wt% SWCNTs or C_{60}/Mg composite has absorbed 6.7 wt% of hydrogen within 2 min. However, the hydrogen sorption capacity is worse when compared to pure Mg in the case of Mg/Non-carbon composites. In the case of Mg/SWCNTs composite, the absorption kinetics is found to be almost one order of magnitude higher than that of pure Mg.

Before the results of Wu *et al.* it was believed that the catalytic effect of carbon on Mg may occur due to the anti-sticking effect of carbon. Wu *et al.* [71] found that Mg/carbon and Mg/non-carbon have similar specific surface area. Thus the effect of grain/particle size and specific surface area are not the responsible factors for the beneficial effect of carbon on Mg. Therefore they speculated that the beneficial effect of carbon materials arises due to some other specific reasons. According to them the incorporation of carbon leads to an increased phase boundary and hydrogen diffusion driven force, which can be explain on the basis of the different hybridization states of carbon-based materials. There are three possible hybridizations of carbon materials: sp, sp^2 and sp^3. During intense milling carbon materials are incorporated into Mg matrix and increase phase boundary. The carbon in Mg matrix are in the sp/sp^2 hybridization and have delocalized Π bond electrons which are responsible for the interaction with H_2 molecule. Thus, adsorbed hydrogen molecule concentrate around the carbon fragments and act as hydrogen source for further chemical absorption which occurs at Mg. Whereas, for the case of non-carbon additives, the above kind of physical adsorption is absent. Moreover, among different carbon materials, SWCNTs showed a beneficial effect because of their novel microstructure which may facilitate the diffusion of hydrogen into Mg grains.

The detailed studies on MgH_2/SWNTs composite were further performed by the same group [85]. The study described the optimum composition and milling parameter for the best result with SWCNTs in MgH_2. They reported that 5 wt% is sufficient for the best results. The addition of too much SWCNTs contrarily degrade the hydrogen storage properties of Mg/SWCNT composite. Wu *et al.* [85] also reported that a longer milling duration of 20 h caused a substantial decrease in the hydrogen storage capacity of about 1.7 wt% lower than that of the 10 h ball-milled sample. This may have occurred because SWCNTs destructed with intense milling through which hydrogen diffusion channels are shortened. Thus, formed carbon lamellas even block the diffusion of hydrogen and produce a negative influence on hydrogen storage of Mg. Thus, the milling time is also a critical parameter for achieving favorable catalytic enhancement in MgH_2/SWCNTs.

4.4.1.3   Carbon-Based Nanostructured Materials on Sorption Behavior of Mg/MgH$_2$

A detailed investigation on the catalytic effect of different carbon materials on MgH$_2$ was performed by Rodenas *et al.* in 2008 [74]. They investigated the mixture of MgH$_2$ and different carbon materials, particularly graphite, activated carbon, MWCNT, CNFs and activated CNFs. All types of carbon materials lowered the decomposition temperature of MgH$_2$, but the best result was found in the mixture containing CNFs and MWCNTs with metallic impurities (Ni and Fe) embedded from the synthesis. The peak desorption temperature of MgH$_2$ was reduced to 322 and 341°C with the presence of CNF and MWCNT, respectively, whereas the decomposition temperature for as-milled MgH$_2$ was found to be 360°C. Also, the desorption kinetics of MgH$_2$ improved with the presence of CNFs. The desorption was completed within 20 min at 300°C with the 5 wt% of CNFs in MgH$_2$. However, the same is true in the case of as-milled MgH$_2$ completed within 240 min under the same conditions of temperature and pressure. They also reported that the positive effect of the addition of the carbon materials is not related to their porosity. The additions of MWCNT and CNF2, both with low porosity but containing appreciable amounts of metal particles (iron and nickel, respectively) from their synthesis, led to relatively large effects on the decomposition of MgH$_2$. Hence, this effective role of MWCNT and, especially, of CNF2, is attributed to the metallic particles that catalyze the decomposition process. Only with 5 wt% of CNF2 does decomposition of MgH$_2$ start below 300°C, and the maximum peak is at around 320°C. On the whole, the incorporation of the carbon materials can offer an additional benefit to the reduction of the MgH$_2$ particle size and stabilization of the powder particle size [69, 86]. They also reported that with only 5 wt% addition of a nanocarbon, the decomposition time at 300°C can be reduced from 240 to 20 min, which makes these materials very promising for the practical kinetic requirement in hydrogen-fuelled vehicles. Thus they concluded that the addition of carbon materials incorporated with metals, such as nickel and iron, enhances the decomposition properties of magnesium hydride.

After the detailed investigation of Rodenas *et al.* [74], recently, our group also synthesized and investigated the catalytic effect of different CNS on sorption behavior of bulk and nano MgH$_2$ with a particular emphasis on carbon nanofibers [87, 88]. The report of

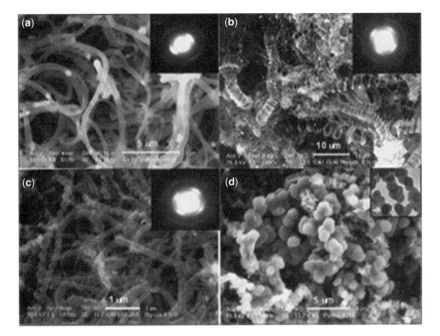

Figure 4.3 SEM images of different carbon structures used as catalyst for received MgH$_2$. (a) planar CNF; (b) helical CNF; (c) twisted CNF, and; (d) spherical carbon particles. Inset of (a), (b) and (c) shows the diffraction patterns obtained by TEM. Reproduced with permission from ref. [87].

Singh *et al.* [87] described the catalytic effect of carbon materials, particularly planer and helical carbon nanofibers and MWCNTs, on sorption behavior of received MgH$_2$. Figure 4.3 represents the morphologies of different synthesized CNS used as the catalyst. Different morphologies of CNFs were synthesized by the CVD technique using LaNi$_5$ as catalyst precursor at different temperature. The planer morphologies of CNFs were synthesized at 550°C, the length of synthesized CNFs have been found to be 6–7 µm, however, helical and twisted morpholiges of CNFs were synthesized at a temperature of 650°C. Spherical carbon particles are synthesized by the ZrFe$_2$ as catalyst precursor at 650°C. Overall all the CNS shown in Figure 4.3 produced a beneficial effect on MgH$_2$.

The SEM micrograph of CNTs are shown in Figure 4.4(a). The synthesized CNTs are multiwalled as confirmed by HRTEM micrograph as shown in Figure 4.2(b). The morphologies of as-received MgH$_2$ shown in Figure 4.4(c), represents that the as-received MgH$_2$ have a particle size in the range 50–100 µm. After 10 min admixing of TCNFs with MgH$_2$ (shown in Figure 4.4[d]),

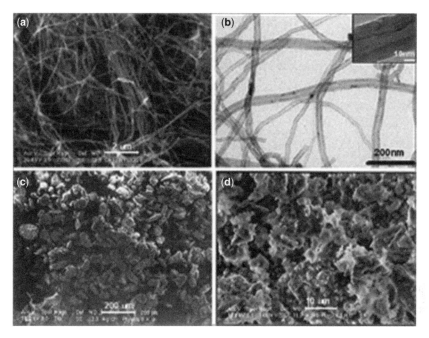

Figure 4.4 (a) SEM images of MWCNT; (b) TEM image of MWCNT (inset shows HRTEM image of MWCNT); (c,d) SEM images of as-received MgH_2 and ball-milled MgH_2. Reproduced with permission from ref. [87].

the particle size has been reduced to 5–10 μm. The temperature programmed desorption (TPD) study revealed that the hydrogen desorption temperature is lowered by ~65 K in the case of TCNFs-catalyzed MgH_2. Also, the absorption kinetics of the received MgH_2 has been increased by the addition of these CNFs. It has been found that the absorption rate for the first 10 min is 0.48 wt%/min. These results are much better as compared to traditional Nb_2O_5 catalyst described by Bhat *et al.* [89]. The improved sorption behavior of MgH_2 was observed mainly because of the size of Ni nanoparticles embedded in CNFs. The larger size Ni particle is expected to have a lower degree of carbon–nickel interaction than that of smaller particles, as in the case of TCNF. This leads to higher catalytic activity of TCNFs.

4.4.1.4   *Catalytic Effect of Different Morphologies of CNFs on Nano MgH_2*

Further detailed investigations have been performed by us with the addition of different morphologies of CNFs in nanocrystalline

MgH$_2$ [88]. The CNFs were synthesized by catalytic thermal decomposition of acetylene (C$_2$H$_2$) gas over LaNi$_5$ alloy. The C$_2$H$_2$ gas was prepared through an inexpensive process which employs reaction of calcium carbide (CaC$_2$) stone and water [90]. The thermal cracking was carried out by filling the gas inside a silica tube. The tube was evacuated and followed by heat treatment of catalyst for 1 h at 500°C under hydrogen and helium ambience. After the activation, the total 450 torr was filled by the 1:4 ratio of H$_2$ and C$_2$H$_2$ gas in the silica tube. The tube was heated at different synthesis temperatures for the synthesis of different morphologies of CNFs. The details of the synthesis process have been published by Raghubanshi *et al.* [90]. At synthesis temperature 650°C, helical CNFs were formed. Figure 4.5(a) represents a typical TEM micrograph of as-synthesized CNF. The as-grown CNFs having the helical morphology and diameter of these CNFs are 150 nm (shown in Figure 4.5[c]). By increasing the synthesis temperature to 750°C,

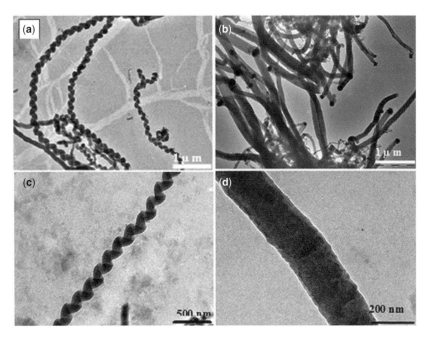

Figure 4.5 TEM micrograph of CNFs synthesized through LaNi$_5$ (a) HCNF synthesized at 650°C and (b) PCNF synthesized at 750°C; (c) magnified micrograph of HCNF; (d) magnified micrograph of PCNF. Reproduced with permission from ref. [28].

planar carbon nanofibers were formed. Figure 4.5(b) represents a typical TEM micrograph of CNFs synthesized at 750°C revealing the as-grown CNFs have planar morphology and diameter of these CNFs are 200 nm as shown in Figure 4.5(d). In both cases the length of these CNFs was found to be 6–8 μm. Admixing of CNFs was performed inside the glove box by using locally fabricated mixer. This is beneficial since it is known that morphology of CNS is important in regard to their catalytic activity, the above-said method of admixing maintains the structural morphology of admixed CNS, which acts as a catalyst for hydrogenation/dehydrogenation of nano MgH_2.

Figure 4.6(a) represents the TPD profile for nano MgH_2 with and without CNFs. It was found that the decomposition temperature was reduced to ~334 and ~300°C for the PCNF- and HCNF-catalyzed nano MgH_2. The effect of ball milling itself reduces the desorption temperature from 422 to 367°C. A further decrease of 67°C takes place due to the catalytic effect of HCNF. The absorption behavior of nano MgH_2 has also been improved by addition of CNFs. The absorption kinetics plots for nano MgH_2 with and without CNF at 300°C and 2 MPa of hydrogen pressure are shown in Figure 4.6(b). Within one hour the nano MgH_2 admixed with PCNF and HCNF reabsorb 5.7 and 5.8 wt% of hydrogen, respectively. Also, it was found that HCNF admixed nano MgH_2 reabsorbs ~5.25 wt% within

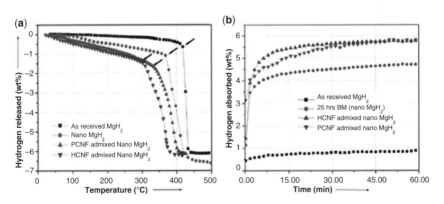

Figure 4.6 (a) TPD profile at heating rate 5°C/min and (b) absorption kinetics plots of nano MgH_2 with and without CNF at 300°C and 20 MPa of hydrogen pressure. Reproduced with permission from ref. [88].

10 min as compared to pristine nano MgH_2, which reabsorbs only ~4.2 wt% of hydrogen.

The morphological changes with admixing of HCNF in nano MgH_2 is also examined by TEM studies of the as-synthesized nano MgH_2 and HCNF-catalyzed nano MgH_2. Figure 4.7(a) and (b) exhibit TEM micrograph of the as-synthesized nano MgH_2. It is evident by the TEM micrograph, the synthesized nano MgH_2 have particles in the range of 10–20 nm. This feature is more clearly

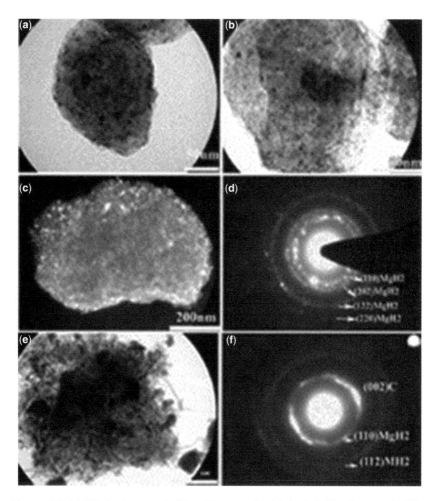

Figure 4.7 (a) TEM micrograph (b) at 50 nm scale; (c) dark field micrograph; (d) SADP of 25 h ball-milled MgH_2 (e) micrograph, and; (f) SADP of HCNF admixed nano MgH_2. Reproduced with permission from ref. [88].

visible in the dark field TEM micrograph (shown in Figure 4.7(c)). The corresponding electron diffraction pattern obtained from the nano MgH_2 is shown in Figure 4.7(d). The spotty rings were indexed based on the tetragonal structure of βMgH_2. Figure 4.7(e) shows representative TEM micrograph of HCNF admixed nano MgH_2. The microstructure also reveals that HCNFs are dispersed homogeneously in nano MgH_2. The presence of diagonally opposite arced diffraction spots in SAED pattern of HCNF admixed nano MgH_2 is the signature of HCNF present in nano MgH_2, other features are similar to those shown in Figure 4.7(d) are representing nano MgH_2.

Since the catalytic activity of these CNFs may also depend on the presence of synthesis-acquired metal nanoparticles, in order to understand the influence of Ni-nanoparticles on the catalytic activities of CNFs, we have further performed investigations on the nano MgH_2 catalyzed separately with Ni-nano particles and purified CNFs. Purification of HCNFs performed through the acid treatment of synthesized HCNFs. After the acid treatment, it has been found that the morphology of the CNF has not changed; only the concentration of metallic particle becomes negligible [90]. From the TPD profile it is found that the admixing of purified CNF and Ni nanoparticle separately show a marginal change in dehydrogenation temperature. Hence for the present case, the combined effect (embedded Ni nanoparticles and fibrous morphologies of CNF) is responsible for the catalytic effect of the as-synthesized CNFs.

4.4.1.5   Catalytic Effect of Graphene on Sorption Behavior of Mg/MgH_2

Among different carbon-based nanostructured materials, graphene has got considerable attention. Because of its notable chemical and physical properties [50] for example large surface area and specific electronic properties, graphene has provided many promising applications in energy storage and catalysis [50]. It is now well known that carbon nanostructures have a beneficial effect on the de/rehydrogenation characteristics of MgH_2 [74]. Therefore it is expected that graphene with high surface area may also have a beneficial effect on the hydrogen sorption properties of MgH_2. It can also act as an inhibitor for preventing the agglomeration of hydride so as to enhance hydrogenation/dehydrogenation performances.

Recently, the catalytic effect of graphene on sorption behavior of MgH_2 was reported by Liu *et al.* [91]. They investigated the catalytic effect of graphene nanosheets (GNS) on the improvement of sorption behavior of MgH_2. The GNS with high surface area were synthesized by thermal exploitation technique and used for the improvement in sorption behavior of MgH_2. The as-synthesized GNS exhibits superior catalytic effect. The study focused on the effect of milling duration on sorption characteristics of GNS-catalyzed MgH_2. The hydrogen sorption kinetics and decomposition behavior improved significantly with increasing milling duration. The 20 h ball-milled material can absorb ~6.6 wt% of hydrogen within one min at 300°C. It can also absorb 6.3 wt% of hydrogen within 180 min even at 150°C. The detail TEM analysis revealed that graphene sheets exist at the edge of the sample, which indicated that GNS dispersed in a disorderly and irregular manner in MgH_2. From HRTEM images they concluded that the crystallites of MgH_2 exist in the range of 5–10 nm. Also, they found that on the edge of the sample 3–5 individual graphene layers with a length less than 5 nm are present. From this it is confirmed that smaller GNS are distributed in an irregular and disorderly manner. It is know well known that the dissociation of hydrogen molecule to hydrogen atom requires 4.5 eV, which indicates that the activation barrier for the rehydrogenation of Mg is quite high [92]. Some recent reports have proved that the presence of carbon on the Mg surface reduced the dissociation barrier of hydrogen molecule to atom. The barrier is lowered to 1.05 ev and the atomic hydrogen is also diffuse easily into the sublayer through the channels [92] due to which hydrogen dissociation/recombination in presence of GNS is facilitated and possible below 300°C. Liu *et al.* [91]. clearly demonstrated that the small GNS dispersed in an irregular and disorderly manner in the composite after ball milling. Which provides more edge sites hydrogen channels, prevent the grain of MgH_2 sintering and agglomeration, thus leading to enhanced hydrogen storage properties.

4.4.2   Sodium Alanate

Sodium aluminum hydride ($NaAlH_4$) would seem to be a possible candidate for application as a practical on-board hydrogen storage material due to the high theoretical reversible hydrogen storage capacity of 5.6 wt%, low cost and its availability in bulk [93]. It decomposes in three steps as follows:

$$3NaAlH_4 \longrightarrow Na_3AlH_6 + 2Al + 3H_2 \quad (3.7\ wt\%\ H_2)\ (4.1)$$

$$Na_3AlH_6 \longleftrightarrow 3NaH + Al + 1.5H_2 \quad (1.8\ wt\%\ H_2) \quad (4.2)$$

$$NaH \longleftrightarrow Na + 1/2H_2 \quad\quad\quad\quad\quad\quad\quad (4.3)$$

Theoretically, $NaAlH_4$ and Na_3AlH_6 contain large amounts of hydrogen, 7.4 and 5.9 wt%, respectively. The reversibility of the above two reactions is a critical factor for the practical applications. The operating temperatures are 185, 230, and 260°C for the first and the second reaction, respectively. Finally, the decomposition of NaH occurs at a much higher temperature, with the total hydrogen release of 7.40 wt%. For hydrogen storage, only the first two reactions need to be considered, because the decomposition of NaH occurs at a too high temperature of 425°C [93]. With 2 mol% TiN as a doping agent, cyclic storage capacity of 5 wt% H_2 is achieved after 17 cycles [94–99]. This clearly indicates that the addition of titanium species enhances not only the dehydrogenation kinetics, but also the rehydrogenation reaction of $NaAlH_4$ [100–102].

4.4.2.1 Catalytic Effect of Carbon Nanotubes on Sorption Behavior of $NaAlH_4$

The effect of CNTs on de-/rehydrogenation characteristics of $NaAlH_4$ was first investigated by our group [103]. This was the first report on improved sorption behavior of $NaAlH_4$ by doping of CNTs. The study demonstrated that among different concentrations of CNTs, 8 mol% of CNTs is optimum to increase the sorption behavior of $NaAlH_4$. It is found that desorption of 3.3 wt% of hydrogen occurs within 2 h at 160°C. Also, the CNTs admixed samples showed good reabsorption kinetics. Figure 4.8 represents the SEM micrograph of CNT-doped $NaAlH_4$. The as-received $NaAlH_4$ have particle size ~40 μm. Figure 4.8(b) describes the morphology of $NaAlH_4$+8 mol% CNT, whereas Figure 4.8(c) shows the microstructural feature of $NaAlH_4$+CNT after complete dehydrogenation. The ball-milled sample has spotty surface features, as shown in Figure 4.8(b), having spot size ~15 μm. The doped CNT generally exists in the bundles and the used CNT also

Figure 4.8 SEM micrographs bring out the microstructure of the (a) $NaAlH_4$; (b) 8 mol% CNT-doped $NaAlH_4$, and; (c) after dehydrogenation of 8 mol% CNT-doped NaAlH4 samples. Reproduced with permission from ref. [103].

have typical strength as typified by 1.8 TPa. Thus, CNTs can work as a needle to penetrate $NaAlH_4$ matrix. Therefore, these beneficial effects of CNTs arise due to their nanosize and strength. The CNTs penetrated the $NaAlH_4$ matrix produce deformed region around them. The insertion thus produces holes surrounded by disc-like regions. Through these the ballistic transfer of the hydrogen occurs and sorption properties of $NaAlH_4$ are enhanced.

Further studies in this sequence have been performed by Chento *et al.* in 2007 [104]. They investigated the catalytic effect of CNTs on sorption behavior of Ti-doped $NaAlH_4$. Since Ti is known as an effective catalyst for $NaAlH_4$ [94–102], they report further improvement in sorption behavior by addition of CNTs in Ti-doped $NaAlH_4$. It is well known that the theoretical storage capacity of $NaAlH_4$ is 5.6 at 256°C. The significant results were obtained with the presence of Ti as catalyst in $NaAlH_4$. With Ti-doped $NaAlH_4$ the material can desorb 3 wt% of hydrogen in 4 h at temperature 125°C, upon further increasing the doping concentration, the desorption becomes faster. The main drawback of Ti as catalyst is the reduction of desorption storage capacity due to the density of Ti which creates dead weight for storage. Thus the addition of carbon in Ti-doped $NaAlH_4$ sample is beneficial with respect to storage capacity of $NaAlH_4$. They also investigated the optimum concentration of CNTs in $NaAlH_4$. They reported that 10 wt% CNTs showed more beneficial results as compared to 20 wt% CNTs. Synergism was reported between carbon and Ti in double-doped $NaAlH_4$.

Recently, the results reported by Chen *et al.* [105] described the beneficial improvement in sorption behavior of $NaAlH_4$ in the presence of MWCNTs. This occurred due to the change in desorption pathway of $NaAlH_4$. The results described that the initial hydrogen desorption temperature is reduced to 80°C from 210°C for pristine $NaAlH_4$ with addition of 50 wt% MWCNTs. The dehydrogenation routes of $NaAlH_4$ were altered substantially. Two metallic hydrides (NaH and AlH_3) and Al were identified at a temperature of as low as 90°C, without the formation of Na_3AlH_4. In addition, Al_4C_3 became noticeable at temperature above 120°C. The formation of Al_4C_3 indicated that MWCNTs not only act as catalyst but also participated in dehydrogenation reaction of $NaAlH_4$. However, such type of behavior has not been observed with the activated carbon.

4.4.2.2   *Metal Decorated Activated Carbon for Enhancing Sorption Behavior of $NaAlH_4$*

A recent report of Lin *et al.* described the effect of the metal-decorated activated carbon as a catalyst for dehydrogenation of $NaAlH_4$ [106]. The report described that the dehydrogenation of $NaAlH_4$ is greatly facilitated by activated carbon (AC) catalyst. This enhancement is further improved by the decoration of carbon with Co, Ni or Cu nanoparticles. As compared to Ti-based catalyst the dehydrogenation was lowered by as much as 100°C using 3 wt% Co- or Ni-decorated carbon. They found that the catalytic activity of Co- and Ni-decorated carbon is similar, whereas Cu-decorated carbon showed less effective catalytic activities. Rehydrogenation reaction is also facilitated by metal-decorated AC. The author proposed that these beneficial effects occurred due to the hydrogen spillover effect from metal to AC. Spillover is a phenomenon in which a hydrogen molecule is dissociatively adsorbed onto a metal surface and the hydrogen atoms then migrate across the metal support interface onto the support. When the process is reversed hydrogen atoms on the support migrate onto the metal where they recombine

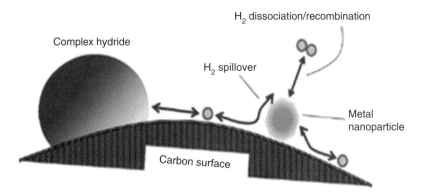

Figure 4.9 Schematic of the proposed model for catalysis by metal-decorated carbon matrix. The carbon component acts as a conduit for transporting hydrogen atoms between hydride and metal catalyst, which is used for hydrogen activation and hydrogen atoms recombination. Reproduced with permission from ref. [106].

and desorbed as hydrogen molecule as shown in Figure 4.9. The effect of this pathway depends on the metal and would be more significant for metal that can dissociate a hydrogen molecule and recombine hydrogen atom at lower energy. This explains why Cu decorated AC has less significant catalytic behavior as compared to Ni or Co decorated AC.

4.4.2.3   Catalytic Activities of Different Carbon Nanostructures for Sorption Behavior of NaAlH₄

Recently Berseth *et al.* (2009) [107] investigated the synergistic effect of zero-, one- and two-dimensional carbon nanostructures (i.e., fullerene, CNTs and graphene, respectively) on sorption behavior of NaAlH$_4$. The study described the theoretical as well as experimental aspect of aforesaid materials as catalyst for NaAlH$_4$. The study of Berseth *et al.* also developed an unambiguous understanding of how the catalyst works. In order to avoid the metallic content in CNS they adopted solvent preparation technique for the synthesis of carbon nanomaterials. Experimentally they have found that C$_{60}$ showed better catalytic effect as compared to other carbon nanomaterials. Also, they proposed that the catalytic effect of these CNSs can be tuned by changing their curvature and dimension. The change in curvature has a different catalytic effect due to the change of the electron localization/delocalization effect.

For NaAlH$_4$ the cluster is stabilized by the formation of the Na$^+$ cation, this charge is balanced by (AlH$_4$) anion. The interaction of

NaAlH$_4$ with a catalyst that is as electronegative as [AlH$_4$]$^{-1}$, is compromised and hence the covalent bond between Al and H is weakened. This weakening of metal hydrogen bond leads to lowering of the hydrogen desorption energy, consequently improving desorption kinetics. The CNSs are formed by the rolling of graphitic carbon sheets, and the accelerated absorption may related to the flat graphite particles, which are responsible for the change in pi and sigma bonding orbital when H$_2$ molecules interact with the surface of the material. The fullerene C$_{60}$ is a 7Å sphere made of graphite carbon that has increased curvature as compared to 50 nm diameter MWCNTs and exhibits the strongest catalytic effect on NaAlH$_4$. Ab initio modeling has also been undertaken to further understand the interactions taking place between NaAlH$_4$ and carbon materials. Figure 4.10 demonstrates that the H removable energy is reduced to 3.6, 2.95 and 2.85 eV, respectively, for graphene, CNTs and fullerene. However, for the pristine NaAlH$_4$ it is found to 3.8 eV. Thus catalytic activity of carbon nanostructures highly depends upon the curvature of the nanostructure.

After the report of Bereth *et al.* the studies on NaAlH$_4$ admixed with various CNSs have also been performed by our group. The study of Qian *et al.* [108] described the beneficial effect of CNS on de/rehydrogenation of NaAlH$_4$. The study particularly focused on the catalytic effect of CNTs and CNFs with two different morphologies,

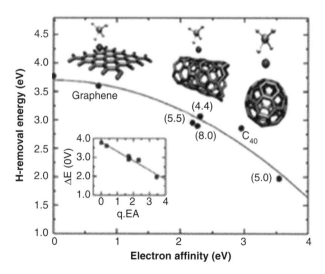

Figure 4.10 Correlation of the electron affinity of carbon substrate and the hydrogen removal energy. Reproduced with permission from ref. [107].

namely HCNFs and PCNFs. It is found that the desorption temperature for Step 1 (Eq. 4.1) and Step 2 (Eq. 4.2) is 170 and 270°C, respectively. The desorption temperature of both steps is reduced due to the presence of CNS. The desorption temperature is lowered to 143°C with the admixing of HCNFs in $NaAlH_4$. Whereas the same for pristine $NaAlH_4$ is ~170°C, as shown in Figure 4.11(i).

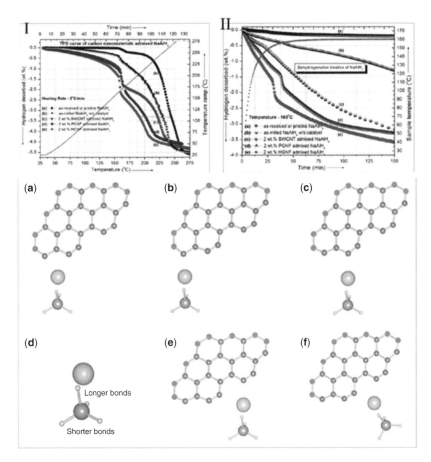

Figure 4.11 (I) TPD curves and (II) desorption kinetic curves of $NaAlH_4$ admixed with 2 wt% carbon nanovariants. Equilibrium configurations of $NaAlH_4$ positions with the Na atom facing various sites of carbon edges (zigzag or armchair edge) and of the isolated $NaAlH_4$ cluster: A–C, sites at zigzag edge; D, isolated one for comparison; E and F, sites at armchair edge. Na atoms are shown in the color of yellow, Al in blue, H in pink, and C in green. The color of "filled" light green in all carbon sheets above means the corresponding carbon atoms are fixed in modeling to simulate the restraining effect that those would experience that are bonded to other carbon atoms within the same graphene layer. Reproduced with permission from ref. [109].

The desorption kinetics plot shown in Figure 4.11(ii) was per-formed at 160°C. It has been found that desorption kinetics of 2 wt% CNFs admixed sample is superior to that of 2 wt% SWCNTs admixed sample. Furthermore, among CNFs, helical CNFs possess better catalytic effect as shown in Figure 4.11(ii). Thus HCNFs possess superior catalytic behavior as compared to other CNSs. The different catalytic activities arose due to the different morphologies of CNFs. The enhanced catalytic behavior of HCNFs is due to the higher curvature. On the other hand, planer CNFs and CNTs possess lower catalytic activity when compared to HCNFs.

The basic constituent of these two types of nanostructures is graphene, rolled up into a tube for SWCNTs and stacked on top of each other in the case of CNFs. The main difference between these types of catalyst is the availability of active sites, which is responsible for the catalytic behavior. The fundamentals behind the catalytic activities of CNFs have also been investigated by the model interaction of $NaAlH_4$ with various edge sites of one single-layer graphene from first principal theory, and the details of theoretical modeling results have been described by Qian *et al.* [109]. Both zig-zag edge and armchair edge are considered for the study (shown in Figure 4.11[a–f]). It is reported that for the pristine $NaAlH_4$ cluster the evaluated Al-H bond length and hydrogen removable energy is ~1.6–1.7 Å and ~3.8 eV, respectively. Both carbon edges decrease the hydrogen removable energy significantly (to less than 2 eV). Also it is noted that the decrease in hydrogen removable energy is lower than those of $NaAlH_4$ mixed with C_{60} and CNTs.

4.4.2.4   *Effect of Graphene on Sorption Behavior of $NaAlH_4$*

Recently, Viswanathan *et al.* [110] investigated the catalytic effect of nitrogen-doped graphene and CNTs as a catalyst for sorption behavior of $NaAlH_4$. The synthesized CNTs exist in bunches have a diameter of 200 nm and length of 60 μm. In the case of graphene and nitrogen-doped graphene, randomly oriented transparent sheet of several microns has been observed in SEM micrograph. These synthesized nanomaterials are used as a catalyst for hydrogen sorption from $NaAlH_4$. They demonstrated that nitrogen-doped carbon materials have superior catalytic effect as compared to pure carbon materials. Desorption of 5.2 wt% of hydrogen occurred within 150 min at 220°C by $NaAlH_4$ catalyzed with nitrogen-doped graphene. Also, the decomposition temperature is decreased for $NaAlH_4$ catalyzed with nitrogen-doped graphene. The total reduction in activation energy for the 1st and 2nd step is found to be ~33.9 and ~77 kJ/mol^{-1}, respectively, for NaAlH$_4$ catalyzed with nitrogen-doped graphene. The rehydrogenation reaction is also facilitated by the deployment of N-doped graphene in NaAlH$_4$. This may occur because of the presence of active centers as the form of nitrogen in nitrogen-doped carbon nanomaterials. Hydrogen molecules get activated on the nitrogen site and then are dissociated into hydrogen atoms. These hydrogen atoms spill over to the nitrogen-doped graphene metal interface and react with metal to form metal hydride [111].

Recently, Lia *et al.* [112] developed an effective strategy for destabilization of NaAlH$_4$ through the morphological tuning of NaAlH$_4$ assisted by CNSs. Through this phenomenon the complex anion interacts with carbon matrix. The study focused on the NaAlH$_4$/graphene (GNs), fullerene and messoporus carbon (MC) selected as model systems to illustrate the effect of morphological tuning assisted by carbon nanomaterials; the dissolution-recrystallization process was used to tune the morphology of the NaAlH$_4$. In this process the desired mixture of NaAlH$_4$ and different carbon nanomaterials were added to dried tetrahydrofuran (THF) solution with vigorous stirring followed by evaporation of THF. The NaAlH$_4$ recrystallized and dispersed homogeneously on the surface and/or the internal pores of carbon material. From the SEM characterization they found that scale-like continuous structure, flower-like structure with diameter ranging from 5 to 10 μm and uniform particle with an average diameter of 2 μm have been formed with GNs-, C$_{60}$- and MC-assisted samples, respectively. The DSC analysis described that the onset desorption temperature is reduced to 188,185 and 160°C for the samples assisted with GNs, C$_{60}$ and MC, respectively. Whereas, the onset desorption temperature for the pristine NaAlH$_4$ is ~210°C. Also, the improved efficiency of the carbon nanomaterials for kinetics is found to be in the order of MC, C$_{60}$, and GNS. The enhanced desorption may occur due to both particle refinement and interaction of [AlH4]$^{-1}$ complex with the carbon matrix induced by the presence of carbon nanomaterials. It has been reported that the hydrogen sorption behavior improved significantly by morphological tuning of NaAlH$_4$ assisted with CNS. These beneficial improvements arose due to the particle refinement and the interaction of complex anion with the carbon matrix.

The recrystallization of $NaAlH_4$ occurred due to the presence of carbon nanomaterials. The nucleation process is affected by the presence of high surface area carbon material which provide more active site and lower energy barrier for nucleation. Also, the recrystallization may altered by the interaction of carbon materials with $[AlH]^{-1}$. As the interaction increases, the nuclei growth is inhibited and grain refinement becomes pronounced. As a consequence, the formation of smaller grains enhance the sorption kinetics of $NaAlH_4$. Also, carbon nanomaterial possess higher electronegativity, and thus the interaction of carbon with $[AlH]^{-1}$ weakens the Al-H bond and reduced the activation energy for the desorption reaction of $NaAlH_4$.

4.4.3   Amides/Imides

The pioneering efforts of Chen *et al.* [113] added a new material in a series of complex hydrides having favorable thermodynamics for on-board storage of hydrogen. Upon heating the $LiNH_2$/LiH mixture to 340–480°C, the amide is converted into imide and ammonia, and the latter immediately combines with the hydride, forming lithium imide and hydrogen. The Li_3N has a high hydrogen storage capacity and the hydrogenation and dehydrogenation follow the following reactions:

$$Li_3N + 2H_2 \longleftrightarrow Li_2NH + LiH + H_2 \longleftrightarrow LiNH_2 + 2LiH \quad (4.4)$$

The sequential dehydrogenation and rehydrogenation reactions of $LiNH_2$ proceed according to Equation 4.4. The first and second reactions release 5.5 and 5.2 wt % hydrogen, respectively [114]. It was proposed [115–117] that partial substitution of Li with more electronegative elements would weaken the bonding between Li^+ and $[NH_2]^-$. Lithium amide with partial Mg substitution was synthesized from the Li-Mg alloys, and the dehydrogenation reactions were investigated [118]. The dehydrogenation reaction from $LiNH_2$ (+LiH) without the partial Mg substitution begins at approximately 277°C upon heating at the rate of 10°C/min under argon. However, the temperature is drastically lowered to around 100°C with increasin;g Mg concentrations. Thus, cation substitution is an effective method for decreasing the dehydrogenation temperatures of $LiNH_2$ [119, 120].

The replacement of LiH by MgH_2 gives rise to a system with better thermodynamics and hydrogen storage characteristics. These systems are one of the hot topics in the hydrogen storage area because of these materials having thermodynamically suitable characteristics for on-board H storage (Eq. 4.5). These systems are composed of $Mg(NH_2)_2$ and LiH in different weight-wise mixing ratios. The different mixing ratios give different dehydrogenated states.

$$Mg(NH_2)_2 + 2LiH \longleftrightarrow Li_2Mg(NH)_2 + 2H_2 \qquad (4.5)$$

It has been found that the $Mg(NH_2)_2$/LiH mixture desorbed ~4.25 wt% within 210 min at 200°C and 0.1 MPa of hydrogen pressure. The slow desorption kinetics prevail even in this system. In some recent studies, absorption/desorption properties have been found to improved through deployment of suitable catalyst [121, 122].

As we have described in the previous section, in several recent studies, different forms of carbon materials have been extensively used as a catalyst for metal (Mg) and complex hydrides $(NaAlH_4)$. For the case of the Li-Mg-N-H system, the effects of carbon nanomaterials on sorption behavior of Li-Mg-N-H are rather sparse. Here we will discuss and describe the results of Wang *et al.* [123] and some of our recent studies on the catalytic activity of CNSs for the de-/rehydrogenation kinetics of 1:2 $Mg(NH_2)_2$/LiH mixture.

4.4.3.1   *Effect of carbon Nanotubes on Sorption Behaviors of Li-Mg-N-H*

Wang *et al.* [123] investigated the catalytic effect of CNSs on sorption behavior of Li-Mg-N-H. This was the initial report on the deployment of CNSs as catalyst for Li-Mg-N-H. They reported that the sluggish dehydriding kinetics of the system is improved by the addition of different carbon materials such as SWCNTs, MWCNTs graphite and activated carbon. It was found that the addition of SWCNTs showed more significant enhancement in dehydriding kinetics of the Li-Mg-N-H system at 200°C. They found that instead of 1 h, SWCNTs-catalyzed mixture showed desorption of 4.5 wt% of hydrogen within 20 min at 200°C with a slight influence on storage capacity. On the other hand, with the addition of other carbon additives the influence is not as significant as with the

SWCNTs. This may occur due to the creation of a large amount of phase boundary between the material matrix and CNSs, and also CNTs provide tunnels for rapid hydrogen diffusion. Moreover, the study also focused on the catalytic effect of embedded metal nanoparticles in SWCNTs. The experiments were also performed for the purified SWCNTs by Wang *et al.* It has been found that purified SWCNTs are less significant as compared to synthesized SWNTs. Hence, it directly meant that the presence of embedded metal nanoparticles in CNSs enhances the catalytic effect of carbon nanomaterials.

4.4.3.2   Comparison of Catalytic Activity of CNSs on Li-Mg-N-H

The detailed investigations on the effect of different CNSs, particularly CNTs and CNFs, on the improvement of the de-/rehydrogenation characteristics of a $Mg(NH_2)_2$/LiH mixture have been performed by us [124]. Amongst CNTs and CNFs, the improvement in the hydrogenation properties for the $Mg(NH_2)_2$/LiH mixture is higher when CNFs are used as a catalyst. In addition, the effect of metallic particles present in carbon nanostructures (acquired from the synthesis process) are also investigated by deployment of two different types of CNFs synthesized through different catalysts: (a) CNF1 (synthesized using a $ZrFe_2$ catalyst), and (b) CNF2 (synthesized using a $LaNi_5$ catalyst).

Different CNSs have been synthesized by chemical vapor deposition. The MWCNTs used in the present study were synthesized by chemical vapor deposition using ferrocene dissolved in benzene [125–135]. Figure 4.12(a) represents the TEM micrograph of the as-synthesized MWCNTs. The diameter of as-synthesized MWCNTs is found to be 20–50 nm. Figure 4.12(b) represents the TEM micrograph of as-received SWCNTs. The diameter of as-received SWCNTs is found to be 10–20 nm. The CNFs were synthesized in our laboratory by catalytic thermal decomposition of hydrocarbon gas acetylene (C_2H_2) over various catalysts (described in Section 4.4.1.4) [36]. Different types of CNFs can be synthesized by using different catalyst precursors. For the present case, planar morphology of CNFs is synthesized by using $ZrFe_2$ alloy as a catalyst at a temperature of 600°C. Planar morphology of CNFs is also synthesized by using

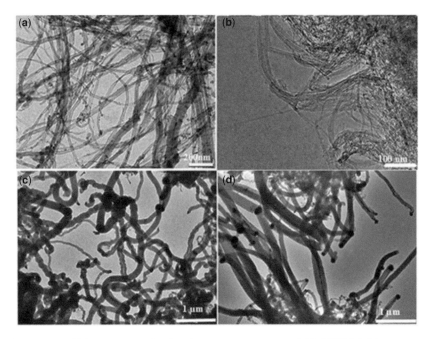

Figure 4.12 TEM micrograph of carbon nanostructures. (a) MWCNT; (b) SWCNT; (c) CNF1 synthesized using ZrFe$_2$ catalyst; (d) CNF2 synthesized using LaNi$_5$ catalyst. Reproduced with permission from ref. [124].

LaNi$_5$ alloy as a catalyst at a temperature of 750°C. From the TEM analysis shown in Figure 4.12(c) and (d), the length and the diameter of these planar CNFs are found to be 5–7 μm and 200 nm, respectively.

The catalytic activity of carbon nanostructures may also depend on the presence of the metallic particle contents. These metallic particles are acquired from the metallic/intermetallic catalysts which are used for the synthesis. The metal elements in synthesized/received and purified CNS were determined by EDX attached with SEM. The as-synthesized MWCNTs, CNF1 and CNF2 have Fe, Fe-Zr, and Ni, respectively. The as-received SWCNTs are free from the metallic catalyst. The purification of CNSs was performed through the acid treatment of the as-synthesized CNSs with concentrated HNO$_3$. After the acid treatment, it was found that only the metallic contents were removed, and also the morphology of the material have not been changed.

The TPD curves described that the decomposition temperature of the mixture was decreased appreciably with the mixing of CNSs.

The initial decomposition temperature for pristine $Mg(NH_2)_2/LiH$ mixture is found to be ~90°C. This is reduced to ~80, 75, 60 and 50°C for SWCNTs, MWCNTs, CNF1 and CNF2 mixed 1:2 $Mg(NH_2)_2/$ LiH, respectively. It was also found that the maximum decomposition temperature for sample without catalyst is ~250°C, which is reduced to ~190, 180, 170 and 150°C for SWCNT, MWCNT, CNF1 and CNF2 mixed $Mg(NH_2)_2/LiH$ respectively. The effectiveness of the catalyst is found to be highest for CNF2 followed by CNF1, MWCNTs and SWCNTs. The hydrogen uptake capacity of pristine and MWCNTs, SWCNTs, CNF1 and CNF2 mixed $Mg(NH_2)_2/LiH$ after 5 h was found to be ~1.8, 2.25, 2.1, 3.25 and 3.5 wt%, respectively. The activation energy for dehydrogenation reaction of the $Mg(NH_2)_2/LiH$ mixture is also reduced from ~97.2 to 88.3, 74.16 and 68.2 kJmol^{-1} with mixing of SWCNT, MWCNT, CNF1 and CNF2, respectively.

The catalytic activity of carbon nanostructures depends on the presence of synthesis-acquired metal nanoparticles [41, 107]. Thus, there are two possible facts which are responsible for catalytic behavior of CNSs and these are: (a) type and morphology of carbon nanostructures, and (b) presence of synthesis-based metallic particles. In order to confirm the effect of metallic particles on the catalytic activity of CNSs. Further TPD experiments were performed for the $Mg(NH_2)_2/LiH$ mixture mixed with purified carbon nanostructures and found that the effect of as synthesized CNSs are more beneficial as compared to purified CNSs. Hence, the morphology of carbon nanostructures is not the only factor for lowering of the desorption temperature. It is the combined effect of both the morphology and the presence of synthesis-acquired metallic particles in CNSs, which play a vital role in lowering the decomposition temperature of the $Mg(NH_2)_2/LiH$ mixture.

It is now well known that the rate of dehydrogenation reaction depends upon the decomposition of $Mg(NH_2)_2$ and the transition metal particles enhance the de-/rehydrogenation characteristics of the $Mg(NH_2)_2/LiH$ mixture [27]. The used CNSs also have Fe and Ni nanoparticles have a beneficial effect on the sorption behavior of Li-Mg-N-H. The possible approach towards mechanism may be dealt with in the presence of unpaired subshell in synthesis-acquired Fe ($3d^6$) and Ni ($3d^8$) nanoparticles which have a tendency to interact with the lone pair of electrons of nitrogen atom, hence weakening the Mg-N bond in the $Mg(NH_2)_2/LiH$ mixture [27]. It may also be pointed out that even without the synthesis-acquired metal particles the catalytic effect of these nanostructures arises due to the curvature effect (electron affinity) of carbon nanostructures [41, 107]. Thus, the decomposition of $Mg(NH_2)_2$ is facilitated by the presence of carbon nanostructures having synthesis-acquired metallic nanoparticles.

It was also found that the catalytic activity of MWCNT is less significant as compared to CNF1 even though the embedded synthesis-acquired metallic particle is the same for both cases, i.e., Fe. These different levels of improvement in hydrogenation properties arise due to the morphological difference of these two carbon nanostructures. The MWCNT is formed through the graphitic carbon sheet rolled into tubes. Thus there are fewer possibilities of active sites. However, in the case of nanofibers, the graphitic carbon sheet becomes aligned and fiber structure gets formed, and the large number of carbon nanosheets and unsaturated bonds are present at the edges of each sheet. Because of these features the numbers of active sites are higher. As also evident by the TEM micrograph of samples mixed with MWCNTs and CNF2 shown in Figure 4.13, parent material agglomerates along with the length of the fiber and provides intimate interaction. This is the most probable reason for better catalytic behaviors of CNF over MWCNTs. Furthermore, we have found that the catalytic activity of CNF1 (having Fe as metallic impurity) is lower as compared to the CNF2 (having Ni as metallic impurity). This may happen because the electron-attracting tendency of Fe ($3d^6$) nanoparticles is less significant as compared to the electron-attracting tendency of Ni ($3d^8$) nanoparticles [28, 29].

4.4.3.3   Effect of Different Morphologies of CNFs on Sorption Behaviour of Li-Mg-N-H

Our investigations have also been focused on the effect of different morphologies of CNFs on sorption behavior of the Li-Mg-N-H system [29]. The two different morphologies, helical and planer, were synthesized by using $LaNi_5$ as catalyst precursor at two different temperatures. The details of the synthesis process and TEM micrographs of these CNFs have already been described in Section 4.4.1.4 and shown in Figure 4.5.

The decomposition characteristics revealed that the initial decomposition temperature of the pristine $Mg(NH_2)_2/LiH$ mixture was reduced to ~70, and 50°C with admixing of PCNF and HCNF,

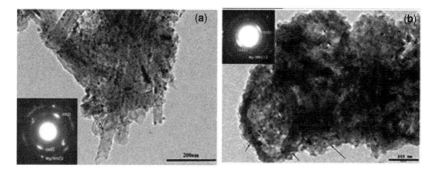

Figure 4.13 TEM micrographs of (a) MWCNTs (mixed); (b) CNF2 mixed parent material. Insets of (a) and (b) represent the selected area diffraction patterns. Reproduced with permission from ref. [124].

respectively. Moreover, the maximum decomposition temperature was reduced to 150°C and 140°C for PCNF and HCNF admixed samples, respectively. Since these CNFs (PCNF and HCNF) have the same synthesis-acquired metallic particles, these different levels of improvement arised due to the morphological differences in CNFs. The hydrogen uptake capacity of pristine and 4 wt% of each PCNF and HCNF admixed $Mg(NH_2)_2/LiH$ mixture after 5 h at 200°C were found to be ~1.8, 3.4 and 3.7 wt%, respectively. This may have occurred due to the presence of carbon fragments having various hybridized orbitals, and these hybridized orbitals can interact with the hydrogen molecule which provides transit site for the hydrogenation. It was found that the activation energy for dehydrogenation reaction of 1:2 $Mg(NH_2)_2/LiH$ mixture was also reduced to 67.8 kJ/mol and 65.2 kJ/mol with admixing of PCNF and HCNF, respectively.

The exact catalytic mechanism for de-/rehydrogenation reaction of the CNF admixed $Mg(NH_2)_2/LiH$ mixture is not yet clear. The CNFs (PCNFs and HCNFs) have similar synthesis-acquired metal nanoparticles (Ni) and have different morphology because they have different levels of improvement in hydrogenation behavior. This improvement may arise due to two possible reasons. Among these one corresponds to the curvature of the HCNFs which may increase the electronegativity of HCNFs as compared to PCNFs [107]. Hence the bonding and lone pair of electrons present in parent material (i.e., $Mg(NH_2)_2/LiH$) become more delocalized. This may result in weakening the bond strength and facilitating the decomposition reaction of $Mg(NH_2)_2/LiH$. The other reason corresponds

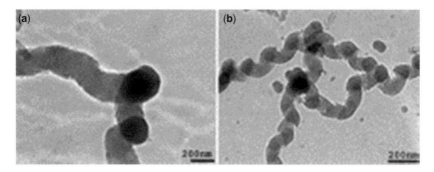

Figure 4.14 TEM micrographs of as-synthesized PCNF and HCNF from the catalyst region. Reproduced with permission from ref. [28].

to the size of the Ni nanoparticles present in CNFs. The HCNFs having smaller (~150 nm) Ni nanoparticles as compared to PCNFs (~200 nm) are shown in Figure 4.14. Ni particles are known to have good catalytic activity for different hydrogen storage materials. The catalytic activities of the Ni metal particles further increase with a decrease in particle size [119].

4.5   Summary

We have reviewed the catalytic effect of different carbon nanostructures on sorption behavior of promising light weight hydrogen storage materials, e.g., MgH_2, $NaAlH_4$ and Li-Mg-N-H systems. Particular attention has been focused on carbon nanotubes, carbon nanofibers and graphene employed as a catalyst for aforesaid hydrogen storage materials. The carbon-based nanostructures have provided innovative opportunities to overcome the high temperature desorption and poor kinetics issues of promising storage materials. Both the theoretical calculation and experimental investigations have described that the presence of carbon in parent materials matrix reduces the kinetic barrier for rehydrogenation. Also, due to the high electronegativity, carbon materials can interact with anion of light metal hydride and weaken the bonds through which desorption temperature is lowered. Additionally, carbon materials are light, high surface area and low cost materials, which makes them an ideal additive for high-capacity storage materials.

Generally, carbon nanomaterials act as an effective catalyst for enhancing the kinetics of the de-/rehydrogenation reaction of hydrides. The mechanism behind the beneficial effect on kinetics is still a matter of debate. Carbon nanostructures often containing metal nanoparticles further improve the kinetics, for example, the better catalytic effect of CNF embedded with Ni nanoparticles has a beneficial effect on sorption behavior and desorption temperature of MgH_2, $NaAlH_4$ and Li-Mg-N-H. Surprisingly, the presence of carbon in hydride matrix reduces the dissociative energy of molecular hydrogen to atomic hydrogen and hence reduces the rehydrogenation barrier of different storage materials. The carbon nanostructures also alter the thermodynamic properties and enhance the thermal conductivity of hydrides. In summary, it is clearly demonstrated that the use of carbon materials offers great opportunities in enhancing the hydrogen storage properties of metal hydrides. Furthermore, the effect of carbon nonmaterial is tuned by morphological changes and the variations in embedded metal nanoparticles.

Acknowledgements

The authors are extremely thankful to Prof. C.N.R. Rao, Prof. R.S. Tiwari, Dr. M.A. Shaz, Dr. K. Awasthi, Dr. H. Ragubanshi and Dr. T.P. Yadav for their encouragement and support. Moreover, thanks are due to Mr. Ashish Bhatanagar for his help in the chapter preparation. The financial support from the Department of Science and Technology, New Delhi, India is gratefully acknowledged.

References

1. L. Schlapbach and A. Züttel, *Nature*, Vol. 414, p. 353, 2001.
2. P.P. Edwards, V.L. Kuznetsov, and W.I.F. David, *Philos. Trans. R. Soc. London, Ser. A*, Vol. 365, p. 1043, 2007.
3. A.W.C. van de Berg and C.O. Areán, *Chem. Commun.*, 2008, 2008.
4. J. Yang, A. Sudik, C. Wolverton, and D.J. Siegel, *Chem. Soc. Rev.*, Vol. 39, p. 656, 2010.
5. E.K. Stefanakos, D.Y. Goswami, S.S. Srinivasan, and J.T. Wolan, Hydrogen energy, In: Myer Kutz, Editor, Environmentally conscious alternative energy production, John Wiley & Sons, Inc., 2007, pp. 165–206.
6. S.M. Acebes, G.D. Berry, J. Martinez-Frias, and F. Espinosa-Loza, *Int. J. Hydrogen Energy*, Vol. 31, p. 2274, 2006.
7. X. Lin, J. Jia, P. Hubberstey, M. Schröder, and N.R. Champness, *Cryst. Eng. Comm.*, Vol. 9, p. 438, 2007.

8. R.E. Morris and P.S. Wheatley, *Angew. Chem. Int. Ed.*, Vol. 47, p. 4966, 2008.

9. K.M. Thomas, *Dalton Trans.*, p. 1487, 2009.

10. Y.H. Hu and L. Zhang, *Adv. Mater.*, Vol. 22, p. E117, 2010.

11. J. Sculley, D.Q. Yuan, and H.C. Zhou, *Energy Environ. Sci.*, 2011, 4, 2721.

12. http://www1.eere.energy.gov/hydrogenandfuelcells/storage/pdfs/targets_onboard_hydro_storage_explanation.pdf, 2009.

13. G.S. Walker, M. Abbas, D.M. Grant, and C. Udeh, *Chem. Commun.*, Vol. 47, p. 8001, 2011.

14. B. Peng, J. Liang, Z.L. Tao, and J. Chen, *J. Mater. Chem.*, Vol. 19, p. 2877, 2009.

15. K.F. Aguey-Zinsou and J.R. Ares-Fernández, *Energy Environ. Sci.*, Vol. 3, p. 526, 2010.

16. R. Bardhan, A.M. Ruminski, A. Brand, and J.J. Urban, *Energy Environ. Sci.*, Vol. 4, p. 4882, 2011.

17. W.Y. Li, C.S. Li, C.Y. Zhou, H. Ma, and J. Chen, *Angew. Chem. Int. Ed.*, Vol. 45, p. 6009, 2006.

18. F. Cheng, Z. Tao, J. Liang, and J. Chen, *Chem.Commun.*, Vol. 48, p. 7334, 2012.

19. W. Grochala and P.P. Edwards, *Chem. Rev.*, Vol. 104, p.1283, 2004.

20. S. Orimo, Y. Nakamori, J.R. Eliseo, A. Züttel, and C.M. Jensen, *Chem. Rev.*, Vol. 107, p. 4111, 2007.

21. V. Bérubé, G. Radtke, M. Dresselhaus, and G. Chen, *Int. J. Energy Res.*, 2007, 31, 637.

22. H. Reardon, J.M. Hanlon, R.W. Hughes, A. Godula-Jopek, T.K. Mandal, and D.H. Gregory, *Energy Environ. Sci.*, Vol. 5, p. 5951, 2012.

23. A. Zaluska, L. Zaluski, and J.O. Strom-Olsen, *Appl. Phys. A: Mater. Sci. Process.*, Vol. 72, p. 157, 2001.

24. J.X. Zhang, F. Cuevas, W. Zaidi, J.P. Bonnet, L. Aymard, J.L. Bobet, and M. Latroche, *J. Phys. Chem. C*, Vol. 115, p. 4971, 2011.

25. R.A. Varin, T. Czujko, C. Chiu, and Z. Wronski, *J. Alloys Compd.*, Vol. 424, p. 356. 2006.

26. R.R. Shahi, T.P. Yadav, M.A. Shaz, and O.N. Srivastava, *Int. J. Hydrogen Energy*, Vol. 33, p. 6188, 2008.

27. R.R. Shahi, T.P. Yadav, M.A. Shaz, and O.N. Srivastava, *Int. J. Hydrogen Energy*, Vol. 35, p. 32, 2010.

28. R.R.Shahi, H. Raghubanshi, M.A. Shaz, and O.N. Srivastava, *Int. J. Hydrogen Energy*, Vol. 37, p. 3705, 2012.

29. Q. Wang, Z. Chen, W. Yu, Y. Chen, and Y. Li, *Ind. Eng. Chem. Res.*, Vol.,*49, p.* 5993, 2010,

30. R.R Shahi, A.P. Tiwari, M.A. Shaz, and O.N. Srivastava, *Int. J. Hydrogen Energy*, Vol. 38, p. 2778, 2013

31. S.K. Pandey, A. Bhatnagar, R.R. Shahi, M.S.L. Hudson, M.K. Singh, and O.N. Srivastava, *J. Nanosc. Nanotech.*, Vol. 13, p. 5493, 2013.

32. J.L. Bobeta, E. Akibab, Y. Nakamurab, and B. Darrieta, *Int. J. Hydrogen Energy*, Vol. 25, p. 987, 2000.

33. M.Y. Song, S.N. Kwon, H.R. Park, and S.H. Hong, *Int. J. Hydrogen Energy*, Vol. 36, p. 3587, 2011.

34. W.N. Yang, C.X. Shang, and Z.X. Guo, *Int. J. Hydrogen Energy*, Vol. 35, p. 4534, 2010.

35. H. Shao, M. Felderhoff, F. Schuth, and C. Weidenthaler, *Nanotechnology*, Vol. 22, p. 235401, 2011.

36. Q. Wang, Y. Chen, X. Zheng, G. Niu, C. Wu, M. Tao, *Physica B*, Vol. 404, p. 3431, 2009.

37. C. Wu, P. Wang, X. Yao, C. Liu, D. Chen, G.Q. Lu, and H. Cheng, *J. Phys. Chem. B*, Vol. 109, p. 22217, 2005.

38. C.Z. Wu, X.D. Yao, and H. Zhang, *Int. J. Hydrogen Energy*, Vol. 35, p. 247, 2010.

39. A. Zaluska, L. Zaluski, J.O. Ström-Olsen, *Applied Physics A*, Vol. 72, p. 157, 2001.

40. Z. Dehouche, L. Lafi, N. Grimard, J. Goyette, and R. Chahine, *Nanotechnology*, Vol. 16, p. 402, 2005.

41. P. Adelhelm and P.E. de Jongh, *J. Mater. Chem.*, Vol. 21, p. 2417, 2011.

42. C. Wu and H.M. Cheng, *J. Mater. Chem.*, Vol. 20, p. 5390, 2010.

43. http://en.wikipedia.org/www/Carbon%nanotube.

44. M.F.L. De Volder, S.H. Tawfick, R.H. Baughman, A.J. Hart, Science, Vol. 339, p. 535, 2013.

45. T. Yamaguchi, S. Bandow, S. Iijima, *Chemical Physics Letters*, Vol. 389, p. 181, 2004.

46. H. Hwang, M. Chhowalla, N. Sano, S. Jia, G.A.J. Amaratunga, *Nanotechnology*, Vol. _, p. 546-550, 2004.

47. K. Anazawa, K. Shimotani, C. Manabe, H. Watanabe, and M. Shimizu, *Applied Physics Letters*, Vol. 81, p. 739, 2002.

48. R.L. Vander Wal, L.J. Hall, and G.M. Berger, *Journal of Physical Chemistry B*, Vol. 106, p. 3564, 2002.

49. R.L. Vander Wal and T.M. Ticich, *Journal of Physical Chemistry B*, Vol. 105, p. 10249–10256, 2001.

50. C.N.R. Rao, A.K. Sood, K.S. Subrahmanyam, and A. Govindaraj, *Angew. Chem. Int. Ed.*, Vol. 48, p. 7752, 2009.

51. K.S. Novoselov, A.K. Geim, S.V. Morozov, D. Jing, Y. Zhang, S.V. Dubonos, I.V. Grigorieva, and A.A. Firsov, *Science*, Vol. 306, p. 666, 2004.

52. M. Becher, M. Haluska, M. Hircher, A. Quintel, V. Skakalova, U. Detlaff-Weglikowska, X. Chen, M. Hulman, Y. Choi, S. Roth, V. Meregalli, *C.R. Physique*, Vol. 4, p. 1055, 2003.

53. M.S. Dresselhaus and G. Dresselhaus, *Adv. Phys.*, Vol. 51, p. 1, 2002.

54. M. Noel and V. Suryanarayanan, *J. Power Sources*, Vol. 111, p. 193, 2002.

55. H. Selig, L.B. Ebert, and A.G. Sharpe, Academic Press, Inc., 1980.

56. P. Adelhelm, K.P. de Jong, and P.E. de Jongh, *Chem. Commun.*, p. 6261, 2009.
57. H. Miyaoka, T. Ichikawa, and Y. Kojima, *Nanotechnology*, Vol. 20, p. 204021 2009.
58. H. Miyaoka, K. Itoh, T. Fukunaga, T. Ichikawa, Y. Kojima, and H. Fuji, *J. Appl. Phys.*, Vol. 104, p. 053511, 2008.
59. X.B. Yu, Z. Wu, Q.R. Chen, Z.L. Li, B.C. Weng, and T.S. Huang, *Appl. Phys. Lett.*, Vol. 90, p. 034106, 2007.
60. Q.H. Yang, W.H. Xu, A. Tomita, and T. Kyotani, *Chem. Mater.*, Vol. 17, p. 2940, 2005.
61. M. Terrones, A.G. Souza, and A.M. Rao, Springer-Verlag, Berlin, p. 531 2008.
62. K.J. Kim, B. Montoya, A. Razani, and K.H. Lee, *Int. J. Hydrogen Energy*, Vol. 26, p. 609, 2001.
63. R. Bogerd, P. Adelhelm, J.H. Meeldijk, K.P. de Jong, and P.E. de Jongh, *Nanotechnology*, Vol. 20, p. 204019, 2009.
64. M.A. Lio-Rodenas, K.F. Aguey-Zinsou, D. Cazorla-Amoros, A. Linares-Solano, and Z.X. Guo, *J. Phys. Chem. C*, Vol. 112, p. 5984, 2008.
65. C.Z. Wu, P. Wang, X.D. Yao, C. Liu, D.M. Chen, G.Q. Lu, and H.M. Cheng, *J. Phys. Chem. B*, Vol. 109, p. 22217, 2005.
66. W. Grochala and P.P. Edwards, *Chem. Rev.*, Vol. 104, p. 1283, 2004.
67. B. Bérubé, G. Radtke, M. Dresselhaus, and G. Chen, *Int. J. Hydrogen Energy*, Vol. 31, p. 637, 2007.
68. I.P. Jain, C. Lal, and A. Jain, *Int. J Hydrogen Energy*, Vol. 35, p. 5133, 2010.
69. M. Güvendiren, E. Bayboru, and T. Ozturk, *Int. J. Hydrogen Energy*, Vol. 29, p. 491, 2004.
70. B. Sakintuna, F. Lamari-Darkrim, and M. Hirscher, *Int. J. Hydrogen Energy*, Vol. 32, p. 1121, 2007.
71. C. Wu, P. Wang, X. Yao, C Liu, D. Chen, G.Q. Lu, and H. Cheng, *J. Alloys and Comp.*, Vol. 414, p. 256, 2006.
72. C. Wu, P. Wang, X. Yao, C Liu, D. Chen, G.Q. Lu, and H. Cheng, *J. Phys. Chem. B*, Vol. 09, p. 22217, 2005.
73. M.A. Lillo- Ródenas, K.F. Aguey-Zinsou, D. Cazorla- Amorós, A. Linares-Solano, and Z.X. Guo, *J. Phys. Chem. C*, Vol. 112, p. 5984, 2008.
74. M.A. Lillo- Ródenas, K.F. Aguey-Zinsou, D. Cazorla- Amorós, A. Linares-Solano, and Z.X. Guo, *Carbon*, Vol. 46, p. 126, 2008.
75. C.Z. Wu, X.D. Yao, and H. Zhang, *Int. J. Hydrogen Energy*, Vol. 35, p. 247, 2010.
76. V. Fuster, F.J. Castro, H. Troiani, and G. Urretavizcaya, *Int. J. Hydrogen Energy*, Vol. 36, p. 9051, 2011.
77. A. Reyhani, S.Z. Mortazavi, S. Mirershadi, A.N. Golikand, and A.Z. Moshfegh, *Int. J. Hydrogen Energy*, 2011, in Press.
78. H. Imamura, M. Kusuhara, S. Minami, M. Matsumoto, K. Masanari, Y. Sakata, K. Itoh, and T. Fukunaga, *Acta Materialia*, Vol. 51, p. 6407, 2003.

79. H. Imamura, N. Sakasai, T. Fujinaga, *J. Alloys Compd.*, Vol. 253, p. 34 1997.

80. H. Imamura, Y. Takesu, T. Akimoto, S. Tabatta, *J. Alloys Compd.*, Vol. 293, p. 568, 1999.

81. H. Imamura, et al., *J. Alloys Compd.*, Vol. 330, p. 579, 2002.

82. H. Imamura, S. Tabatta, Y. Takesue, Y. Sakata, S. Kamazaki, *Int. J. Hydrogen Energy*, Vol. 25, p. 837, 2000.

83. C.X. Shang and Z.X. Guo, *J. Power Sources*, Vol. 129, p. 732004.

84. D. Chen, L. Chen, S. Liu, C.X. Ma, D.M. Chen, L.B. Wang, *J. Alloys Compd.*, Vol. 372, p. 231-237, 2004.

85. C. Wu, P. Wang, X. Yao, C. Liu, D. Chen, G.Q. Lu, and H. Cheng, *J. Alloys Compd.*, Vol. 420, p. 278, 2006.

86. G. Liang, J. Huot, S. Boily, A. van Neste, and R. Schulz, *J. Alloys Compd.*, Vol. 292, p. 247, 1999.

87. R. Singh, H. Raghubanshi, S.K. Pandey, and O.N. Srivastava, *Int. J. Hydrogen Energy*, Vol. 35, p. 4131, 2010.

88. R.R. Shahi, H. Raghubanshi, M.A. Shaz, and O.N. Srivastava, *J. Appl. Nanosci.*, Vol. 2, p. 195, 2012.

89. V.V. Bhat, A. Rougier, L. Aymard, G.A. Nazri, and J.M. Tarascon, *J. Alloys Compd.*, Vol. 460, p. 507, 2008.

90. H. Raghubanshi, M.S.L. Hudson, and O.N. Srivastava, *Int. J. Hydrogen Energy*, Vol. 36, p. 4482-4490, 2011.

91. G. Liu, Y. Wang, C. Xu, F. Qiu, C. An, L. Li, L. Jiao, and H. Yuan, *Nanoscale*, Vol. 5, p. 1074, 2013.

92. A.J. Du, S.C. Smith, X.D. Yao, and G.Q. Lu, *J. Phys. Chem. B*, Vol. 110, p. 1814, 2006.

93. C.M. Jensen and K.J. Gross, *Appl. Phys. A*, Vol. 72, p. 213, 2001.

94. G. Sandrock, K.J. Gross, G. Thomas, C. Jensen, D. Meeker, S. Takara, *J. Alloys Compd.*, Vol. 330–332, p. 696, 2002.

95. G. Sandrock and K.J. Gross, *J. Alloys Compd.*, Vol. 339, p. 299, 2002.

96. M. Fichtner, O. Fuhr, O. Kircher, and J. Rothe, *J. Alloys Compd.*, Vol. 14, p. 778, 2003.

97. P. Wang and C.M. Jensen, *J. Alloys Compd.*, Vol. 379, p. 99, 2004.

98. O. Kircher and M. Fichtner, *J. Alloys Compd.*, Vol. 404–406, p, 339–42, 2005.

99. B. Bogdanovic, M. Felderhoff, S. Kaskel, A. Pommerin, K. Sclichte, and F. Schüth, *Adv. Mater.*, Vol. 15, p. 1012, 2003.

100. D. Sun, T. Kiyobayashi, H.T. Takeshita, N. Kuriyama, and C.M. Jensen, *J. Alloys Compd.*, Vol. 337, p. L8, 2002.

101. D. Sun, S.S. Srinivasan, T. Kiyobayashi, N. Kuriyama, and C.M. Jensen, *J. Phys. Chem. B*, Vol. 107, p. 101769, 2003.

102. S.S. Srinivasan, H.W. Brinks, B.C. Hauback, D. Sun, and C.M. Jensen, *J. Alloys Compd.*, Vol. 377, p. 283, 2004.

103. D. Pukazhselvan, B.K. Gupta, A. Srivastava, and O.N. Srivastava, *J. Alloys Compd.*, Vol. 403, p.312, 2005.

104. C. Cento, P. Gislon, M. Bilgili, A. Masci, Q. Zheng, and P.P. Prosini, *J. Alloys Compd.*, Vol. 437, p.360, 2007.
105. T.T. Chen, C.H. Yang, and W.T. Tsai, *Int. J. Hydrogen Energy*, Vol. 37, p.14285, 2012.
106. S.S.-Y. Lin, J. Yang, and H.H. Kung, *Int. J. Hydrogen Energy*, Vol. 37, p. 2737, 2012.
107. P.A. Berseth, A.G. Harter, R. Zidan, A. Blomqvist, C.M. Araujo, and R.H. Scheicher, *Nano Lett.*, Vol. 9, p. 1501, 2009.
108. M.S.L. Hudson, H. Raghubanshi, D. Pukazhselvan, and O.N. Srivastava, *Int. J. Hydrogen Energy*, Vol. 37, p. 2750, 2012.
109. Z. Qian, M.S.L. Hudson, H. Raghubanshi, R.H. Scheicher, B. Pethak, C.M. Aroujo, A. Blomqvist, B. Johansson, O.N. Srivastava, and R. Auja, *J. Phys. Chem. C*, Vol. 116, p. 10861, 2012.
110. L.H. Kumar, C.V. Rao, and B. Viswaanathan, *J. Mater. Chem. A*, Vol. 1, p. 3355, 2013.
111. B. Chen and J.L. Falconer, *J. Catal.*, Vol. 134, p. 737, 1998.
112. Y. Li, F. Fang, H. Fu, J. Qiu, Y. Song, Y. Li, D. Sun, Q. Zhang, L. Ouyang, and M. Zhu, *J. Mater. Chem. A*, Vol. 1, p. 5238, 2013.
113. P. Chen, Z. Xiong, J. Luo, J. Tan, and K.L. Lin, *Nature*, Vol. 420, p. 302, 2002.
114. S. Nayebossadri, *Int. J. Hydrogen Energy*, Vol. 36, p. 8335, 2011.
115. M. Aoki, K. Miwa, T. Noritake, G. Kitahara, Y. Nakamori, S. Orimo, and S. Towata, *Appl. Phys. A*, Vol 80, p. 1409, 2005.
116. T. Ichikawa, N. Hanada, S. Isobe, and H. Len, *J. Phys. Chem. B*, Vol. 108, p. 7887, 2004.
117. T. Ichikawa, S Isobe, N. Hanada, H. Fujii, *J. Alloys Compd.*, Vol. 365, p. 271, 2004.
118. S. Orimo, Y. Nakamori, G. Kitahara, K. Miwa, N. Ohba, T. Noritake, S. Towata, *Appl. Phys. A*, Vol. 79, p. 1765, 2004.
119. S.S. Srinivasan, M.U. Niemann, J.R. Hattrick-Simpers, K. McGrath, P.C. Sharma, and D.Y. Goswami, *Int. J. Hydrogen Energy*, Vol. 35, p. 9649, 2010.
120. X. Zhang, Z. Li, F. Lv, H. Li, J. Mi, S. Wang, X. Liu, and L. Jiang, *Int. J. Hydrogen Energy*, Vol. 35, p. 7809, 2010.
121. C. Liang, Y. Liu, H. Fu, Y. Ding, M. Gao, and H. Pan, *J. Alloys Compd.*, Vol. 509, p. 7844, 2011.
122. L.P. Ma, H.B. Dai, Z.Z. Fang, X.D. Kang, Y. Liang, and P.J. Wang, *J. Phys. Chem. C*, Vol. 113, p. 9944, 2009.
123. Y. Chen, P. Wang, C. Liu, and H.M. Cheng, *Int. J. Hydrogen Energy*, Vol. 32, p. 1262, 2007.
124. R.R. Shahi, H. Raghubanshi, M.A. Shaz, and O.N. Srivastava, *Int. J. Hydrogen Energy*, article in press Vol. 38, p.8863, 2013.
125. A. Srivastava, O.N. Srivastava, S. Talapatra, R. Vajtai, P.M. Ajayan, *Nature Mat.*, Vol. 3, p. 610, 2004.

126. K. Awasthi, A.K. Srivastava, and O.N. Srivastava, *J. Nanoscience Tech.*, Vol. 5, p. 1616, 2005.

127. K. Awasthi, S. Awasthi, A. Srivastava, R. Kamalakaran, S. Talpatra, P.M. Ajayan, and O.N. Srivastava, *Nanotechnology*, Vol. 17, p. 5417, 2006.

128. S. Awasthi, K. Awasthi, R. Kumar, and O.N. Srivastava, *J. Nanosci. Nanotech.*, Vol. 9, p. 5455, 2009.

129. K. Awasthi, R. Kumar, H. Ragubanshi, S. Awasthi, R. Pandey, D. Singh, T.P. Yadav, and O.N. Srivastava, *Bull. Mater. Sci.*, Vol. 34, p. 607, 2011.

130. A. Srivastava, A.K. Srivastava, and O.N. Srivastava, *Appl. Phys. Lett.*, Vol. 72, p.1685, 1998.

131. A. Srivastava, A.K. Srivastava, and O.N. Srivastava, *Carbon*, Vol. 39, p. 201, 2001.

132. K. Awasthi and O.N. Srivastava, American Scientific Publisher U.S.A., pp 4-5, 2008.

133. K. Awasthi, R. Kumar, R.S. Tiwari, and O.N. Srivastava, *J. Expt. Nanosci.*, Vol. 5, p. 498, 2010.

134. K. Awasthi, R. Kamarakaram, A.K. Singh, and O.N. Srivastava, *Int. J. Hydrogen Energy*, Vol. 27, p. 425, 2002.

135. K. Awasthi, S. Awasthi, A. Srivastava, S. Talpatra, R. Kamarakaram, P.M. Ajayan, and O.N. Srivastava, *Nanotechnology*, Vol. 17, p. 5417, 2006.

5

Carbon Nanotubes and Their Applications

Mohan Raja[1,*] and J. Subha[2]

[1]*Amity Institute of Nanotechnology, Amity University, Noida, India*
[2]*Central Institute of Plastics Engineering & Technology (CIPET), Bhopal, India*

Abstract

Carbon nanotubes (CNTs), which consist of rolled graphene sheets built from sp2-hybridized carbon atoms, are nowadays attracting scientists from various disciplines due to their attractive physical and chemical properties. In this account, we will describe the recent progress in the advance of synthetic techniques for the large-scale production of carbon nanotubes, purification, and chemical modification that are enabling the integration of CNTs in thin-film electronics and large-area coatings. Although providing strong mechanical strength in polymer composites and electrical and thermal conductivities for many electronics applications, CNTs sheets have already shown promising performance for use in applications including supercapacitors, actuators, lightweight electromagnetic shields,in biomedical areas ranging from biosensing, etc. We predict that carbon nanotubes will find numerous applications and continue to take an important place in the development of emerging technologies in the near future.

Keywords: Carbon nanotubes, synthesis, composite, microelectronics, coatings, energy storage, biosensing

5.1 Introduction

Elemental carbon can form a diversity of remarkable structures. Apart from the renowned graphite, carbon can construct closed and open cages with honeycomb atomic arrangement. Kroto *et al* [1]

Corresponding author: mohanraja27@yahoo.com

Ashutosh Tiwari and S.K. Shukla (eds.) Advanced Carbon Materials and Technology, (173–192)
2014 © Scrivener Publishing LLC

discovered the structure of C_{60} molecules. Carbon nanotubes can be thought of as rolled-up graphene sheets with no overlapping edges and were first isolated by Iijima in 1991 [2]. Their diameters typically vary from 1 to 100 nm and their lengths can be several orders of magnitude larger, up to millimeters, even centimeters long [3]. The nanotubes consist of several types such as single-wall carbon nanotubes (SWCNTs) [4], and multi-wall carbon nanotubes (MWCNTs) [2]. The beginning of widespread CNTs research in the early 1990s was preceded in the 1980s by the first industrial synthesis of what are now known as MCWNTs, and documented observations of hollow carbon nanofibers as early as the 1950s.

However, CNTs-related commercial activity has grown most substantially during the past decade. Since 2006, worldwide CNTs production capacity has increased at least ten-fold, and the annual number of CNTs-related journal publications and issued patents continues to grow. Most CNTs production today is used in bulk composite materials and thin films, which rely on unorganized CNTs architectures having limited properties. Organized architectures of CNTs such as vertically aligned forms and sheets show promise to scale up the properties of individual CNTs and also to realize new functionalities. However, presently realized mechanical, thermal and electrical properties of CNTs macrostructures such as yarns and sheets remain significantly lower than those of individual CNTs. Worldwide commercial interest in carbon nanotubes (CNTs) is reflected in a production capacity that presently exceeds several thousand tons per year. Currently, bulk CNTs powders are incorporated in diverse commercial products ranging from rechargeable batteries, automotive parts, boat hulls, sporting goods, water filters and antimicrobial coatings. Advances in CNTs synthesis, purification, and chemical modification are enabling integration of CNTs in thin-film electronics, large-area coatings and biosensors, etc. Although not yet providing compelling mechanical strength or electrical or thermal conductivities for many applications, CNTs yarns and sheets already show promising performance for use in applications including supercapacitors, actuators, and lightweight electromagnetic shields.

5.2   Carbon Nanotubes Structure

The binding in carbon nanotubes is sp2 with each carbon atom joined to three neighbor carbon atoms, as in graphite. Graphite

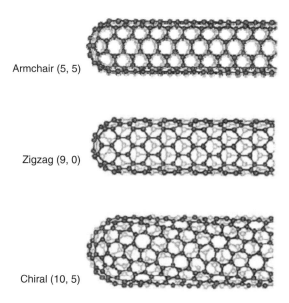

Armchair (5, 5)

Zigzag (9, 0)

Chiral (10, 5)

Figure 5.1 Carbon nanotube structures.

has a sheet-like structure where carbon atoms all lie at the corners of hexagons in a plane and are only weakly bonded to the graphite sheets above and below with 0.34 nm of interlayer distance. Carbon nanotubes can be considered as rolled-up graphene sheets (graphene is the term to describe an individual graphite layer) [5]. Three types of nanotubes are possible, namely armchair, zigzag and chiral nanotubes, depending on how the two-dimensional graphene sheet is rolled up, as shown in Figure 5.1. The primary symmetry classification of a carbon nanotube is either achiral or chiral. Both armchair and zigzag nanotubes are achiral since their mirror image has an identical structure to the original one. Chiral nanotubes exhibit a spiral symmetry whose mirror image cannot be superimposed on the original one.

The different structure can be most easily explained in terms of the unit cell of carbon nanotubes in Figure 5.2. The so-called chiral vector of the nanotube, C_h, is defined by $C_h = na_1 + ma_2$, where a_1 and a_2 are unit vectors in the two-dimensional hexagonal lattice, and n and m are integers.

Another important parameter is the chiral angle, which is the angle between C_h and a_1. An armchair nanotube corresponds to the case of m = n and a zigzag nanotube corresponds to the case of m = 0. All other (n, m) chiral vectors correspond to chiral nanotubes.

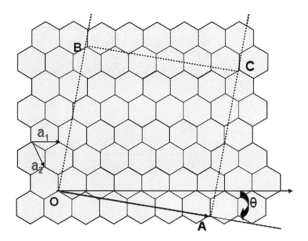

Figure 5.2 Carbon nanotube lattice.

The diameter of carbon nanotube, D_t, can be calculated by L/π, in which L is the circumference of the carbon nanotube;

$$D_t = \frac{L}{\pi} = a\sqrt{n^2 + m^2 + nm}$$

(5.1)

Where the lattice constant a = 0.249 nm. For example, the diameter of the zigzag nanotube (9, 0) in the Figure 5.1 is D_t = 0.882 nm.

The chiral angle θ denotes the tilt angle of the hexagons with respect to the direction of carbon nanotube axis. It can be calculated by the following equation:

$$\tan\theta = \frac{\sqrt{3m}}{2n + m}$$

(5.2)

5.3   Carbon Nanotube Physical Properties

The special nature of carbon combines with the molecular perfection of nanotube structures to endow them with exceptional material properties, such as very high strength, stiffness, toughness, and electrical and thermal conductivity. There is no other element in the periodic table that bonds to itself in an extended network structure with the strength of the carbon-carbon bond. The delocalized π-electron donated by each carbon atom is free to move along the complete structure, moderately, than stays with its donor atom, giving rise to the first identified molecule with metallic-type electrical conductivity. In addition, the high-frequency carbon-carbon bond vibrations provide an intrinsic thermal conductivity higher than even diamond. Carbon nanotubes represent a mixture of molecules with various diameters, length and chirality.

The reported Young's modulus of CNTs is in the order of 1–1.2 TPa, tensile strength in the order of 36GPa at a failure strain of 6% [6,7]. However, there are discrepancies in the values of reported tensile modulus and strength for CNTs as a result of different testing methods, calculations and system errors. Electronic conductivity of CNTs was predicted to depend sensitively on tube diameter and chiral angle, with only a slight difference in one parameter causing the change from a metallic to semiconductor state [8–10]

5.4   Carbon Nanotube Synthesis and Processing

In order to use carbon nanotubes in many devices, it is essential to produce these materials with a high crystallinity economically on a large scale. In this context, the catalytic chemical vapor deposition (CVD) method is considered optimum for producing large amounts of carbon nanotubes, mostly with the use of a floating-catalyst method [11–13]. This technique is more controllable and cost efficient when compared with arc-discharge and laser ablation methods [14–17]. Various groups have been able to produce SWCNTs for laboratory experiments [18,19]. More recently, we reported an alternative route for the large-scale synthesis of SWCNTs by combining the use of catalytic substrates and floating methods [20]. The template (substrate) prevents metal particle aggregation, thus resulting in the formation of high-purity SWCNTs. The method is able to produce either individual nanotubes or nanotube bundles, exhibiting a wide range of diameters (0.4–4 nm). In particular, nanosized zeolites were impregnated with Fe-containing compounds (seeding method) and placed inside a furnace (*ca.* 1000°C) together with vapor of benzene, and H_2 as the carrier gas. The resulting material was purified via immersion in hydrofluoric acid (HF).

Recently, some new synthetic methods were introduced for the production of carbon nanotubes at low temperature [21,22] along with the aligned growth of single-walled carbon nanotubes (SWCNTs) [23] which opened a possibility for the cost-effective synthesis of carbon nanotubes with controlled structures for practical applications. The CNTs were produced from the mixture solution of dichlorobenzene along with $ZnCl_2$ particles, which act as catalyst and nucleation site for CNTs growth. This solution is sonicated under ambient condition using ultrasonic water bath [24]. Alternatively, the synthesis of aligned CNTs that can be processed without the need for dispersion in a liquid offers the promise for cost-effective realization of compelling large properties. These methods include self-aligned growth of horizontal [25] and vertical [26] CNTs on substrates coated with catalyst particles, and production of CNTs sheets and yarns directly from floating-catalyst CVD systems [27]. CNTs forests can be manipulated into dense solids [28], aligned thin films [29], and intricate three-dimensional (3D) microarchitectures [30], and can be directly spun or drawn into long yarns and sheets [31,32].

5.5   Carbon Nanotube Surface Modification

Carbon nanotubes are observed in bundles due to the substantial van der Walls attraction. In order to manipulate the methods for CNTs, it is attractive to functionalize the sidewall of CNTs, thereby generating CNTs derivatives that are compatible with solvent as well as organic matrix materials. Both surface modification techniques and non-covalent wrapping methods have been reported [33]. In chemical surface modification, functional groups are covalently linked to the CNTs surface; this is also referred to as the covalent functionalization method [34] Based on the reaction, two approaches have been explored: the first approach involves direct attaching of functional groups to the graphitic surface, and in the second approach, the functional groups are linked to the CNTs-bound carboxylic acids, which are created during the CNTs synthesis, or during post treatment of CNTs for the purification purpose.

These carboxylic acids are considered the defect sites on the CNTs' surface and the method is also known as "defect chemistry" [35,36]. The advantage of the chemical functionalization method is that the functional groups are covalently linked on the CNTs surface; the linkage is permanent and mechanically stable. However, reaction with the graphitic sheet also results in breaking of the sp2 conformation of the carbon atom. The conjugation of the CNTs wall is therefore disrupted, and it was observed that the electrical and mechanical properties of the chemically functionalized CNTs decreased dramatically as compared to the pristine tubes [37,38]. The surface of single-wall carbon nanotubes was modified by the addition of photoinitiator under UV irradiation with hexylamine. The effects of the CNTs surface modification by ionic functionalization enhance their degree of dispersion in polymer matrices [39]. The larger surface area of CNTs can be used as templates to prepare nanoparticulate hybrid systems consisting of silver and copper nanoparticles for metal-functionalized CNTs [40].

5.6   Applications of Carbon Nanotubes

5.6.1   Composite Materials

Carbon nanotubes were initially used as electrically conductive fillers in plastics, due to the attractive benefit of their high aspect ratio to form a percolation network at concentrations as low as 0.01 weight percent (wt%). Disordered CNTs/polymer composites reach conductivities as high as 10,000 S m−1 at 10 wt% loading [41]. The inclusion of CNTs into polymer holds the potential to improve the mechanical, electrical, or thermal properties compared to traditional fillers [42], and significant efforts have gone into fabricating CNTs/polymer composites for high performance [43–45] and multifunction [46–50]. It is known that the homogeneous dispersion of CNTs in polymer matrix is difficult due to the tendency for formation of CNTs bundles. The chemical functionalization of CNTs is considered as an effective way to achieve homogenous dispersion of CNTs in polymer matrices [51–53]. Surfactants have also been used to improve the dispersion and strengthen the interactions between CNTs and polymer matrix [54]. Several methods have been developed to prepare CNTs/polymer composites, for example, solution casting [55–58], melt mixing [36], and *in situ* polymerization [59,60]. In the automotive industries, CNTs-filled plastics have enabled electrostatic-assisted painting of mirror housings, in addition to fuel lines and filters that dissipate electrostatic charge. Further products consist of electromagnetic interference (EMI)-shielding packages and wafer carriers for the microelectronics industry. For load-bearing applications, CNTs powders mixed with polymers or precursor resins can enhance stiffness, strength, and toughness [61]. Adding ~1 wt% MWCNTs to epoxy resin improves stiffness and fracture toughness by 6 and 23%, respectively, without compromising other mechanical properties [62]. These enhancements depend on CNTs diameter, aspect ratio, dispersion, alignment and interfacial interaction with the matrix. Many manufacturers of CNTs sell premixed resins and master batches with CNTs loadings from 0.1 to 20 wt%. Additionally, engineering nanoscale stick-slip among CNTs and CNTs/polymer contacts can increase material damping [63], which is used to enhance sporting goods, such as baseball bats, tennis racquets and bicycle frames (Figure 5.3).

Carbon nanotube resins are also used to enhance fiber composites [64]. The latest good examples such as strong, lightweight wind turbine blades and hulls for maritime security boats are made by using carbon fiber composite with CNTs-enhanced resin (Figure 5.4) and wind turbine composite blades. The CNTs can also be deployed as additives in the organic precursors used to form carbon fibers.

Figure 5.3 CNTs composite bicycle frame (Photo courtesy of BMC Switzerla nd AG).

Figure 5.4 Carbon fiber laminate with CNTs dispersed in the epoxy resin (inset), and a lightweight CNT-fiber composite boat hull for maritime security boats. (Images courtesy of Zyvex Technologies).

The carbon nanotubes influence the arrangement of carbon in the pyrolyzed fiber, making possible the fabrication of 1-mm diameter carbon fibers with over 35% increase in strength (4.5 GPa) and stiffness (463 GPa) compared with control samples without CNTs [65]. The electroactive shape memory behavior was observed in polymer nanocomposites based on PU/M-CNT nanocomposites prepared using polyurethane and metal nanoparticles-decorated MWCNTs through the melt mixing process [66]. The mechanical properties such as Young's modulus and yield stress increased significantly for PU/metal nanoparticles-decorated CNT composites in comparison to the pristine CNT nanocomposites. The PU/M-CNT composites have excellent shape recoverability compared to PU/pristine CNT composites. The PU/metal-decorated CNTs showed high strain recovery ability after several cycles of training compared to PU/pristine carbon nanotube composites (Figure 5.5).

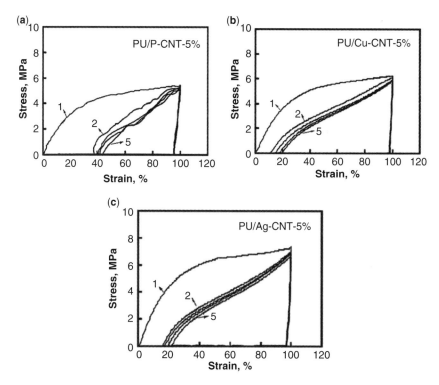

Figure 5.5 Stress vs strain plots of (a) PU/pristine CNT nanocomposites, (b) PU/Cu-CNT nanocomposites, and (c) PU/Ag-CNT nanocomposites. (Reprinted with permission from Elsevier).

Besides polymer composites, the addition of small amounts of CNTs to metals has provided amplified tensile strength and modulus [67] that may find application in aerospace and automotive structures. Commercial Al-CNTs composites have strengths comparable to stainless steel (0.7 to 1 GPa) at one-third the density (2.6 g cm–3). This strength is also comparable to Al-Li alloys, yet the Al-CNTs composites are reportedly less expensive

Carbon nanotubes can also be used as a flame retardant additive in plastics; this effect is mainly predictable to changes in rheology by CNTs loading [68]. These CNTs additives are commercially smart as alternatives to halogenated flame retardants, which have restricted use due to environmental regulations. Last, DuPont intends to launch new CNTs-based materials for armor that improve the durability, strength and performance of body and vehicle armor and helmets (Figure 5.6). Additionally, this technology platform will enable DuPont to provide new materials for other industrial uses such as ropes and cables that offer strong and durable support for off-shore oil and gas platforms.

5.6.2 Nano Coatings – Antimicrobials and Microelectronics

Carbon nanotubes are promising as a multifunctional coating material. For example, the metal nanoparticle-decorated CNTs hybrid systems can be synthesized using CNTs consisting of carboxyl groups in the backbone, which can bind ions of transition metals easily (Ag^+ and Cu^{2+}). These ions are well known for their

Figure 5.6 Carbon nanotubes ballistic coupon. (Image courtesy of DuPont).

Figure 5.7 Antifouling "Green Ocean Coating Heavy Duty" coatings used on the hull of a ship (Image courtesy of Bayer Material Science).

broad-spectrum antimicrobial activity against bacterial and fungal agents together with their lack of cross resistance with antibiotics [40]. The MWCNTs-containing paints reduce biofouling of ship hulls (Figure 5.7) by discouraging attachment of algae and barnacles [69]. They are a possible alternative to environmentally hazardous biocide-containing paints. Inclusion of CNTs in anticorrosion coatings for metals can enhance coating stiffness and strength while providing an electric pathway for cathodic protection.

Widespread development continues on CNTs-based flexible transparent conducting films [70] as an alternative to indium tin oxide (ITO). A concern is that ITO is becoming more expensive because of the shortage of indium, compounded by mounting demand for displays, touchscreen devices, and photovoltaics. Besides cost, the flexibility of CNTs transparent conductors is a major advantage over brittle ITO coatings for flexible displays. Further, transparent CNTs conductors can be deposited from solution (e.g., slot-die coating, ultrasonic spraying) and patterned by cost-effective nonlithographic methods (e.g., screen printing, microplotting). The latest profitable development effort has resulted in CNTs films with 90% transparency and a sheet resistivity of 100 ohm per square (Figure 5.8). This surface resistivity is adequate for some applications but still substantially higher than for equally transparent, optimally doped ITO coatings [71]. Related applications that have less stringent requirements include CNTs thin-film heaters, such as for use in defrosting

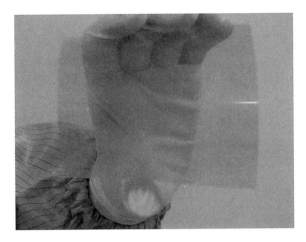

Figure 5.8 Carbon nanotubes flexible transparent conducting film. (Image courtesy of *Printed Electronics World*).

automobile windows or sidewalks. All of the above coatings are being employed industrially.

In recent years, the nanothickness CNTs films used are transparent, flexible and stretchable, can be tailored into many shapes and sizes, and are freestanding or placed on a variety of rigid or flexible insulating surfaces. A piece of carbon nanotube (CNTs) thin film can be a practical magnet-free loudspeaker simply by applying an audio frequency current through it (Figure 5.9). This CNTs film loudspeaker can produce sound with wide frequency range, high sound pressure level and low total harmonic distortion [72]. The CNTs thin-film transistors (TFTs) are particularly attractive for driving organic light-emitting diode (OLED) displays, for the reason that they have shown higher mobility than amorphous silicon ($\sim$1 cm^2 V^{-1} s^{-1}) (56) and can be deposited by low temperature, non-vacuum methods. Recently, flexible CNTs TFTs with a mobility of 35 cm2 V–1 s–1 and an on/off ratio of 6 × 106 were demonstrated (Fig. 5.9a) [73].

5.6.3   Biosensors

Sensors are devices that detect a change in physical quantity or event. There are many studies that have reported the use of CNTs-based pressure, flow, thermal, gas, chemical and biological sensors. Carbon nanotubes can be used as flow sensors [74,75]. The flow of a liquid on bundles of SWCNTs induces a voltage in the direction of

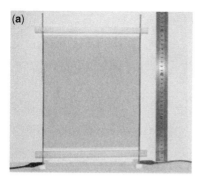
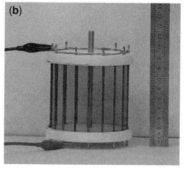

Figure 5.9 Carbon nanotubes thin film loudspeakers. (a) A4 paper size CNTs thin film loudspeaker . (b) The cylindrical cage shape CNTs thin film loudspeaker can emit sounds in all directions; diameter 9 cm, height 8.5 cm. (Reprinted by permission from the American Chemical Society).

the flow. In the future, this finding can be used in micro machines that work in a fluid environment, such as heart pacemakers, that need neither heavy battery packs nor recharging [74]. Piezoresistive pressure sensors can be made with the help of CNTs. Single-wall carbon nanotubes were grown on suspended square polysilicon membranes [76]. When uniform pressure was applied to the membranes, a change in resistance in the SWCNTs was observed. According to Caldwell *et al.* [77], fabrication of piezoresistive pressure sensors that incorporate CNTs can bring dramatic changes to the biomedical industry, as many piezoresistance-based diagnostic and therapeutic devices are currently in use there. The CNTs-based nanobiosensors may be used to detect deoxyribonucleic acid (DNA) sequences in the body [78,79]. These instruments detect a very specific piece of DNA that may be related to a particular disease [80]. Such sensors enable detection of only a few DNA molecules that contain specific sequences, and thus possibly diagnose patients as having specific sequences related to a cancer gene. Biosensors can also be used for glucose sensing. Carbon nanotube chemical sensors for liquids can be used for blood analysis, for example, to detect sodium or find pH value [81].

5.6.4   Energy Storages

Carbon nanotubes are being conceived for energy production and storage. Graphite, carbonaceous-based electrodes have been used for decades in fuel cells, batteries and several electrochemical

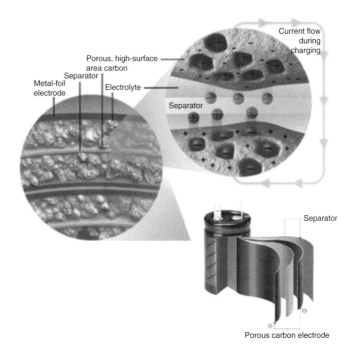

Current flow during charging

Porous, high-surface area carbon
Separator
Metal-foil electrode
Electrolyte
Separator
Separator
Porous carbon electrode

Figure 5.10 Concept for CNTs-based supercapacitors. (Image courtesy of YEG Ultracapacitors).

storage applications (Figure 5.10) [82]. Carbon nanotubes are special because they have small dimensions, perfect surface specificity and smooth surface topology, since only the basal graphite planes are exposed in their structure. The rate of electron transfer in carbon-based electrodes ultimately determines the efficiency of fuel cells and this depends on various factors, such as the structural morphology of the carbon-based material used as electrodes. Pure MWCNTs and MWCNTs deposited with metal catalysts (Pd, Pt, Ag) have been used to electrocatalyze an oxygen reduction reaction, which is important for fuel cells [82]. It is seen from several studies that nanotubes could be excellent replacements for conventional carbon-based electrodes. Similarly, the improved selectivity of carbon nanotube-based catalysts has been demonstrated in heterogeneous catalysis. Ru-supported carbon nanotubes were found to be superior to the same metal on graphite and on other carbons in the liquid phase hydrogenation reaction of cinnamaldehyde [83]. The properties of catalytically grown carbon nanofibers (which are basically imperfect carbon nanotubes) have been found to be desirable for high power electrochemical capacitors [84].

5.7 Conclusion

This chapter has described several possible applications of carbon nanotubes, with an emphasis on materials science and electronics applications. What we would like to express through this chapter is that the unique structure, topology and dimensions of carbon nanotubes have created an excellent carbon material, which can be considered as the most perfect fiber that has ever been fabricated. The extraordinary physical and chemical properties of carbon nanotubes create a host of application possibilities, some derived as an extension of traditional carbon fiber applications, but many are new possibilities, based on the novel electronic and mechanical behavior of carbon nanotubes. It needs to be said that the excitement in this field arises due to the versatility of this material and the possibility to predict properties based on its well-defined perfect crystal lattice. Nanotubes truly bridge the gap between the molecular realm and the macro world, and are destined to be a star in future technology. According to reports, many companies are investing in diverse applications of CNTs such as transparent conductors, thermal interfaces, antiballistic vests, and wind turbine blades. However, often few technical details are released, and companies are likely to keep technical details hidden for a very long time after commercialization, which makes it challenging to predict market success. Hence, the increases in carbon nanotube production capacity and sales are an especially important metric for emerging CNTs applications.

References

1. H.W. Kroto, J.R. Heath, S.C. O'Brien, R.F. Curl, R.E. Smalley, *Nature*, 318, 162, 1985.
2. S. Iijima, *Nature*, 354, 56, 1991.
3. S. Chakrabarti, H. Kume, L. Pan, T. Nagasaka, and Y. Nakayama, *Journal of Physical Chemistry C*, 111, 1929, 2007.
4. S. Iijima, T. Ichihashi, *Nature*, 363, 603, 1993.
5. R. Saito, G. Dresselhaus, M.S. Dresselhaus, *Physical Properties of Carbon Nanotubes*. Imperial College Press: London, 1998.
6. M.F. Yu, B.S. Files, S. Arepalli, R.S. Ruoff, *Physical Review Letters*, 84, 24, 5552, 2000.
7. M.F. Yu, O. Lourie, M.J. Dyer, K. Moloni, T.F. Kelly, R.S. Ruoff, *Science*, 287, 5453, 637, 2000.

8. R. Saito, M. Fujita, G. Dresselhaus, M.S. Dresselhaus, *Applied Physics Letters*, 60, 18, 2204, 1992.

9. N. Hamada, S. Sawada, A. Oshiyama, *Physical Review Letters*, 68, 10, 1579, 1992.

10. J.W. Mintmire, B.I. Dunlap, C.T. White, *Physical Review Letters*, 68, 5, 631, 1992.

11. M. Endo, *Chemtech*, 18, 568, 1988.

12. G.G Tibbetts, C.A. Bernnardo, D.W. Gorkiewicz, and R.L. Alig, *Carbon*, 32, 569, 1994.

13. M.L. Lake, *Applied Science, Series E*, 372, 187, 2001.

14. T.W. Ebbesen and P.M. Ajayan, *Nature*, 358, 220, 1992.

15. C. Journet, W.K. Maser, P. Bernier, A. Loiseau, M.L. Delachapelle, S. Lefrant, P. Deniard, R. Lee, and J.E. Fisher, *Nature*, 388, 756, 1997.

16. J. Liu, A.G. Rinzler, H. Dai, J.H. Hafner, R.K. Bradley, P.J. Boul, A. Lu, T. Iverson, K.Shelimov, C.B. Huffman, F.R.Macias, Y.S. Shon, T.R. Lee, D.T. Colbert, and R.E. Smalley, *Science*, 280, 1253, 1998.

17. A. Thess, R. Lee, P. Nikolaev, H. Dai, P. Petit, J. Robert, C. Xu, Y. Lee, S. Kim, A. Rinzler, D.T. Colbert, G.E. Scuseria, D. Tomanek, J.E. Fischer, and R.E. Smalley, *Science*, 273, 483, 1996.

18. H. Cheng, F. Li, G. Su, H. Pan, and M.S. Dresselhaus, *Appl. Phys. Lett.*, 72, 3282, 1999.

19. J.F. Colomer, C. Stephan, S. Lefrant, G. Van Tendeloo, I. Willems, Z. Konya, A. Fonseca, Ch. Laurent, and J.B. Nagy, *Chem. Phys. Lett.*, 317, 83, 2000.

20. T. Hayashi, Y.A. Kim, T. Matoba, M. Ezaka, K. Nishimura, T. Tsukada, M. Endo, and M.S. Dresselhaus, *Nano Lett.*, 3, 887, 2003.

21. J.K. Vohs, J.J. Brege, J.E. Raymond, A.E. Brown, A.E. Williams, and B.D. Fahlman, *J. Am. Chem. Soc.*, 126, 9936, 2004.

22. S.H. Jeong, J.H. Ko, J.B. Park, and W. Park, *J. Am. Chem. Soc.*, 126, 15982, 2004.

23. Y. Murakami, S. Chiashi, Y. Miyauchi, M. Hu, M. Ogura, T. Okubo, and S. Maruyama, *Chem. Phys. Lett.*, 385, 298, 2004.

24. M. Raja and S.H. Ryu, *J. Nanosci. Nanotechnol.*, 9, 5940, 2009

25. Q. Cao and J.A. Rogers, *Adv. Mater.*, 21, 29, 2009.

26. K. Hata, D.N. Futaba, K. Mizuno, T. Namai, M. Yumura, and S. Iijima, *Science*, 306, 1362 2004.

27. K. Koziol, J. Vilatela, A. Moisala, M. Motta, P. Cunniff, M. Sennett, and A. Windle, *Science*, 318, 1892, 2007.

28. D.N. Futaba, K. Hata, T. Yamada, T. Hiraoka, Y. Hayamizu, Y. Kakudate, O. Tanaike, H. Hatori, M. Yumura, and S. Iijima, *Nat. Mater.*, 5, 987, 2006.

29. Y. Hayamizu, T. Yamada, K. Mizuno, R.C. Davis, D.N. Futaba, M. Yumura, and K. Hata, *Nat. Nanotechnol.*, 3, 289, 2008.

30. M. De Volder, S.H. Tawfick, S.J. Park, D. Copic, Z. Zhao, W. Lu, A.J. Hart, *Adv. Mater.*, 22, 4384, 2010.
31. M. Zhang, K.R. Atkinson, and R.H. Baughman, *Science*, 306, 1358, 2004.
32. K.L. Jiang, Q.Q. Li, and S.S. Fan, *Nature*, 419, 801, 2002.
33. A. Hirsch, *Chem. Int. Ed.*, 41, 1853, 2002.
34. S. Banerjee, T. Hemraj-Benny, and S.S. Wong, *Adv. Mater.*, 17, 17, 2005.
35. J.L. Bahr and J.M. Tour, *J. Mat. Chem.*, 12, 7, 1952, 2002.
36. Y. Sun, K. Fu, Y. Lin, and W. Huang, *Acc. Chem. Res.*, 35, 1096, 2002.
37. E. Bekyarova, M.E. Itkis, N. Cabrera, B. Zhao, A.P. Yu, J.B. Gao, and R.C. Haddon, *J. Am. Chem.Soc.*, 127, 16, 5990, 2005.
38. A. Garg and S.B. Sinnott, *Chem. Phy. Lett.*, 295, 4, 273, 1998.
39. M. Raja, A.M. Shanmugharaj, and S.H. Ryu, *Soft Materials*, 2, 65, 2008.
40. M. Raja, A.M. Shanmugharaj, and S.H. Ryu, *J. Biomedical Materials Research B: Applied Biomaterials*, 96b, 1, 119, 2011.
41. W. Bauhofer, J.Z. Kovacs, *Compos. Sci. Technol.*, 69, 1486, 2009.
42. M. Dresselhaus and P. Avouris, *Carbon Nanotubes: Synthesis, Structure Properties and Applications*; Springer: Berlin 2001.
43. P.M Ajayan, O. Stephan, C. Colliex, and D. Trauth, *Science*, 265, 1212, 1994.
44. P.M. Ajayan, L.S. Schadler, C. Giannaris, and A. Rubio, *Adv. Mater.*, 12, 750, 2000.
45. A.B. Dalton, S. Collins, E. Munoz, J.M. Razal, V.H. Ebron, and J.P. Ferraris, *Nature*, 423, 703, 2003.
46. H. Ago, K. Petritsch, M.S.P. Shaffer, A.H. Windle, and R.H. Friend, *Adv. Mater.*, 11, 1281, 1999.
47. E. Kymakis, I. Alexandrou, and G.A.J. Amaratunga, *J. Appl. Phys.*, 93, 1764, 2003.
48. H.W. Goh, S.H. Goh, G.Q. Xu, K.Y. Lee, G.Y. Yang, Y.W. Lee, and W.D. Zhang, *J. Phys. Chem. B*, 107, 6056, 2003.
49. P.C.P. Watts, W.K. Hsu, H.W. Kroto, and D.R.M. Walton, *Nano Lett.*, 3, 549, 2003.
50. S. Barrau, P. Demont, A. Peigney, C. Laurent, and C. Lacabanne, *Macromolecules*, 36, 5187, 2003.
51. D.E. Hill, Y. Lin, A.M. Rao, L.F. Allard, and Y.P. Sun, *Macromolecules*, 35, 9466, 2002.
52. B.Z. Tang and H. Xu, *Macromolecules*, 32, 2569, 1999.
53. M.J. O'Connell, P. Boul, L.M. Ericson, C. Huffman, Y.H. Wang, E. Haroz, and C. Kuper, *Chem. Phys. Lett.*, 342, 265, 2001.
54. H.J. Barraza, F. Pompeo, and E.A. O'Rear, *Nano Lett.*, 2, 797, 2003.
55. M.S.P. Shaffer and A.H. Windle, *Adv. Mater.*, 11, 937, 1999.
56. D. Qian, E.C. Dickey, and R. Andrews, *Appl. Phys. Lett.*, 76, 2868, 2000.
57. B. Safadi, R. Andrews, and E.A. Grulke, *J. Appl. Polym. Sci.*, 84, 2660, 2002.

58. C. Pirlot, I. Willems, A. Fonseca, and J.B. Nagy, *Adv. Eng. Mater.*, 4, 109, 2002.
59. G. Viswanathan, N. Chakrapani, H. Yang, B. Wei, C.Y. Ryu, and P.M. Ajayan, *J. Amer. Chem. Soc.*, 125, 9258, 2003.
60. W. Wu, S. Zhang, and D. Zhu, *Macromolecules*, 36, 6286, 2003.
61. T.W. Chou, L. Gao, E.T. Thostenson, Z. Zhang, J.-H. Byun, *Compos. Sci. Technol.*, 70, 1, 2010.
62. F.H. Gojny, M.H.G. Wichmann, U. Kopke, B. Fiedler, K. Schulte, *Compos. Sci. Technol.*, 64, 2363, 2004.
63. J. Suhr, N. Koratkar, P. Keblinski, P. Ajayan, *Nat. Mater.*, 4, 134, 2005.
64. J.N. Coleman, U. Khan, W.J. Blau, Y.K. Gun'ko, *Carbon*, 44, 1624, 2006.
65. H.G. Chae, Y.H. Choi, M.L. Minus, S. Kumar, *Compos. Sci. Technol.*, 69, 406, 2009.
66. M. Raja, A.M. Shanmugharaj, and S.H. Ryu, *Mater. Chem. Phys.*, 129, 925, 2011.
67. S.R. Bakshi, A. Agarwal, *Carbon*, 49, 533, 2011.
68. T. Kashiwagi, F.M. Du, J.F. Douglas, K.I. Winey, R.H. Harris, and J.R. Shields, *Nat. Mater.*, 4, 928, 2005.
69. A. Beigbeder, P. Degee, S.L. Conlan, R.J. Mutton, A.S. Clare, M.E. Pettitt, M.E. Callow, J.A. Callow, and P. Dubois, *Biofouling*, 24, 291, 2008.
70. Z. Wu, Z. Chen, X. Du, J.M. Logan, J. Sippel, M. Nikolou, K. Kamaras, J.R. Reynolds, D.B. Tanner, A.F. Hebard, A.G. Rinzler, *Science*, 305, 1273, 2004.
71. S. De and J.N. Coleman, *MRS Bull.*, 36, 774, 2011.
72. L. Xiao, Z. Chen, C. Feng, L. Liu, Z.Q. Bai, Y. Wang, L. Qian, Y. Zhang, Q. Li, K. Jiang, and S. Fan, *Nano Letters*, 8, 12, 4539, 2008.
73. D.M. Sun, M.Y. Timmermans, Y. Tian, A.G. Nasibulin, E.I. Kauppinen, S. Kishimoto, T. Mizutani, Y. Ohno. *Nat. Nanotechnol.*, 6, 156, 2011.
74. S. Ghosh, A.K. Sood, and N. Kumar, *Science*, 299, 5609, 1042, 2003.
75. K.J. Liao, W.L. Wang, Y. Zhang, L.H. Duan, and Y. Ma, *Microfab. Technol.*, 4, 57, 2003.
76. J. Liu and H. Dai, Design, fabrication, and testing of piezoresistive pressure sensors using carbon nanotubes, 2002, http://www.nnf.cornell.edu/2002re u/ Liu.pdf,
77. R. Caldwell, H. Dai, Q. Wang, and R. Grow, Carbon nanotubes as piezoresistors for a pressure sensor, 2002, http://www.nnf.cornell.edu/2002re u/Caldwell.pdf
78. J. Wang, G. Liu, and M.R. Jan, *J. Amer. Chem. Soc.*, 126, 3010, 2004.
79. Y. Xu, Y. Jiang, H. Cai, P.G. He, and Y.Z. Fang, *Anal. Chim. Acta*, 516, 19, 2004.
80. P. He and L. Dai, *Chem. Commun.*, 3, 348, 2004.
81. P. Adrian, Nanosensors targeted at the right markets could generate big business opportunities. *Sens. Bus. Dig.* 2003, http://www.sensorsmag.com/resources /businessdigest/sbd0703.shtml

82. R.L. McCreery, *Electroanal. Chem.*, 17, Marcel Dekker, New York, 401, 1991.
83. G. Che, B.B. Lakshmi, E.R. Fisher, and C.R. Martin, *Nature*, 393, 346, 402. 1998.
84. J.M. Planeix, N. Coustel, B. Coq, V. Brotons, P.S. Kumbhar, R. Dutartre, P. Geneste, P. Bernier, and P.M. Ajayan, *J. Am. Chem. Soc.*, 116, 7935 402, 1994.
85. C. Niu, E.K. Sichel, R. Hoch, D. Moy, and D.H. Tennet, *Appl. Phys. Lett.*, **7**, 1480, 402, 1997.

6

Bioimpact of Carbon Nanomaterials

A. Djordjevic[1,*], R. Injac[2], D. Jovic[1], J. Mrdjanovic[3] and M. Seke[4]

[1]Faculty of Sciences, Department of Chemistry, Biochemistry and Environmental Protection, University of Novi Sad, Novi Sad, Serbia
[2]Faculty of Pharmacy, Institute of Pharmaceutical Biology, University of Ljubljana, Ljubljana, Slovenia
[3]Oncology Institute of Vojvodina, Experimental Oncology Department, Sremska Kamenica, Serbia
[4]Institute of Nuclear Sciences "Vinca," University of Belgrade, Vinca, Serbia

Abstract

The unique size-dependent properties of carbon nanomaterials (CNMs)—graphene, nanotubes (CNTs) and fullerenes— make them very attractive for diagnostic and therapeutic application. This chapter presents possible application of CNMs. Graphene with extraordinary chemical and physical properties has already revealed a great number of potential applications such as environmental toxic material removal, drug delivery, tissue engineering, and fluorescence-based biomolecular sensing. The CNT derivatives have many interesting properties which make them potentially useful in a living system as biosensors, bioelectronic devices based on enzyme–nanotube or antibody–nanotube conjugates, chemotherapeutic agents, hyperthermia therapy and immunotherapy agents, agent in treatment of central nervous system disorders, and tissue engineering agent. Fullerene C_{60} derivatives have been used as: drug and gene delivery vectors, magnetic resonance imaging agents, radio protectors, antioxidants, HIV-1 protease inhibitors, antigenotoxic agents, and phototherapy agents. Fullerenols are polyhydroxylated derivatives of fullerene ($C_{60}(OH)n$) with remarkable antioxidant, xenobiotic-protective, radioprotective, nanodrug and endohedral gadolinium carrier properties.

Corresponding author: aleksandar.djordjevic@dh.uns.ac.rs

Ashutosh Tiwari and S.K. Shukla (eds.) Advanced Carbon Materials and Technology, (193–272)
2014 © Scrivener Publishing LLC

Keywords: Engineered nanomaterials, nanomedicine, graphene, carbon nanotubes, fullerene derivatives, fullerenol, nanotoxicity, drug delivery, diagnostic, genotoxicity, malignancy, nanoecotoxicity

6.1 Biologically Active Fullerene Derivatives

6.1.1 Introduction

Fullerenes are cluster structures made entirely of carbon atoms. The whole family of cluster carbon structure, Cn 20 ≤ n ≤ 5 40, was named after Richard Buckminster Fuller, an American architect who projected geodesic homes which resemble spherical fullerenes. The most prominent member of the fullerene family is undoubtedly C_{60} (Figure 6.1), a perfectly symmetrical molecule that has a shape of a soccer ball, which is why the molecule has also been called "buckyball" or "footballene." Fullerene C_{60} was accidentally disovered in 1985 by Robert F. Curl, Jr., Richard E. Smalley, and Sir Harold W. Kroto [1]. The Nobel Prize in Chemistry in 1996 was awarded to this research team for the discovery and structure determination of the third allotropic modification of carbon. Fullerene can be found in nature in meteor rocks. Nowadays, several processes are being used for the synthesis of carbon clusters [2, 3]. In comparison with the quantity of fullerene produced at the end of the previous century in grams, today, the annual production of fullerene materials reaches tones. Fullerenes are defined as polyhedral closed cages made up entirely of three-coordinate carbon atoms and having 12 pentagons and different number of hexagons; for instance, molecule C_{60} possesses 12 pentagons and 20 hexagons and has a geometry of truncated icosahedron (fullerene-60-Ih), a shape well-known in art ever since the 16th century. All carbon atoms in molecule are equivalent, which was proved by ^{13}C NMR spectrosopy. Diameter of the molecule is 0.710 ± 0.007 nm and inner diameters of π electron cloud are of estimated size 0.34 nm–0.35 nm. Pure fullerenes are black crystals, or powder, and practically insoluble in water and proton-acceptor solvents. They are, however, soluble in halogen and alkyl-substituted benzene, CS_2, 1,2-dichlorobenzene, and naphthalene. The structural and electronic characteristics of fullerene C_{60} provided the possibility for conducting various types of chemical transformations [4] that have resulted in a wide variety of biologically active water-soluble fullerenes.

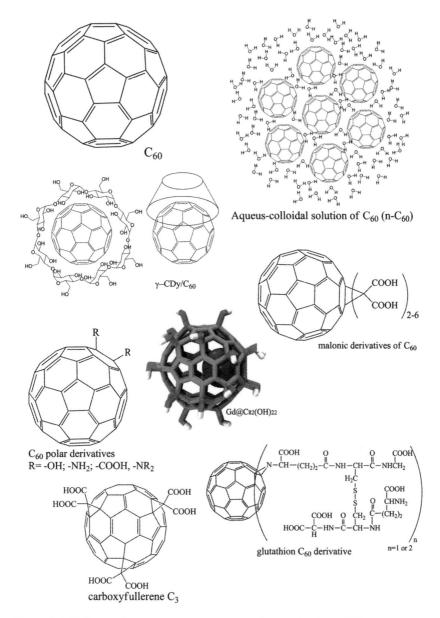

Figure 6.1 Fullerene C_{60}, non-covalent and covalent derivatives of C_{60}.

By combination of nucleophilic and electrophilic additions, cyclo-additions and radical additions, it is possible to covalently attach a large nuber of organic compounds to a fullerene cage. Since the discovery of fullerene, many biological features of non-derivatized and derivatized fullerenes have been unraveled, some of which are anti-viral and antibacterial properties, antioxidative, neuroprotective and prooxidative activities, cell signaling and apoptosis, cytotoxic and cytoprotective properties, radio protective activities, anticancer drugs, etc. [5]. Fullerenes are used in nanomedicine technologies due to their size and possibilities for use in diverse surface modifications as nano-structured materials. This part of the chapter will specifically deal with the biological activity of derivatized fullerene C_{60}, with a particular accent on water-soluble, polyhydroxylated derivatives—fullerenols.

6.1.2 Functionalization/Derivatization of Fullerene C_{60}

Fullerenes have a wide range of applications in biomaterials, medicine, electronics, and new materials. Nanomedicine has become one of the most promising areas of nanotechnology. Water-soluble fullerene derivatives that come from chemical transformation largely enhance the biological efficacy in *in vitro* and *in vivo* models. Major challenges face neuropharmacology, antineoplastics, nanodrug cariers and nanodiagnostic. Poor solubility of fullerene in polar solvents is one of the limiting factors for application in *in vitro* and *in vivo* models. Numerous examples of chemical transformations aimed at increasing solubility of fullerene C_{60} have been developed and can be classified into four basic groups:

1. Chemical modification of fullerene cage by attaching various polar functional groups (-OH, -NH$_2$, -COOH) [6–10].
2. Fullerene incorporation in water-soluble supramolecular structures by calixarene and cyclodextrin [11, 12].
3. Synthesis of stable water suspensions of fullerene by using organic solvents and intensive sonication [13].
4. Method of long-term stirring of C_{60} in water [14]. Despite the fact that this method allows examination of native C_{60} in water, drawbacks of this method are low reproducibility, production of large aggregates and low concentration of fullerene in water.

6.1.3 Biological Activity of Non-Derivatized Fullerene C_{60}

The unique π-electron system of fullerene C_{60} makes it highly photosensitive in VIS and UV range of spectrum. Partial solubility of fullerene C_{60} (fullerene nanoaggregats n-C_{60}) in water can be achieved by preparing several different compositions (Fig. 6.1): long-term sonicated toluene or tetrahydrofuran (THF) solution of C_{60} in water (THF/nC_{60}); monomeric $(\gamma\text{-CyD})_2/C_{60}$, ($\gamma$-cyklodextrin bicapped C_{60}) and γ-CyD/nC_{60} (obtained by heating the $(\gamma\text{-CyD})_2/C_{60}$ aqueous solution). Transmission Electron Microsopy (TEM) results have demonstrated that dimensions of water-soluble nano-aggregates C_{60} in water stand within 20 to 140 nm [15]. *In vitro* research on phototoxicity of $(\gamma\text{-CyD})_2/C_{60}$, THF/n$C_{60}$, sonicated/ n$C_{60}$ and γ-CyD/nC_{60} conducted on keratinocyte cell lines (HaCaT) showed that their phototoxicity was mainly mediated via singlet oxygen production. Cell viability analysis of HaCaT cell line has shown that $(\gamma\text{-CyD})_2/C_{60}$ exhibits up to 60 times higher phototoxicity than polyhydroxylated derivative fullerenol, $C_{60}(OH)_{24}$.

Upon UVA irradiation of human keration cytes in the presence of $(\gamma\text{-CyD})_2/C_{60}$, a significant increase in concentration of intracellular peroxides from proteins was noted. Sayes *et al.* [16] have found that n-C_{60} showed cytotoxic effect on human dermal fibroblasts, human liver carcinoma cells (HepG2), and neuronal human astrocytes with LC_{50} from 2 ppb to 50 ppb (depending on the cell type), after 48 h exposure. L-Ascorbic acid was shown to have a protective effect against oxidative damage caused by the toxic effects of n-C_{60} [17, 16]. Isakovic *et al.* [18] have published that higly prooxidative capacity of n-C_{60} is responsible for rapid cell death in mouse L929 fibrosarcoma, rat C6 glioma, and U251 human glioma cell lines. Contrary to *in vitro* results, aqueous C_{60} suspension effects on acute intoxication in rats induced by carbon tetrachloride showed that n-C_{60} suspension had no acute or subacute toxicity [19]. On the contrary, the group claimed that n-C_{60} protected the rats' livers from free-radical damage. In low concentration range, n-C_{60} has been found to inhibit growth of gram-positive (Bacillus subtilis CB315, JH642 derivative) and gram-negative (Echerichia coli DH5) cultures in a wide range of conditions [20, 13]. Based on toxicological studies, it was established that residual quantities of organic solvents in nano-aggregates were responsible for the toxic effects of n-C_{60} [20, 18].

6.1.4   Biological Activity of Derivatized Fullerene C_{60}

By adding polar functional groups on C_{60}, many water-soluble derivatives have been synthesized (Fig. 6.1). These derivatives present highly attractive nanomaterials for various biomedical applications

[7, 21–27]. In some earlier conducted studies [28], researchers concluded that functionalization of fullerene C_{60} under photosensitive conditions decreases its prooxidative properties and makes it less toxic than non-functionalized C_{60} [28]. In the absence of photosensitive reaction source, most water-soluble fullerenes express characteristics of powerful deactivators of different types of reactive oxygen species (ROS) [29]. Electron spin resonance (ESR) study of fullerenol encapsulated gadolinium, $Gd@C_{82}(OH)_{22}$ (Fig. 6.1), fullerenol $C_{60}(OH)_{22}$, and $C_{60}(C(COOH)_2)_2$ (Fig. 6.1) confirmed the ability of scavenging following reactive oxygen species: singlet oxygen, superoxide anion radical, hydroxyl radical and stable DPPH radical [30]. Dynamic light scattering (DLS) analysis has shown that the size of derivatized fullerene particles ranges from 78 to 170 nm. The antioxidative properties of fullerene derivatives weaken with the increase in size of nanoparticles, which is in accordance with the decrease in reactive surface area. In addition, nanoparticle size affects distribution in cells, cytoprotective and antioxidative properties in biological systems (adenocarcinomic human alveolar basal epithelial cells, A549 and rat brain capillary endothelial cell cultures, rBCEC). Malonic derivative of fullerene, $C_{60}(C(COOH)_2)_2$, with average size of particles around 170 nm and zeta-potential -22 mV at pH 7, exists as a stable form in physiological conditions. Derivatives of fullerene have exhibited a protective effect on the primary culture of microvascular endothelial cells (CMECs) under conditions of sodium nitroprusside (SNP)-induced cell apoptosis. Besides, it was founded that nanoparticles can improve reparation of endothelial cells and inhibit polimerization of actin, induced by toxic effects of reactive nitrogen species (RNS) and ROS [31]. The proposed mechanism of fullerene derivative action is based on catalytic activity with electron-deficit regions on derivatized spheres of C_{60}.

The trimalonic acid derivative of fullerene has been shown to be effective in the treatment of both gram-positive and gram-negative infections, where a far better outcome was achieved in the case of gram-positive bacteria [32].

Tris(dicarboxymethyl)-fullerene C3 (Fig. 6.1) isomer expressed cytoprotective effect on adrenal gland cells through increase in survival of cells and cell death prevention, including apoptosis, which is of great significance in the treatment of Parkinson's disease.[33]. Carboxy derivatives of fullerene isomer C3 inhibited excitotoxic death of cultured cortical neurons [34] and had a protective effect on the cerebellar granule cells apoptosis induced by oxidative stress [35]. It was found that tris(dicarboxymethyl) fullerenes form aggregates in aqueous solution, and those within 40–80 nm are responsible for the neuroprotective action in cells [36]. Carboxyfullerenes proved effective in the treatment of gram-positive and gram-negative infections [37]. Carboxyfullerene (C3) was proven to have a radioprotective effect on B lymphoblastoid cell lines because of its strong free radical scavenging capacity, as well as its ability to decrease apoptosis of human intestinal crypt epithelial cells (HIECs). Carboxyfullerene was also shown to decrease DNA damage 8 hours after irradiation. In *in vivo* investigations on mice, after 7,2 Gy γ-irradiation with doses of 100 mg/kg, a 75% survival rate was noted [38].

Photosensitizer tris-cationic-buckminsterfullerene, irradiated with VIS light, expressed an antimicrobial effect against gram-positive and gram-negative bacteria by producing reactive oxygen species [39]. Stable nanoparticle formed between human serum albumin and C3 isomer presents a stable protein-fullerene complex, and it showed better water solubility and biocompatibility than C3 isomer [40].

N-ethyl-polyamino C_{60} (Fig. 6.1) possesses potential activity towards allergic response via inhibition of free radicals [41]. Fullerene derivative $C_{60}(ONO_2)_{7+2}$ showed effective antioxidative protection on *in vivo* model in oxidative stress induced by ischemia-reperfusion-induced lung injury [42]. Glutathione C_{60} derivative (Fig. 6.1) was tested on a cell line of rat adrenal gland (PC12) and proved to scavenge ROS and RNS, without causing necrosis and apoptosis of cells [43].

Monomethoxy triethylene glycol substituted fulleropyrollidines completely inhibit *Mycobacterium avium* at a dose of 260 μg/ml and *Mycobacterium tuberculosis* at a dose of 50μg/ml [37]. Trans-2 and trans-4 C_{60}-bis (N,N dimethylpyrrolidinium iodide) (Figure 6.2) express bacteriostatic effect against *Escherichia coli* [44].

Cationic quinazolinone conjugated fullerene (Fig. 6.2) has a proven antibacterial activity against both gram-positive and gram-negative bacteria, and also antimicotic activity against some fungi (Candida albicans, Aspergillus clavatus, and Aspergillus niger), with lower minimal inhibitory concentrations (MIC) than some commercially used drugs [45].

A fullerene-isoniazid conjugate forms stable nanosize aggregates in water suspension with successfull antimycobacterial activity against *Mycobacterium avium* and strains of *Mycobacterium tuberculosis-H(37)* even with low concentration doses [46].

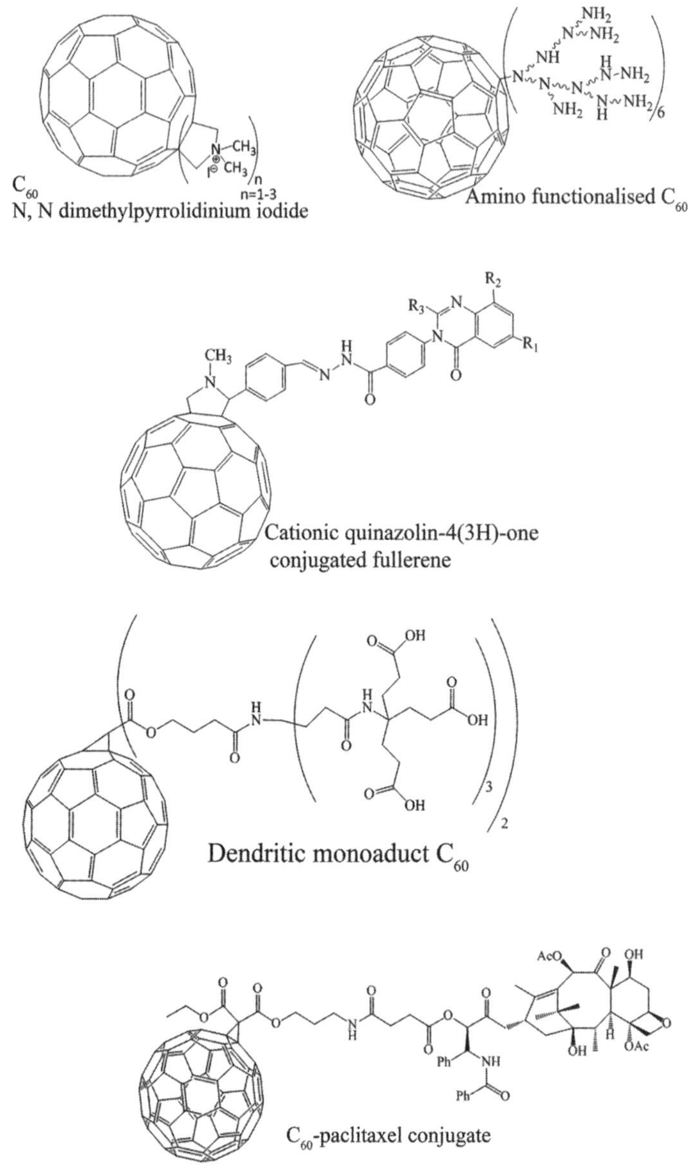

Figure 6.2 Covalent derivatives of C_{60}.

Bis-methanophosphonate C_{60} derivative displayed photo-induced cytotoxicity on HeLa cells [47].

Dendritic mono-adduct C_{60} (Fig. 6.2) inhibits the growth of Jurkat cells, and it was also proven that, in presence of UVA and UVB light, the cytotoxic effects are even more intensive [48].

Antineoplastic C_{60}-paclitaxel (Fig. 6.2) conjugate was purposely synthesized as a slow-release drug-delivery system for liposome aerosol delivery to lungs [49]. Tetraamino fullerene proved to be a good solution for gene delivery in mammalian cells [50]. It binds to DNA, delivers the fullerene/DNA complex into the cytoplasm and allows the encoded gene to be expressed either transiently or steadily with efficiency comparable or higher to that achieved by the lipofection. Fullerene immunoconjugates have also been synthesized and characterized [51].

Monoclonal fullerene specific antibody [52] could be applied in assays for the measurement of dosage and serum levels of fullerene derivatives and for the detection of preferential intracellular localization of fullerenes as drug delivery agents [53]. Kim *et al.* [54] have published that 1,2-(dimethoxymethano)fullerene inhibits amyloid peptide aggregation by binding to the central hydrophobic part of the peptide, which indicates its potential use in Alzheimer's disease therapy.

Amine-functionalized C_{60} (Fig. 6.2) was synthesized by attaching ethylenediamine to fullerene, after which it was derivatized with polyethyleneimine (PEI), (C_{60}-PEI), via cationic polymerization of aziridine on the surface of C_{60}-NH_2. The C_{60}-PEI was encapsulated with folic acid (for the purpose of targeting) and conjugated with new antineoplastic docetaxel, thus producing a novel drug carrier. By comparing docetaxel and fullerene conjugate results, it was established that the tumor targeting drug delivery system could efficiently cross cell membrane, as well as exhibit higher antitumor efficacy in cultured prostate cancer cells. Additionally, experiments conducted on *in vivo* murine tumor model showed that fullerene conjugate afforded higher antitumor efficacy without obvious toxic effects to normal organs [55].

6.1.5   Chemical Synthesis of Fullerenol $C_{60}(OH)_n$

Fullerenols or fullerols, ($C_{60}(OH)_n$ ($2 \leq n \leq 44$), are polyhydroxylated derivatives of fullerene C_{60} (Fig. 6.2). The first synthesis of fullerenol was done in 1993 by Li *et al.* [56] out of toluen solution of C_{60} with sodium hydroxide in presence of catalytic quantities of quaternary ammonium hydroxides [56]. Fullerenols can be synthesized in some of the following ways: with oleum [57], by hydroboration [58], in synthesis of highly-functionalized $C_{60}(OH)_{44} \cdot 8H_2O$ with toluen solution of C_{60} and 30% hydrogen peroxide with catalytic

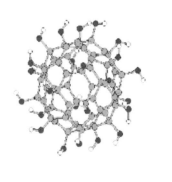

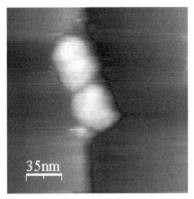

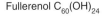

Fullerenol $C_{60}(OH)_{24}$

AFM Fullerenol nano particles

Figure 6.3 Fullerenol $C_{60}(OH)_{24}$ and AFM fullerenol nano particles.

quantities of quaternary ammonium base [59]. Fullerenol is possible to synthesize by direct oxidation, as well [9]. Yao *et al.* [60] have published the synthesis and separation of fullerenols by dialysis. Sequential synthesis of fullerenol $C_{60}(OH)_{24}$ intermediated via polybromic derivative $C_{60}Br_{24}$ [61] results in a single reaction product with significant yield [62]. Fullerenols are highly soluble in water, and even more soluble in dimethyl sulfoxide. Water solubility increases with the number of hydroxyl groups. In aqueous solution at different pH values, fullerenols are prone to forming stable nanoaggregates (Fig. 6.3) [6, 63].

6.1.6　Fullerenol and Biosystems

The potential diagnostic, analytical, industrial, environmental, scientific, and health effects of different fullerenes, including fullerenols, have attracted increasing attention in the past decade, especially with the improvement of water-soluble forms that alleviate their use in biological systems [64]. These compounds have potential application in many difficult diseases. So far, they have been investigated for suppression of metastases, treatment of cerebral conditions such as Alzheimer's and Parkinson's diseases, type-C hepatitis therapy, and HIV treatment.

The chemical adjustment of fullerenes by adding the -OH groups to their carbon surface yields a variety of polyhydroxylated structures $C_{60}(OH)_x$ exhibiting different degrees of solubility and antioxidant activity [12, 65]. The crucial advantage of fullerenol, according to previous investigations, compared to other branded antioxidants or cytoprotectors, is its potential dual function as radio-protector and organo-protector for the period of the anticancer therapy [66, 21, 67–71, 23, 72–77]. Nevertheless, before a definite statement concerning the potential efficacy of fullerenol as an adjunct to anticancer therapy, there is a need for further studies, including clinical trials in humans.

6.1.6.1 Fullerenol In Vitro Activity

The antioxidative and positive toxicity effectiveness of fullerenol is being broadly evaluated in both *in vitro* and *in vivo* models. One of the most studied antioxidative effects of fullerenol is its protective influence against doxorubicin-induced cell death. Fullerenol modulated doxorubicin-induced cytotoxicity on human breast cancer cell lines and human hepatocellular carcinoma cells (HepG2) [78, 71]. When speaking about the biological antioxidative activity of different fullerenols $C_{60}(OH)_x$ *in vitro*, what should be mentioned is their protective effect against oxidative stress in RAW 264.7 macrophage cell line, ischemia-reperfused rat lungs [79], and their ability to extensively reduce doxorubicin toxicity against human breast cancer cell lines and Hepg2 cells [80, 71].

It was shown that fullerenols have protective effects on cytotoxicity in three human breast cancer cell lines (T47D, MDA-MB-231, and MCF). Polyhydroxylated fullerene derivatives in combination with doxorubicin significantly inhibited doxorubicin cytotoxicity by more than 50% for most concentrations at each time point. Cells had the best protection if fullerenol was added to culture one hour prior to doxorubicin. Accordingly, the inhibition of doxorubicin cytotoxicity with fullerenol in estrogen receptor positive cell lines was the most prominent after two days. The concentration 0.44 µmol/l of fullerenol was found to be the most effective. The rates of inhibition by the fullerenol alone were similar to the doxorubicin-fullerenol ones. Assuming that doxorubicin cytotoxicity can be explained by induction of oxygen free-radical formation, and proposed antioxidative activity of the fullerenols, the effect of fullerenol was compared with the effect of some natural antioxidant. In the same investigation, a comparison was done with proanthocyanidine using the same experimental system. The results show that both molecules added as single agents to MCF-7 cell culture for 48 h, induce low growth inhibition. In combination with doxorubicin, both agents powerfully reduce doxorubicin cytotoxicity. Doxorubicin cytotoxicity was decreased by 80% (fullerenol) and 87% (proanthocyanidin) if the agents were added to the culture at least one hour before doxorubicin [80]. In the same way, fullerenol showed protection of cell death in human hepatocellular carcinoma cells (HepG2) induced by doxorubicin. Doxorubicin-induced cytotoxicity in the HepG2 cell line is dose- and time-dependent. Fullerenol in two different concentrations (8.8 µmol/l and 39 µmol/l) was co-added with doxorubicin (5 µmol/l) and significantly increased the cell survival rate almost-two fold within the first 24 h. After 24 h of treatment, the survival fractions of the HepG2 cells were as follows: 29, 61 and 72% (doxorubicine alone, doxorubicine/fullerenol 8.8 µmol/l, doxorubicine/fullerenol 39 µmol/l, respectively). After four days, the survival rate of the HepG2 cells was almost 7 times higher among the doxorubicine/fullerenol-treated cells compared to doxorubicine alone. The level of cytotoxicity of the doxorubicine/fullerenol combination was similar to the cytotoxicity induced by the fullerenol alone [71]. Localization of fullerenol close to the mitochondria [81], its free radical scavenger activity, and the ability to act as an artificial electron acceptor, could explain the potential protective effect of fullerenol in the model of doxorubicin-induced cytotoxicity in different cell lines [80].

Bogdanović *et al.* [82] investigated and confirmed fullerenol effects on antioxidative enzymes activity in irradiated human erythroleukemia cell lines. Investigations have provided facts of reactive nitrogen and oxygen species (RNS/ROS) scavenger properties *in vitro* as well as radio protective efficiency *in vivo*. Pre-treatment of irradiated human erythroleukemia K562 cells by fullerenol exerted the statistically significant effects on cell number and response of antioxidative enzymes to X-ray irradiation-induced oxidative stress in cells. The enzyme activities of γ–glutamyl transferase (GGT), total superoxide dismutase (SOD) and glutathione peroxidase (GSH-Px) were determined by kinetic method [83–85]. The number of K562 cells was increased 48 h after irradiation, indicating that the surviving fraction of cells continued to proliferate. This effect was more evident in fullerenol pre-treated cells. Activity of GGT in irradiated K562 cells was significantly increased in a time-related manner (4-fold at 24 h and 2.8-fold at 48 h post-irradiation time, respectively). The fullerenol pre-treatment significantly reduced the levels of GGT in irradiated cells. Published data are showing the major increase of total SOD levels in irradiated cells during 48 h post-irradiation time. More fascinatingly, 4-fold elevated SOD levels were observed at the 1 h time point in fullerenol pre-treated non-irradiated cells, and a 5.5-fold elevated level in fullerenol pre-treated irradiated cells. These results are in accordance with the fullerenol ROS scavenging activity observed in both *in vivo* and *in vitro* systems, indicating the important role of fullerenol for maintenance of redox homeostasis [86–88]. The hypothesis concerning the mechanisms responsible for the increased activity of SOD in K562 cells pre-treated by fullerenol, may be viewed in light of the possibility that the NO-scavenging fullerenol activity may prevent superoxide consummation in the reaction of formation of peroxynitrite anion, along with increasing the O_2^- concentration, and increasing SOD activity as a consequence of superoxide excess. It was hypothesized that the fullerenol pre-treatment prevented the toxic effects of ROS directly by increasing the antioxidant enzyme activities. Fullerenol pre-treatment did not have an influence on the control levels of GSH-Px activity, while the levels of GSH-Px were higher in both irradiated cellular sets. Following fullerenol pre-treatment and 24Gy irradiation, the GSH-Px level was significantly elevated at 1 h post-irradiation time. This effect of fullerenol pre-treatment in irradiated cells may be partially explained by consecutive activation of GSH-Px due to SOD increased activity. On the other hand, level of substrate accumulation in cell cytosole, which inhibits the GSH-Px enzyme active center, could also have an influence on the previously described effect [82]. The fullerenol capacity to form clusters in water, at physiological conditions, may be to some extent responsible for the biological effects that were observed [63].

The assessment of fullerenol activity including toxicity is an unconditional and noticeable precondition for its potential use in biomedicine, cosmetics and consumer products. Recent studies describe lower toxicity of fullerenol and other water-soluble fullerenes against dermal fibroblasts and liver carcinoma HepG2 cells, in comparison to pure fullerenes (C_{60}/C_{70}). It was confirmed that the cytotoxic activity of C_{60} colloid was caused by ROS-mediated cell membrane lipid peroxidation [16]. In addition, Isakovic *et al.* [18] demonstrated that pure C_{60} and its polyhydroxylated derivatives $C_{60}(OH)_x$ in fact employ different cytotoxic mechanisms resulting in superior induction of caspase-independent necrosis and caspase-dependent apoptosis, respectively. From a molecular point of view, C_{60} induces production of oxygen radicals, which are involved in lipid peroxidation, and consequently stimulate necrotic cell death. In contrast, $C_{60}(OH)_x$-triggered apoptosis in some cell types seems to be ROS independent. The pro-oxidant activity of pure C_{60} is connected to its chemical structure and characteristics. It was recognized that derivatization of C_{60} decreases ROS generation and cytotoxicity [17].

The very first scientific literature with a lot of optimism and positive results supports a protective role of fullerenol in biological systems. Results of numerous investigations in this field are evidence of the cytotoxic effects of this nanoparticle [89, 17, 90, 91]. Fullerenol's mechanism of cell death appears to be cell-type-specific. Both apoptotic and non-apoptotic mechanisms have been investigated and the results published in the literature [90, 92]. Water soluble fullerene derivatives have been reported to cause cell cycle arrest at the G1 phase in Chinese hamster lung and ovary cells [91], as well as suppressed proliferation of human breast cancer cells in a cell line-dependent manner [80]. Derivatized fullerenes have also been reported to demonstrate differential cytotoxicity in liver carcinoma cell lines, human dermal fibroblasts and colon adenocarcinoma cells, with the more water-soluble derivatives demonstrating minor adverse effects in culture [71, 17].

To illustrate, fullerenol selectively inhibits proliferation of the human colon adenocarcinoma cell line (Caco-2) and hepatocellular carcinoma cells (HepG2) in a concentration- and time-dependent manner, with greater influence on Caco-2 cells. Knowing that low levels of free oxygen species promote cell proliferation and that redox alterations play a considerable role in a signal transduction pathway, vital for cell growth regulation [93, 94], it is sound to propose that fullerenol might influence the HepG2 and Caco-2 cell redox state, leading to decreased cell proliferation [71]. Such an observation could clarify the selectivity of fullerenol on growth inhibition for the HepG2 cells in contrast to Caco-2 cells. For cells to regenerate after injury and protect themselves against oxidative stress, it is a high requirement for enzymes to regulate cellular processes [95]. Fullerenols have been confirmed as good free radical trappers in biological systems [89, 96–98]. Our research group confirmed that fullerenol does disturb the mitochondrial activity in view of the fact that a significant dose-dependent drop of mitochondrial potential incubation with fullerenol was shown. The effect was more pronounced in Caco-2 cells than in HepG2 cells, which can result from the higher regeneration properties and mitochondrial capacity of the HepG2 cells [71].

Loss of mitochondrial membrane potential after addition of fullerenol to cell culture was also observed in kidney cells [99], with proof that fullerenol-induced cell death was associated with cytoskeleton disruption and autophagic vacuole accumulation. Treatment of cell culture with fullerenol also resulted in simultaneous loss of cellular mitochondrial membrane potential and depletion of ATP.

Since many of the fullerene applications are based on intravascular administration, it is of great import to identify the effects of fullerenes on endothelial cells. Yamawaki and Iwai [90] have reported that the addition of fullerenol to endothelial cells decreases cell density, cell proliferation and cell attachment, promotes LDH release, and increases accumulation of polyubiquitinated proteins. These authors concluded that fullerenol did not induce apoptosis based on the failure, but relatively, similarly as in kidney cells [99] induced atophagy. Further study by Gelderman *et al.* [92] on human epithelial HUVEC cells revealed that the fullerenol had both pro-apoptotic and pro-inflammatory effects on cells, illustrated by fullerenol-induced DNA fragmentation. Additionally, cell cycle capture and increase of intracellular Ca^{2+} were observed in cells treated with $C_{60}(OH)_{24}$. For that reason it is reasonable to hypothesize that both apoptotic and autophagic cell death may occur in fullerenol-treated endothelium.

The FNP-treated irradiated cells and significant overexpression of anti-apoptotic Bcl-2 and Bcl-xL and cytoprotective genes such as GSTA4, MnSOD, NOS, CAT and HO-1 genes, may indicate that FNP exerts cytoprotective function in K562 leukemic cells, rendering K562 cells more tolerant to radiotherapy [100].

Recently, a few new papers have been published using different *in vitro* models as well as different fullerenols. The outcomes of the latest studies have not been much different than those already reported. Firstly, our research group did an investigation regarding the efficacy of nano-fullerenol $C_{60}(OH)_{24}$ form on micronuclei and chromosomal aberration frequency in peripheral blood lymphocytes [101]. Both genotoxic and antigenotoxic effects of fullerenol nanoparticles have been evaluated. Nanoparticle number distribution in a culture medium with serum showed that predominant particles were about 180 nm and 90 nm connected to chromosomal aberrations and micronuclei, respectively. Furthermore, cytogenetic assay showed that fullerenol decreased chromosomal abbreviations and micronucleus frequency on the undamaged and the mitomycin C-damaged human peripheral blood lymphocytes at concentrations ranging from 5.5.to 221.6 µmol/l. It was found that fullerenol in nano form did not demonstrate genotoxic but induced antigenotoxic effects at subcytotoxic concentrations on isolated human lymphocytes [101]. With a similar aim, Zha *et al.* [102] have used water-soluble polyhydroxyfullerene $C_{60}(OH)_x(ONa)_y$ (y = 6–8; x + y = 24) for investigating the influence of its different concentrations on cultured rat hippocampal neurons. Once again, it was confirmed that fullerenol can be deemed a promoter of cell death as well as protector against oxidative risk, depending on the concentrations used. This kind of dual effect was not anything new, but much still remains unknown about its dual behavior from a mechanism point of view.

With broad potential use of fullerenes in mind, the safety hazards of these materials need to be systematically evaluated and confirmed.

6.1.6.2   *Fullerenol* In Vivo *Activity*

Many research works published in the past two decades have included different *in vivo* and *in vitro* topics about fullerenes (C_{60} and C_{70}), but a very limited number of investigations have included fullerenols, in particular, $C_{60}(OH)_{18}$, $C_{60}(OH)_{22}$ and $C_{60}(OH)_{24}$. The situation in regards to evidence about the *in vivo* toxicity of fullerenols is even worse in comparison to published antioxidative studies and fullerenes toxicity testing. In most cases, toxicity was presented as a potential cytostatic application in different tumor cell lines with a tendency to be used in anticancer therapy [23].

The antioxidant properties of fullerenol were tested by measuring their ability to scavenge stable 2, 2-diphenyl-1-picrylhydrazyl (DPPH) free radical and reactive hydroxyl radical (OH). The addition of fullerenol to the reaction system (0.18–0.88 mmol/l) resulted in a dose-dependent inhibition of the DPPH radical. The concentration of 0.88 mmol/l of fullerenol was more effective in inhibition of hydroxyl radical (inhibition ratio 83%) in comparison with inhibition of DPPH radical. Accordingly, the possible mechanism of antioxidative activity of fullerenol is the radical-addition reaction of 2n·OH radicals to remaining olefinic double bonds of fullerenol core to yield $C_{60}(OH)_{24}+2n\cdot OH$ (n=1–12). The other proposed mechanism is the possibility of hydroxyl radical interacting with a hydrogen from fullerenol, which results in the formation of a stable form such as fullerenol radical $C_{60}(OH)_{23}O$ [66].

The prevention of lipid peroxidation in liposome model system by fullerenol was also confirmed. Briefly, tested substances (including fullerenol) and liposome were added to $FeSO_4$/ascorbic acid system and incubated 1 h at room temperature [103]. Commercial antioxidant butylated hydroxy toluen (BHT) was used for comparison of fullerenol antioxidant potential toward lipid peroxides. Treatment of liposome with $FeSO_4$ and ascorbic acid leads to oxidation of polyunsaturated fatty acid in liposome. As a result of this kind of reaction, formation of TBA-RS can be detected. Results confirmed that fullerenol induced dose-dependent inhibition of $FeSO_4$/ascorbic acid-stimulated formation of TBA-RS. In parallel, it was examined and shown that the effect of BHT on lipid peroxidation was similar to the efficiency of fullerenol under the same conditions. Calculated ED_{50} for $C_{60}(OH)_{24}$ and BHT were 2.5 mmol/l, and 1.2 mmol/l, respectively. Additionally, superoxide radical-scavenging activity of fullerenol in xanthine/xanthine oxidase system was used for testing. Xanthine oxidase reduces cytochrome C generated through enzymatic activity of superoxide radicals. Removing superoxide radicals leads to decreased rate of cytochrome C reduction. Application of $C_{60}(OH)_{24}$ into xanthine/xanthine oxidase system effected the decrease in reduction rate of cytochrome C compared to control. Published results showed that $C_{60}(OH)_{24}$ in the range of nmol/l and μmol/l concentrations decreased reduction of cytochrome C up to 20%, while concentration of 1 mmol/l decreased reduction of cytochrome C for 40% [62].

In addition to previous investigations, nitric oxide scavenging in chemical model system was also used to prove a statement concerning antioxidative potential of fullerenol. Examination of the possible nitric oxide-scavenging activity fullerenol confirmed that it expressed direct scavenging activity toward nitric oxide radical (NO) liberated within solution of sodium nitroprusside (SNP), which is a well known NO donor. Different concentrations of SNP (1, 2 or 5 mmol/l) and of fullerenol (0.05 or 0.10 mmol/l) were incubated alone or in combination to estimate possible NO-scavenging activity of fullerenol. The NO release on light and room temperature from SNP dissolved in phosphate buffer was time- and dose- dependent. In comparison to the nitrite levels obtained when SNP was dissolved alone, co-incubation of SNP with fullerenol resulted in the dose-dependent decrease of the level of nitrite. On the other hand, the percentage of the decrease was the same regardless of the concentration of SNP present in the solution. The average ED_{50} for fullerenol calculated for all dose-dependent curves was 10^{-4} mol/l [62].

The radioprotective effect of fullerenol is connected to its capability to react with free radicals in *in vivo* systems, which is the same mechanism for antioxidant organo-protection. Ionizing radiation produces harmful effects on living organisms by inducing enhanced production of free radical species. Exposure of healthy tissue to ionizing radiation leads to the formation of ROS that are related to radiation-induced toxicity. The antioxidative enzyme SOD catalyzes the dismutation of superoxide anions into hydrogen peroxide and has a very important role in increasing protection from lethal irradiation to haematopoietic cells. An investigation of the potential radioprotective effects of fullerenol in a dose of 100 mg/kg i.p. given 30 min prior to X-irradiation was published by Trajkovic *et al.* [104]. Previous reports about the radioprotective effect of fullerenol established that i.p. administration of 100 mg/kg of fullerenol 30 minutes before irradiation is suitable in comparison to a smaller dose of 10 mg/kg [80], but probably the most suitable dose from the effectiveness and safety point of view would be something in between. In relation to fullerenol radioprotectivity, importance must be placed on its hemato- and tissue-protective effects. So far, the tissue protective effect has mainly been confirmed on the spleen, small intestine and lung tissue [105].

Some polyxydroxylated fullerenes have been more favorable in recent investigations according to their previous successful results as potential antioxidants for different medical needs. One of them is fullerenol $C_{60}(OH)_n$ (n=18,22,24). It is suggested that fullerenols may act as free radical scavengers in biological systems, in xenobiotics, on the skin surface, as well as on radioactive irradiation-induced oxidative stress. They have repeatedly demonstrated protective effects against different cytotoxicities of doxorubicin in animal models, especially by fullerenol $C_{60}(OH)_{24}$ [62, 66, 21, 75, 67–71, 23, 72–74, 76, 77].

Sprague Dawley outbred rats with chemically-induced mammary carcinomas were used to study a potential protective role of fullerenol against major organo-toxicity induced by doxorubicin [67, 70, 71, 74]. According to preliminary studies on healthy adult Wistar rats [75, 72], it was established that 100 mg/kg (i.p.) of fullerenol administered 30 min before doxorubicin has a potential protective influence on heart and liver tissue. Therefore, in further investigations that dose was chosen as the effective one, but not as a final one.

Nephrotoxicity caused by doxorubicin has been well-known in the literature for a few decades. It was shown that pre-treatment with fullerenol prevented oxidative stress, lipid peroxidation and a disbalance of GSH/GSSG level in kidney tissue caused by doxorubicin. Most recent results approve the suitable nephroprotective efficacy of fullerenol in the acute phase of toxicity and support further studies of it as a possible nephroprotector [67]. An *in vivo* study was examined to confirm the potential protective role of fullerenol $C_{60}(OH)_{24}$ (25, 50, and 100 mg/kg/week) on doxorubicin-induced (1.5 mg/kg/week) liver and heart toxicity within three weeks of treatment, using female Sprague Dawley rats with chemically-induced colorectal carcinoma [71]. The outcome for the treatment with doxorubicine alone had significant variations in the relevant serum levels (ALT, AST, LDH, -HBDH) within three weeks of treatment, as well as in the levels of MDA, GSH, GSH-Px, TAS, GR, CAT, and SOD in both liver and heart tissue at the end of the treatment period. These toxicity effects were significantly reduced for all measured parameters by pre-treatment with fullerenol. Chronic cardio- and hepato-toxicity was confirmed and fullerenol antioxidative, or better said, the organoprotective effect, was compared with the well-known antioxidant vitamin C in the same dose (100 mg/kg/week). According to macroscopic, microscopic, hematological, biochemical, physiological, pharmacological, and pharmacokinetic results, it was confirmed that fullerenol exhibits a protective role in the heart and liver tissue against chronic toxicity induced by doxorubicine at all examined doses. In contrast, vitamin C in the investigated dose was not a sufficient organo-protector [71].

Tissue injury to the heart muscle after doxorubicin administration was also confirmed by changes in the ultra-structural pathology results and different levels of oxidative stress parameters. In addition, the same parameters have been used for confirmation of potential cardioprotective influence of fullerenol as a pre-treatment agent for doxorubicin therapy in the acute phase. Fullerenol itself did not have an effect on heart injury in rats with breast cancer in a dose of 100 mg/kg, but very comparable results to the control group. The presented results suggested that fullerenol might be a potential cardioprotector in doxorubicin-treated *in vivo* systems, but only after progress in the preferred physical properties of fullerenol and its improvement. In particular, improvement is needed regarding solubility of the solid-state form in water or some other cell acceptable solvent [70].

In one recently published paper the intent was to investigate the nephroprotective effects of fullerenol $C_{60}(OH)_{24}$, on doxorubicin-induced nephrotoxicity. The study was conducted on adult female Sprague Dawley outbred rats with chemically-induced breast cancer (1-methyl-1-nitrosourea; 50 mg/kg; ip). Animals were divided into control healthy, control cancer, doxorubicin alone (8 mg/kg; ip; cancer), doxorubicin plus fullerenol as pre-treatment agent (8 mg/kg and 100 mg/kg, respectively; ip; cancer), and fullerenol alone (100 mg/kg; ip; cancer) groups with eight subjects per investigated group. Two days after the treatment, bio-samples (blood and kidney tissues) were taken for analysis. The activity of LHD and -HBDH as serum enzymes, as well as level of MDA, GSH/GSSG, GSH-Px, GR, CAT and SOD, were measured. It was once again confirmed that doxorubicin caused nephrotoxicity on the acute level, but fullerenol pre-treatment prevented oxidative stress and lipid peroxidation in kidney tissue caused by doxorubicin [67]. The same model was used for the first time in pulmoprotective investigation of fullerenol $C_{60}(OH)_{24}$ on doxorubicin-induced lung toxicity using a biochemical and histopahological approach. The levels of MDA and oxidized GSH in the lung tissue were higher in the group treated with doxorubicin alone than in both control groups. On the other hand, activities of CAT, GR, SOD, and LDH were found to be increased in the lung tissue of the rats in the anticancer agent group over all the other groups. The GSH-Px significantly decreased in the activity compared with the control and fullerenol groups. There was no significant difference in any measured parameters in either control or fullerenol groups. The acute macroscopic and microscopic change found in the doxorubicin group was subpleural edema typical in anticancer therapy. On the other hand, these histopathological changes have been normal in the groups treated with fullerenol. In conclusion, this study clearly indicated that doxorubicin treatment, which obviously harmed pulmonal function and pre-treatment with fullerenol, might avoid toxicity in rats through inhibition of oxidative stress [74]. The overview of doxorubicin toxicity and potential protection is summarized in Figure 6.4.

Recently, Icevic *et al.* [106] investigated the protective effect of fullerenol after oral administration in healthy rats in an animal

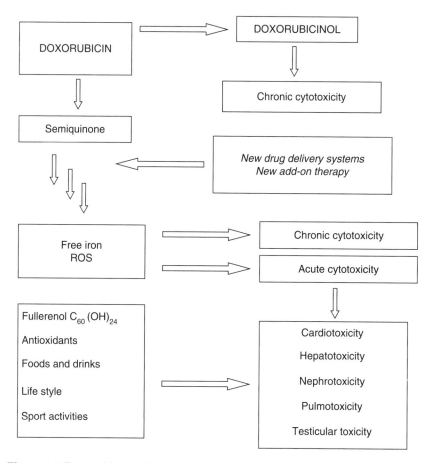

Figure 6.4 Doxorubicin toxicity and potential protection scheme.

model. The goal of that work was to investigate the potential protective effects of orally applied fullerenol *in vivo* after a single dose of doxorubicin (8 mg/kg (i.p.)) 6 h after the last application of fullerenol. The biochemical and pathological showed that fullerenol in solution form H2O:DMSO (80:20, w/w) given orally in final doses of 10.0, 14.4, and 21.2 mg/kg three days consecutively, has the protective (hepato and nephro) outcome against doxorubicin-induced toxicity through its antioxidant properties.

Polyhidroxilated, water soluble fullerenol $C_{60}(OH)_{22}$, used in the studies of Maksim *et al.* [107] was synthesized by a procedure of complete substitution of bromine atoms from polybromine derivatives $C_{60}Br_{24}$. With the aim of investigation the pharmacological behavior and metabolism of the fullerenol, they presented possibilities for its labeling with ^{99m}Tc ($T_{1/2}$= 6.01 h, E = 141 keV). Two different labeling approaches with technetium-99m were executed: directly by tin (II) chloride method and with ^{99m}Tc(I) using $[^{99m}Tc(CO)_3(H_2O)_3]^+$ precursor. The results have shown that the amount of free ^{99}mTcO$_4$ in the samples was dependent on the fullerenol-stannous chloride ratio, and it increased if the ratio was higher. The biological performance of labeled fullerenol was different depending on the labeling method and the time when the animals were sacrificed. From the latest investigations it can be concluded that the oxidation state in technetium coordination complexes has a great impact on *in vitro* and *in vivo* behavior of these complexes [107].

Fullerenol $C_{60}(OH)_{24}$ is a promising candidate for many biomedical applications due to its strong free-radical scavenging and antioxidative potential. Using the carrageenan-induced rat footpad edema test, the anti-inflammatory effect of fullerenol has been estimated in comparison with those of amifostine and indomethacin. Fullerenol (12.5–75 mg/kg, i.p.) and indomethacin (3–10 mg/kg, i.p.) were dissolved in DMSO, and amifostine (50–300 mg/kg, i.p.) was prepared in saline. The control groups were given identical vehicles (DMSO + saline). The drugs or vehicles were given half an hour prior to carrageenan injection. Footpad swelling was measured three hours after, followed by calculation of a percent of inhibition derived through comparison with the control groups. Histopathological assessment of the inflamed foot skin biopsies was also performed. Fullerenol dose-dependently and significantly reduced the extent of footpad edema, comparable to that of both indometacin and amifostine. The largest therapeutic index of fullerenol suggests its safety for potential future use in humans. Presented results support the hypothesis that fullernol produces a strong acute, anti-inflammatory activity [77].

There are many ongoing studies with fullerenols as potential strong antioxidative agents in different medical, therapeutical and pathological cases. Nevertheless, molecule(s) still have to pass many present challenges until their future use in *in vivo* human studies.

One recently published paper [108] demonstrated the positive tumor-inhibitory effect and immunomodulatory activity of $C_{60}(OH)_x$, which significantly inhibits the growth of murine H22 hepatocarcinoma through a 17-day treatment after tumor inoculation, and reduces the damnification of liver function in tumor-bearing mice.

The $C_{60}(OH)_x$ therefore exhibits effective tumor therapeutic activity by improving the innate immunity. It was shown that macrophages are efficiently activated by $C_{60}(OH)_x$. They contributed to the building and consolidation of the immunological defence systems against malignancies during the different stages of cancer tissue growth. A very similar result was also observed in one of our investigation with $C_{60}(OH)_{24}$. This kind of finding is of great importance, because it opens new possibilities for the advanced therapeutic usage of the cytostatic properties of fullerenol [71].

The potential anti-tumor and antimetastatic activities of fullerenol and their linked mechanisms were the purpose of a Jiao et al. [109] study with $C_{60}(OH)_{20}$. Thirty EMT-6 tumor-bearing mice were injected intraperitoneally with 0.1 ml saline or 0.1 ml saline containing fullerenol $C_{60}(OH)_{20}$ in different concentrations (0.08 and 0.4 mg/ml) on a daily basis for 16 days. Imbalances in the oxidative defense system have been investigated as well as the expression of several angiogenesis factors in the tumor tissue samples. Fullerenol exhibits anti-tumor and antimetastatic activities in EMT-6 breast cancer metastasis animal model, and it was confirmed to change oxidative stress significantly. Conversely, the expression of some angiogenesis factors was reduced in tumor tissues after 16 days of treatment with fullerenol $C_{60}(OH)_{20}$. Exceptionally, CD31 expression and vessel density were evidently reduced in tumors from fullerenol-treated mice compared with tissue from control animals. All these findings could be part of an important mechanism by which fullerenol aggregates inhibit tumor growth and suppress carcinoma metastasis in vivo [109].

Even though many research and development groups around the globe have been fascinated by different fullerenols for their radical scavenging and antioxidant capacity in vivo and in vitro and their promising use as a drug, drug carrier or diagnostic/analytical tool, there is little information on their organ specific toxicological properties. One unexplored areas is, for example, their pulmonary toxicological properties. The aim of a study performed by Xu et al. [110] was to examine the effect of fullerenols on the Sprague Dawley rat model after intratracheal instillation. Fullerenols [$C_{60}(OH)_x$, x = 22, 24; combination of two forms] were administered intratracheally (1, 5 or 10 mg per animal). After three-day exposures, the lungs of the rats were assessed. Bronchoalveolar lavage fluid biomarkers and a pathological evaluation of tissue have been investigated. Introductions to 1 mg dose did not induce adverse pulmonary toxicity, while the two other doses induced a cell injury effect. This was confirmed by oxidative/nitrosative stress and inflammation parameters. Consequently, results showed and confirmed once again that fullerenols produced reactions in a dose-dependent way. The dosage of $C_{60}(OH)_x$ retained in the lung and the ensuing aggregation might be the main factor in the process, which is very similar to the aggregation in the fatty liver tissue after i.p. administration and connected to very low solubility of polyhydroxylated form [70, 71, 73]. Existing results might be in the inconsistency of the possible use of fullerene and its derivates as inhaled drugs or carriers [110].

Polyhydroxylated fullerenes as part of novel applications in cancer drug delivery systems were part of a recently presented investigation [11]. Doxorubicin was conjugated to fullerenols through a carbamate linker, achieving ultrahigh loading efficiency. It was found that fullerenol-doxorubicin conjugate strongly suppressed the proliferation of cancer cell lines *in vitro* through a G2-M cell cycle block in comparison to each of them separately. The new system resulted in apoptosis. In addition, in an *in vivo* murine tumor model, conjugate exhibited comparable antitumor efficacy as free drug without the systemic toxicity of free doxorubicin. Furthermore, it was also shown that the fullerenol anticancer use can be extended to other chemotherapeutic agents, such as cisplatin. Anyhow, as a potential anticancer agent, fullerenol has to be tested in patients with different kinds of tumors, and before that, some investigations on bigger animal models should be performed.

Nanotechnology and nanoscience, including nanomaterials development with biomedical application, is a very progressive area nowadays. The cytotoxic effects of certain engineered nanomaterials towards malignant cells form the basis for one aspect of nanomedicine as well as its potential use as delivery system for highly toxic drugs, with the goal of reducing side effects [112]. Injection of specific gadolinium-fullerenol form of $[Gd@C_{82}(OH)_{22}]_n$ nanoparticles could express reduction of enzyme activities connected with the metabolism of reactive oxygen species (ROS) in tumor-bearing mice [113]. Its application can powerfully repair the functions of injured tissues such as liver and kidney. It has been presented that investigated metallo-form nanoparticles containing fullerenol were delivered to almost all types of tissues in mice after i.p. administration, which is in very good correlation with fullerenol $C_{60}(OH)_{24}$ distribution in rats [73]. The strong and stable fullerene cage cannot be changed during metabolism in organisms, which means that the presence of Gd *in vivo* represents the exact distribution of used nanoparticles complex. In one of the published papers, it was found and confirmed that the accumulation of Gd in liver was higher than that in tumor when they examined the accumulation of Gd-incorporating nanoemulsion in mice tumor [114]. Fullerenols could be entrapped in the reticuloendothelial system and distributed to all organs by bloodstream *in vivo* [115]. In contrast, the intravenous administration of $Gd@C_{82}(OH)_{40}$ to mice essentially caused delivery to some organs such as lung, liver, spleen and kidney [115]. The thrombin and fibrinogen related parameters in plasma were measured after $[Gd@C_{82}(OH)_{22}]_n$ administration with clear increase of fibrinogen, and shorten protrombin time. Surprisingly, in the $[Gd@C_{82}(OH)_{22}]_n$-treated nude mice, splenomegaly and hepatomegaly were less, which confirmed inhibition of the liver injury. These kinds of results are very well in alignment with the conclusion about the serum enzyme levels reported before. After i.p. injection of $[Gd@C_{82}(OH)_{22}]_n$, the serum AST and ALT activities were significantly decreased. Presented results are not in good correlation with the ALT and AST level in rats treated with fullerenol $C_{60}(OH)_{24}$ by i.p. way of administration in rats with different carcinomas [71, 73]. Indicators of kidney injury were repaired to the typical level by nanoparticle-treatments well-known on renal function tests. The activities of oxidative stress parameters like hepatic SOD, GSH-Px, GST and CAT were measured in mice administered with $[Gd@C_{82}(OH)_{22}]_n$ nanoparticles, to verify the molecular level of interaction on the fullerenol-cell level. The GSH-Px, CAT, GSH, protein bound thiols, MDA, and SOD activities and concentrations were significantly lower after treatment with $[Gd@C_{82}(OH)_{22}]_n$ nanoparticles. With all these tests, it was confirmed that polyhydroxylated fullerene derivatives might be new antioxidants, with very high potential in different therapeutic areas [116].

The acute toxicity of fullerenol was published using mice pretreated i.p. with polyhydroxylated C_{60} derivatives. The LD_{50} value of fullrenol was projected to be around 1.2 g/kg. Pretreatments with 0.5 and 1.0 g/kg fullerenol significantly reduced cytochromes P450 and b(5) levels, as well as NADPH-cytochrome P450 reductase, benzo[a]pyrene hydroxylase, 7-ethoxycoumarin O-deethylase, aniline hydroxylase, and erythromycin N-demethylase activities in liver tissue. Stressing cells with 0.01 and 0.1 g/kg fullerenol had no influence on these monooxygenases levels. On the other hand, added extras of fullerenol to mouse liver microsomes inhibited monooxygenases activities toward benzo[a]pyrene, 7-ethoxycoumarin, aniline, and erythromycin with IC_{50} values of 42, 94, 102 and 349 µM, respectively. Fullerenol presented noncompetitive and mixed-type of inhibition in investigated molecules. Additions of fullerenol to rat liver mitochondria caused an inhibition of ADP-induced uncoupling and markedly repressed mitochondrial Mg^{2+}-ATPase activity with an IC_{50} value of 7.1 µM, which was confirmed to be dose-dependent. These results demonstrate that fullerenol can reduce the activities of P450-dependent monooxygenase and mitochondrial oxidative phosphorylation *in vitro*, as well as suppress the levels of the microsomal enzymes *in vivo* [89].

The effect of fullerenol $C_{60}(OH)_{24}$ on lipid peroxidation in the kidneys, testes and lungs of rats treated with doxorubicin has been recently published by Vapa *et al.* [117]. The experiment was conducted with healthy male Wistar rats divided into five investigation groups. The control was saline treated while the other four groups received doxorubicin alone, doxorubicin/fullerenol in two different concentrations and fullerenol alone in higher concentration. Tissue samples were taken after 2 and 14 days of the treatment. Once again, results confirmed previously published evidence about fullerenol suppression of doxorubicin toxicity in different organs when used as pretreatment agents. A dose of 100 mg/kg i.p. (higher dose) in combined therapy exhibited a better protective effect. On the other hand, fullerenol applied alone at the same dose did not show significant influence on lipid peroxidation in all tested organs [117].

Magic bullet or useful cell killer are still not synonyms for fullerenol, since there still is a long way to go before it can be confirmed for ethical, beneficial and pharmacological use because of the toxicological and safety issues discovered in numerous investigations. However, with all its advantages and disadvantages, it has big potential to be the molecule of future biomedical interest.

Dimethyl sulfoxide (DMSO) has been described as a universal cure, but there is no adequate modulator of DMSO activity that can minimize its side effects. The properties of FNP make this nanoparticle a good candidate to be a modulator of DMSO activity that could minimize its side effects. Isolated spontaneous and calcium-induced contractile active rat uteri (Wistar, *virgo intacta*), were treated with DMSO and FNP in DMSO. The activity of GR is modulated by FNP and is of interest in cryopreservation to maintain the GSH level in medium [118].

6.2 Biologically Active Graphene Materials

6.2.1 Chemical Synthesis and Characterization of Important Biologically Active Graphene Materials

Novoselov *et al.* [119] identified single layers of graphene and other two-dimensional crystals. Graphene and its derivatives are two-dimensional (2D) hexagonal single-layer structures with sp^2-hybridized carbon atoms. In-plane carbons exist between strong C-C bonds, while out of plane (the angle of 90°) a delocalized π network of electrons allows weak interactions between graphene sheets and other materials. The graphene material family (graphene oxide [GO], graphene sheets [GS], reduced graphene [rGO], exfoliated expandable graphite [EG]) has recently attracted attention mostly for its unique physical properties such as heat stability, electrical conductability, mechanical strength and chemical reactivity [120–123]. Because of their high chemical reactivity, it is to be expected that derivatized graphene materials will find a purpose in numerous novel, potentially applicable nanomaterials in photonic devices, electronics, clean energy, cell cultures, memory devices, electron acceptors, light absorbers, electrochemistry, fluorescence and sensors, matrices for mass spectrometry, and nano drug delivery [124]. Pristine graphene is hydrophobic, water insoluble and purely soluble in most organic solvents. This issue can be overcome by chemical covalent and non-covalent derivatization (Table 6.1). So far, numerous methods for the production of graphene and its derivatives have been developed: chemical vapor deposition, [125, 126], re-intercalation and ultrasonication of thermally exfoliated expandable graphite (EG) [127], epitaxial growth [128], non-covalent functionalization of reduced graphene oxide [129], chemical reduction of graphene oxide [130] and Scotch tape method [131]. There are a great number of 2D graphene-based sheet products; zero dimensional (0D) graphene quantum dots [132], 1D graphene nanoribbons [133] and graphene nanomeshes [134]. These materials possess different optical, electrical and biological propreties.

Graphene oxide (GO) and derivatized graphenes are precursors for chemical synthesis of stable colloid systems of thin platelets. By using aqueous solutions with ethanol, dimethyl sulfoxide, acetonitrile, dimethyformamide, tetrahydrofurane, highly reduced and derivatized GO sheets can be obtained [135]. A method for synthesis was developed by Hummers and Offeman [136] who started

Table 6.1 Covalently and non-covalently functionalized nanographene.

Type of functionalization	References
Covalent functionalization	
Polyethylene glycol	[154–156]
Polyacrylic acid	[157]
Azomethine ylides	[152]
Poly-L-lysine	[158]
Sulfonic acid	[159]
Dextran	[160]
Chitosan	[161, 162]
Non-covalent functionalization	
Polyethylenimine	[163]
PEGylated phospholipid	[164]
Pyrene terminated poly(N-isopropylacrylamide)	[165]
Protein	[166]
Pluronic F127	[167]
Poly (maleic anhydride-alt-1-octadecene)	[164, 168]
DNA	[169]
Tween	[170]
Sliver nanoparticles	[171]
Iron oxide nanoparticles	[168, 172]

by treating graphite with strong oxidizing agents (H_2SO_4, $KMnO_4$ and $NaNO_3$) in order to get graphite oxide, which was exfoliated to graphene oxide sheets with intensive long-term sonication until stable aqueous suspension was obtained. After modification of this method by including phosphoric acid, it was possible to reach a higher degree of oxidation and to decrease the nitrogen oxides produced. Separation of GO is a complex process that starts with dialysis, which helps in removing all the inorganic inpurities [137]. It is also possible to synthesize GO by using benzoyl peroxide at 110°C [138].

Reduced graphene oxide (rGO) is of particular interest for research. A number of different methods for GO reduction have been developed: synthesis with hydrazine [139] by using strong bases [140], electrochemical reduction [141], photochemical [142] and thermal reduction [143] by using sodium borohydride [144] and hydroquinone [145]. Solubility of rGO in water is lower than 0,5 mg/ml [146]. The most common reduction process for GO is by using hydrazine monohydrate. With reduction comes the elimination of oxygen functional groups (hydroxyl, carboxyl and etheric) and founding of π electron conjugated system, which results in an increase in hydrophobicity, which is why the reduced product precipitates. When reduction with strong bases is used, oxygen functional groups in rGO are ionized and product synthesized in that manner has better water solubility and greater stability without any sign of agglomeration within four months [135]. Chemical reduction of stable suspension of GO with hydrazine monohydrate at 80°C and 12-hour stirring results in a homogeneous black suspension of highly reduced graphene oxide with few aggregates that are easy to break by sonication. Highly reduced graphene oxide in N,N-dimethylformamide/H_2O (9:1) exists in the form of individual sheets, which were determined by atomic force microscopy and transmission electron microscopy analysis. A combination of organic solvents with suspension of GO in N,N-dimethylformamide/H_2O in ethanol, dimethylsulfoxide, N-methylpyrrolidone, acetonitrile in volume ratio 90:9:1 gives stable suspensions or black agglomerated powders with addition of 1,2-dichlorobenzene, diethyleter, toluen. After adding tetrahydrofurane and acetone, suspensions are being formed only after 24 hours. Brief sonication results in stable suspension forming. After stirring 1 ml of highly reduced graphene oxide solution in DMF/H_2O (9:1) agglomerates are formed, while addition of 9 ml of pyridine results in the formation of stable suspension, after which follows filtration. Solvothermal reduction of GO (DMF, 180°C, with hydrazine monohydrate) only produces particles with an average size of 300 nm of single sheets of reduced graphene. This type of synthesis conducted in H_2O/N-methylpyrrolidone at 200°C was proven to obtain a low yield of soluble graphene (0.001 mg/ml) [147]. The structure of GO can be best described as graphene sheets with undefined positions and number of oxygen functional groups (-OH, -COOH, -C=O, -O-). X-ray photoelectron spectroscopy, ^{13}C-nuclear magnetic resonance spectroscopy and scanning transmission electron microscopy have confirmed the existence of approx. 40% of sp^3 C–O bonds [148, 149]. A high percentage of sp^3-hybridized carbon atoms leads to structural distortion of GO sheets. These areas are characterized as isolated pentagonal and heptagonal pairs, which present quasi-amorphous single-layer carbon structure. The dimensions of topological distortions are within 1–2 nm and occupy roughly 5%. These defects cause both, in-plane and out-of-plane deformations in GO structure.

Synthesis of lightly sulphonated graphene (G-SO$_3$H) can be summarized in three main steps: 1) pre-reduction of graphene with sodium borohydride at 80°C for 1 h in order to remove the majority of functionalized oxygens, 2) sulfonation with aryl diazonium salt for 2 h, and 3) post-reduction with hydrazine (100°C for 24 h). Sulfonated graphene can be dissolved in water at pH within 3–10 at a concentration of 2 mg/ml [150].

Derivatized graphene sheets are able to dissolve in a mixture of water and organic solvents such as methanol, acetone, acetonitrile, which makes it possible to further modify the surface.

Sulphonated graphene remains stable for more than a month in the form of single carbon layer. Like graphene oxide, sulphonated graphene is also negatively charged (55–60 mV) at pH 6, which is an ideal charge for the stabilization of nanoparticles. Derivatized graphenes with anionic functional groups can undergo stabilization in aqueous solutions by using positively charged surfactants, some of which is didodecyldimethylammonium bromide [151]. Quintana *et al.* [152] succedeed in modifying graphene through 1,3-dipolar cycloaddition with azomethine ylides.

Amino modified graphene can be synthesized from GO with SOCl$_2$ and sodium azide, used to convert carboxyl groups on GO into amino groups [153].

6.2.2 Biologically Active Graphene Materials

The toxicology assessment of graphene family materials and their influence on the environment are of great importance. Recently, a great number of scientific studies conducted on *in vivo* and *in vitro* models have been published. Studies point to interaction of nanographene sheets and GO with primary biomolecules, single-stranded or double-stranded DNA, and hydrophobic core of lipid bilayers.

Besides that, they also show high cargo carrying capacity for conjugated small molecule drugs [173]. Mostly, studies are focused on GO in adsorption of enzyme, cell imaging, drug delivery, biosensors, molecular imaging, drug/gene delivery, and cancer therapy [174]. The biomedical application of graphene family materials is not complete without a detailed toxicological study on different biological models.

6.2.2.1   In Vitro Toxicity

Many scientific groups have been working on the investigation of toxic effects of graphene nanomaterials on *in vitro* and *in vivo* models [175, 176]. Schinwald *et al.* [177] have found that layers of approx. 15 µm were not completely uptaken by immortalized human monocytic (THP-1) cells, which led to inhibition of the phagocytosis process and frustrated phagocytosis occurrence. A significant decrease in lactat dehydrogenase results in the loss of membrane integrity and decreases the viability of macrophages in concentrations higher than 5 µg/ml, which could be a direct consequence of reactive oxygen species production. Graphene exhibits a cytotoxic effect on human neuronal cells and mitochondrial injury 4 and 24 h after treatment with a concentration of 10 µg/ml [178]. Graphene was determined to have IC_{50} cytotoxic activity on human cervical cancer cells (HeLa) in a concentration of approx. 100 µg /ml [157]. Being biocompatible and of low cytotoxicity, thin graphene and its derivative sheets are good candidates for mediums for cell growth. Ruiz *et al.* [179] have discovered that a proliferation of mammalian colorectal adenocarcinoma HT-29 cells without any morphological enlargement occurs on GO film (glass slides coated with 10 µg of graphene oxide). It was also established that mouse neuronal cells could grow on the surface coated with poly-L-lysine [180]. Good biocompatibility and quite an increase in number of cells, as well as an increase in the avarage length, wihin 2–7 days suggest that GO as surface efficiently promotes neuronal cell growth. Surfaces coated with rGO film showed good biocompatibility and had no cytotoxic effect on mouse pheochromocytoma cells, human oligodendroglia cells, or human fetal osteoblasts [181]. Wang *et al.* [174] have proven that GO can have dose- and time-dependent cytotoxicy on human lung fibroblasts cytoplasm and nucleus by disabling cell adhesion and by inducing cell floating and apoptosis at doses above 20 µg/ml. In a concentration lower than 20 µg/ml GO is cytotoxic to human fibroblast cells, while in doses higher than 50 µg/ml it influences cell adhesion and induces apoptosis in a 1–5 day period. Chang *et al.* [175] investigated the influence of GO on morphology, viability, mortality, and membrane integrity of human lung epithelial cells. The GO did not enter the cells, neither did it express any cytotoxicity. Concentrations of GO higher than 50 µg/ml cause oxidative stress with increased loss of cell viability. Hu *et al.* [166] confirmed concentration-dependent cytotoxicity of GO on alveolar basal epithelial cells (A549). The influence of graphene to efflux hemoglobin from suspended red blood cells was also investigated, where smaller particles proved to be more harmful and where chitosan film almost completely eliminated hemolytic activity. Results have shown that particle size, oxygen functional groups and graphene charge have a great impact on biological/toxicological responses to red blood cells. Experiments with mammalian fibroblasts proved that cytotoxicity depends on exposure environment (i.e., whether or not aggregation occurs) and mode of interaction with cells (i.e., suspension versus adherent cell types). Graphene sheets were more harmful than GO to mammalian fibroblasts and human skin fibroblast cells in terms of ROS production [161]. Wojtoniszak *et al.* [182] have tested L929 mouse fibroblasts for the biological effects of functionalized GO and rGO with polyethylene glycol, polyethylene glycol–polypropylene, glycol–polyethylene glycol (Pluronic P123), and sodium deoxycholate as dispersants. It was established that toxic effects depend on type of dispersant and concentration of nanomaterial in suspension. The most promising was GO functionalized with polyethylene glycol. GO, in comparison to rGO, has better biocompatibility, especially at higher concentrations (50 to 100 mg/ml). Ultra-small reduced nano-GO with high NIR absorbance could find its purpose in target photothermal therapy on *in vitro* cancer cell ablation [164].

6.2.2.2   Microbiological Toxicity

Some of the first published results concerning the antibacterial activity of GO and rGO investigated on *Escherichia coli* and *Staphylococcus aureus* were announced in 2010 by Akhavan and Ghaderi [183]. They concluded that rGO reduced with hydrazine was more toxic than non-reduced GO because between rGO and bacteria a better charge transfer during contact was established. Based on efflux of cytoplasm monitoring, it was possible to detect the mechanism of rGO antibacterial activity, which was conditioned with physical interaction between extremely sharp edges of nanowalls with bacteria. Gram negative *E. coli* has a more resistant outer membrane than *S. aureus;* therefore it is more resistant to antibacterial activity of rGO. This was similarly confirmed by Hu *et al.* [184] who showed that cell metabolic activity is GO dose-dependent and decreased to approx. 70%–13% at GO concentration of 20–85 µg/ml, respectively. The rGO also exhibited high antibacterial effects; the metabolic activity of *E. coli* was reduced to approx. 24% at rGO concentration of 85 µg/ml. The loss of membrane integrity of *E. coli* was confirmed to be responsible for this outcome. S. Liu *et al.* [185] have proposed the mechanism of antibacterial activity of graphene materials (GMs) against *E. coli*, which starts with initial cell deposition of GMs and membrane stress caused by the attack of sharp nanosheets. Akhavan and Ghaderi [186] investigated the interaction between chemically exfoliated graphene oxide nanosheets and *E. coli* that live in acid medium and anaerobic conditions. By using XPS, it was concluded that *E. coli* reduces GO. The GO sheets turned out to be a potent biocompatible surface for bacterial growth, unlike rGO, which possesses inhibitory effects to bacteria proliferation. Graphene was proven to accumulate in monkey renal cell membranes causing severe damage caused by oxidative stress that results in apoptosis, while carboxylate derivative expressed no toxicity in the same model [187].

In vitro **models.** Polyethyleneimine (1.2 kDa and 10 kDa) functionalized GO formed stable complexes in physiological conditions and proved not to be toxic against HeLa cell lines at a concentration of 300 mg/l. Besides, it appeared to be an effective gene transfection nanocarrier [188]. Induced by photothermal activation in near-infrared spectrum, graphene nanoparticles have shown significant anticancer activity against human glioma (U251) cells [189]. Thew cytotoxic mechanism is based on oxidative stress and depolarization of mitochondria that causes apoptosis and cell necrosis characterized by caspase activation/DNA fragmentation and cell membrane damage, respectively. Coating GO with branched polyethylene glycol resulted in stable conjugate in many biological surroundings. Thusly prepared GO was used for the attachment of aromatic compound, camptothecin analog, and SN38 via π-π interactions. Polyethylene glycol GO-SN38 complex expressed excellent water solubility and high cytotoxic activity against human colon cancer cells. The single-layered rGO sheets functionalized with polyethylene glycol polymer chains have shown mild cyto-toxicity against human epithelial breast cancer cells at concentra-tions of 80 mg/ml and higher [164]. Recent studies have shown that GMs could serve as carriers of commercial water-insoluble antineoplastics [190]. When GO dextrane coating was applied to cervical cancer HeLa cells remarkably reduced cellular toxicity. After intravenous application in mice, it was found that obvious clearance without any clear signs of toxicity within seven days was achieved. PEGylated single-layer nanographene oxide sheets have shown excellent water solubility and stability in buffers and serum without agglomeration.

In vivo **models.** Zanni *et al.* [191] have conducted research on toxicity of graphite nanoplatelets on *Caenorhabditis elegans* where no signs of acute toxicity were observed. The influence on life expectancy, as well as reproductive abilities were investigated [191]. Gollavelli and Ling [157] examined the toxicity of gra-phene coated with polylactic acid and fluorescein o-methacrylate on zebrafish embryo and proved biocompatibility without nega-tive effects. Confocal and electron microscopy have revealed that GMs were localized in the embryo cytoplasm region with head to tail distibution. Direct administration of pure graphene (Pluronic-block copolymer-dispersed) and GO in mouse lung resulted in severe and persistent lung injury, which was proved to originate from the increase in mitochondria respiration and ROS produc-tion which consequently activated inflammatory and apoptotic processes [167]. Several studies dealing with toxicity and biodis-tibution after intravenous and intratracheal application of GO in mice have been published [174]. Experiments were conducted with three doses of GO 0.1, 0.25, and 0.4 mg/kg, where the first two con-centrations had no toxicity, while the highest concentration caused chronic toxicity induced by granuloma formation in kidney, liver, spleen and lung without the possibility to be renally cleared. Zhan *et al.* [192] established that concentration of GO 1 mg/kg within a 14-day period causes no pathological changes in organs; however the increase in concentration (10 mg/kg) leads to inflammation, cell infiltrations, and lung edema and liver granuloma formation. Polyethylene glycol coating of nanographene oxide sheets presents an important process for achieving better biocompatibility of GMs [193]. In an *in vivo* farmacological study conducted by K. Yang *et al.* [194, 195], pharmacokinetics, biodistribution studies and system-atic toxicity were investigated by using ^{125}I-labeled nanographene sheets functionalized with polyethylene glycol. Polyethylene glycol nanographene sheets were mostly accumulated in a reticuloendothelial system including liver and spleen after intravenous administration. Three months after the treatment of female mice with polyethylene glycol nanographene sheets at 20 mg/kg neither biochemical nor histological analysis showed significant toxicity of this material. Also measured was the major gradual clearance of graphene from the mice, which probably happens via both renal and fecal clearance. The polyethylene glycol-coated GO sheets intravenously applied have proven to accumulate in the liver and spleen, after which a gradual clearance takes three to five days [194]. By modification of graphene with polyethylene glycol and doxorubicin, after photothermal activation, the complete loss of viability of murine mammary tumor (line EMT6) cells and complete destruction of solid tumor (EMT6 tumor-bearing mice were used) were successfully achieved without any loss in mice body weight [159]. Investigation of polyethylene glycol nanographene sheets in tumor-bearing mice by *in vivo* fluorescence imaging showed that intravenous application of graphene at 20 mg/kg results in highly efficient tumor passive targeting of graphene sheets in several different tumor models [196]. Nanocomposite obtained from transferrin-functionalized gold nanoclusters/graphene oxide was synthesized in order to be used in the bioimaging of cancer cells and small animals [197]. It proved to be water soluble, biocompatible and of low cytotoxicity when under near-infrared fluorescent conditions. K. Yang *et al.* [198] have compared covalently conjugated PEGylated GO and covalently conjugated PEGylated highly-reduced GO to more condensed non-covalent PEGylated conjugates and deduced that covalent derivatives have longer blood circulation and remarkably increased tumor passive uptake. Besides, it was established that smaller particles, unlike larger ones of approx. 65 nm, after intravenous application in mice showed decreased accumulation in the liver and spleen. Graphene nanoplatelets were also proven to cause severe lung injuries after being inhaled [177].

6.2.2.3   Nanographene Imaging

By functionalization of GO with fluorescent colors and magnetic nanoparticles it is possible to create nanocomposites potentially applicable in biological imaging [199, 198]. Positron emission tomography, computed tomography imaging of radiolabeled (^{64}Cu)

PEGylated GO conjugate with anti-CD105 antibody and 1,4,7-tri-azacyclononane-1,4,7-triacetic acid exhibited excellent stability and target specificity, as well as significantly improved tumor uptake of thus functionalized GO in mice bearing 4T1 tumors [200]. They have also published positron emission tomography/computed tomography imaging, with PEGylated GO labeled with [66]Ga [201]. By covalent bonding of GO and iron oxide nanoparticle a product was obtained that can be used for multimodal fluorescence and magnetic resonance imaging of cells [202]. Cellular imaging research has revealed that multifunctionalized magnetic graphene is being localized in cytoplasm [157].

6.2.2.4   Delivery of Anticancer Drugs to the Cancer Cells By Nanographene

GO was widely investigated as a nanocarrier with numerous chemo-therapy drugs: doxorubicin [203, 204, 154, 205], camptothecin [206, 207], ellagic acid [208], 1,3-bis(2-chloroethyl)-1-nitrosourea [209], and b-lapachone [210]. Sun *et al.* [193] have synthesized a conjugate of PEGylated GO with an anti-CD20 antibody, rutixan and doxorubicin. Thus synthesized conjugate possesses selective toxic activity against B cell lymphoma [193]. Wen *et al.* [154] showed that a poly-ethylene glycol shell of GO can be applied as a smart drug delivery system in doxorubicin delivery which would drastically enhance the efficacy of this drug in *in vivo* research. Experiments including thermo-responsive poly(N-isopropylacrylamide) and GO proved physiological stability of nano complex without any significant sign of toxicity [206]. After camptothecin attachment, high cyto-toxicity against malignant cell lines was achieved. Nanocomposite GO–iron oxide nanoparticle is also able to serve as doxorubicin carrier with possibility of pH-dependent controlled release [211]. This nanocomposite possesses a dual targeting delivery system due to the magnetic properties of material and conjugated folic acid [212]. The GO–iron oxide nanocomposite synthesized by hydrothermal method and functionalized with polyethylene glycol expresses physiological stability and biocompatibility with the possibility of magnetically targeted drug delivery, as well as photothermal treatment of cancer in *in vitro* models [172]. TiO_2-modified GO has proved itself as an effective system in the photodynamic killing of cancer cells [213]. By functionalization of GO with sulfonic acid groups and covalent binding to folic acid, it is possible to obtain a novel nanocarrier for loading and targeted delivery of anticancer drugs such as doxorubicin and camptothecin, where interactions with anticancer drugs are achieved via π-π stacking [159]. Folic acid allows specific targeting of human breast cancer cells, exhibiting folic acid receptors. S.A. Zhang *et al.* [160] have revealed that covalent conjugate of GO and dextrane had great biocompatibility as well as stability in physiological conditions. Based on photosensitivity, PEGylated GO can be useful for cell imaging in near infrared spectrum. Furtheremore, it was established that by simple physisorption via π-stacking doxorubicin can be easily binded [193].

6.2.2.5   Gene Delivery by Nanographene

Several different studies have proved that surface functionalized GO can find its purpose in gene delivery in *in vitro* and *in vivo* conditions. By derivatization of carboxylic groups on GO with polyethylenimine, chitosan or 1-pyrenemethylamine, amidic bonds were established [214, 162] that allow electrostatic [188] or non-polar π-π stacking interactions [215] and show efficacy with plasmid DNA or small interfering RNA delivery. Functionalized GO could have effective delivery of molecular beacons [216] and aptamers [217] to cells for *in situ* specific molecule detection. L. Zhang *et al.* [163] prepared amide derivative of GO with polyethyleniminom in order to load small interfering RNA and doxorubicin. X. Yang *et al.* [215] revealed similar results with polyethylene glycol and 1-pyrenemethylamine-functionalized GO. After further conjugation with folic acid, the complexes were able to selectively deliver small interfering RNA to cells and could effectively knockdown the targeted gene. Graphene materials can be successfully used as graphene-based DNA sensors [218]. Conjugation of low-molecular weight branched polyethylenimine with GO can serve as a nonviral gene delivery vector, as well as a small interfering RNA delivery system and photothermal therapy agent [214]. Polyethyleneimine conjugated GO was proven to be non-toxic and effective as nanocarrier for gene transfection [188].

6.2.2.6   Mechanisms of Action

Unlike spherical fullerene and carbon nanotube mechanisms, mechanisms of graphene biological activity are yet to be completely described [219]. So far, it has been thought that the biological effects of graphene materials are the consequences of reactive oxygen species (ROS). It is well-known that the level of ROS is controlled by cellular antioxidative enzymes. If ROS are being overproduced, destruction of macromolecules of great importance can occur starting with unsaturated fatty acids, proteins and DNA. For their hydrophobicity, graphene materials achieve interactions with hydrophobic lipid layers of membranes. Some studies emphasized that after exposition to graphene materials, ROS production is time- and dose-dependent [178]. Wang *et al.* [174] have found that GO can reach the nucleus of human lung epithelial cells or fibroblasts where it induces cell floating and apoptosis. On the other hand, Chang *et al.* [175] published that GO had no toxic effects on alveolar basal epithelial cells. According to the research of Hu *et al.* [184], it was concluded that the cytotoxic activity of GO nanosheets is based on direct physical destruction of cell membranes. It was also established that *E. coli* bacterial membranes damage caused by GO and rGO was exclusively the result of physical contact [184]. Yuan *et al.* [220] have investigated the expression of 37 proteins involved in metabolism and cell growth. The authors have identified key enzymes for the redox processes regulation of the cell and deduced that graphene did not influence upregulation of the thioredoxin-peroxiredoxin system in order to overcome ROS stress nor did it induce apoptosis. A paper published by Li *et al.* [176] confirmed that graphene induces cytotoxic effects via depletion of mitochondrial membranes and potentially increased oxidative stress. This study also proved that graphene could trigger apoptosis via the mitochondrial activation pathway. Consequently, the caspase 3 and its downstream effector proteins were activated and the execution of apoptosis was initiated. Hu *et al.* [166] have investigated the influence of fetal bovine serum (FBS) on GO nanosheet toxicity. In 10% FBS solution, a significant decrease in toxicity on human cells was noticed.

6.3   Bioimpact of Carbon Nanotubes

6.3.1   Introduction

The first appearance of carbon nanotubes (CNTs) in papers is connected with scientists Lukyanovich and Radushkevich [221], but usually a Japanese electron microscopist Sumio Iijima is credited for the discovery of CNTs. Since his discoveries in 1991 and 1993

[222, 223], CNTs have become an even more interesting topic for various investigations, some of which are in the chemical, biomedical and ecotoxicological field of research.

6.3.2 Properties of CNTs

Carbon nanotubes (CNTs) are carbon-made cylindrical tube-like structures, with sp^2-hybridized carbon atoms that form hexagonal structural units. Their prominent characteristic is large length-to-diameter ratio [224]. The degree of chirality in CNTs is a measure of their electrical and conductivity properties, the diameter of nanotubes, and metallic or semimetallic characteristics [225, 226]. Being hydrophobic in nature, CNTs are subjected to van der Waals interactions, which results in them being found in the form of self-assembled aggregates, which makes them quite insoluble in water. Nithiyasri et al. [227] have investigated the possibility of self-assembling CNTs in aqueous dispersions using a water-soluble protein; bovine serum albumin (BSA). They concluded that CNTs self-assemble in the intermediate temperature range of 45–65°C in both native BSA at 35°C, as well as denatured BSA solution at 80°C [227].

6.3.3 Classification of CNTs

Depending on the number of walls, CNTs can be classified as single-walled carbon nanotubes (SWNTs) or multi-walled carbon nanotubes (MWNTs). Single-walled carbon nanotubes are made of only one sheet of graphene in the form of a cylindrical structure with a radius up to 1 nm, and provide outstanding mechanical properties, for example, high tensile strength [228, 229]. Multi-walled carbon nanotubes consist of 2–10 layers of graphene sheets. Nanostructures with an outer diameter less than 15 nm are classified as MWNTs, and nanostructures above that size as nanofibers; unlike nanofibers, nanotubes are hollow structures. Depending on their shapes, CNTs can be divided into three different structures: carbon nanohorns (CNHs), nanobuds and nanotorus [230].

6.3.4 Synthesis of CNTs

Nowadays, CNTs are synthesized by artificially developed methods by using one of three main methods: arc-discharge, laser ablation, and chemical vapor deposition (CVD) [223, 231–233].

6.3.5   Functionalization of CNTs

Low water solubility can be overcome by different functionalization of CNTs [234], which can also expand their potential field of application. Georgakilas *et al.* [235] and Pantarotto *et al.* [236] have chemically functionalized CNTs with biomolecules such as peptides; Peng *et al.* [237] did functionalization of CNTs with carboxylic acid, while Zhang *et al.* [238] functionalized CNTs with poly-L-lysine. W. Yang *et al.* [168] have shown the possibility of non-chemical functionalization of CNTs using hydrophobin coatings, which made hydrophobic substrates wettable and soluble. In this process, no organic solvents or surfactants were used.

6.3.6   Drug (Molecule/Gene/Antibody) Delivery, Targeting, Drug Release

Carbon nanotubes (CNTs) have attracted much attention for their possibility of controlled drug delivery of molecules [239, 240]. Carbon nanotubes as hydrophobic structures are able to deliver hydrophobic biomolecules [241], while particular covalent functionalization of the sidewalls makes them hydrophilic [242, 243]. Drugs or biomolecules can be loaded inside the tube, or can be directly attached to the walls of CNTs or, for example, particular drug molecule can be loaded inside the CNTs, while the outer surface can be modified, thus achieving biocompatibility and biodegradation. Since when applied along with CNTs, chemotherapeutics have been shown to have better uptake by malignant cells, nanotubes can potentially lower the dose of drug. Besides possessing target selectivity, cancer treatments should also enable controlled release of the therapeutics. The most prominent drug-delivery system (out of chemotherapeutic) of CNMs is probably CNTs-doxorubicin complex based on π-π interactions [244–249]. Arlt *et al.* [250] have reported that carboplatin, delivered in the form of CNT–CP, exhibited higher anticancer activity than free carboplatin.

R. Li *et al.* [251] have shown that SWNTs functionalized with p-glycoprotein antibodies and loaded with the anticancer agent doxorubicin, in comparison to free doxorubicin, demonstrated higher cytotoxicity by 2.4-fold against K562R leukemia cells. Luksirikul *et al.* [252] have used C_{60} fullerenes to close uranyl acetate-filled SWNTs and demonstrated pH-triggered release of the encapsulated material. Bhirde *et al.* [253] functionalized SWNTs with the anticancer drug cisplatin, with epidermal growth factor as targeting agent and quantum dots as imaging agents for the treatment of head and neck carcinoma tumors, and successfully achieved a decrease in tumor size. Chaudhuri *et al.* [254] conjugated SWNTs to doxorubicin via an enzymatically cleavable carbamate bond that allowed sustained release of the active drug.

In order to increase aqueous solubility and biocompatibility, spacers between CNTs and a drug molecule are usually used, some of which are polyethylene glycol, PEG [255], and polyvinyl alcohol [207]. Z. Liu *et al.* [256] conducted an *in vivo* study using PEGylated SWNTs conjugated to the anticancer drug paclitaxel. The PEGylated nanotubes were able to prolong circulation and enhance cellular uptake of the drug by the cancer cells. Z. Liu *et al.* applied PEGylated SWNT loaded with doxorubicin *in vivo* [257] and came to the conclusion that a CNT-based drug delivery system expressed reduction in toxicity compared with free doxorubicin, or a commercially available product. Beside doxorubicin, camphotecin was also examined [258].

In vivo studies revealed that the CNT-chitosan complexes had better release and targeting properties than pure CNTs [259]. F. Yang *et al.* suggested that by controlling the size of CNTs the effective uptake into the lymphatics could be allowed. They functionalized CNTs with folic acid and loaded magnetic nanoparticles along with drugs, where such a complex proved to have better delivery to cancer cells in the lymph nodes; the MWNTs could be retained in the targeted lymph nodes for several days continuously releasing chemotherapeutic drugs by using an externally placed magnet [260].

Because of their ability to cross the BBB, CNTs have attracted much attention for delivery of drug molecules to the brain [240]. Besides the blood-brain barrier (BBB), the obstacle in delivering therapeutics to the brain is the presence of degradable enzymes [261, 262].

In order to overcome the problem of limitations in the use of amphotericin B caused by its adverse effects, Benincasa *et al.* [263] tested the antifungal activity of CNT-amphotericin B (AMB) conjugates, where they found that minimum inhibitory concentration (MIC) values for CNT-AMB conjugates were either comparable to or better than those for AMB.

Carbon nanotubes are used in delivery of DNA and antibodies. The fact that nanotubes do not alter the specificity of attached antibodies enables them to deliver the drug at the targeted site. McDevit *et al.* [264] completed the research in which CNT was functionalized with antibody to effect specific targeting, metal-ion chelate, and fluorescent chromophore moieties to identify location. Carbon nanotubes functionalized in a previous manner were found to be specifically reactive against the human cancer cells *in vivo* in a model of disseminated human lymphoma and *in vitro* by flow cytometry and cell-based immunoreactivity assays.

Since genetic materials have a poor ability to cross the biological membranes, the use of viral or nonviral vectors to carry the gene and internalize it into the cell is necessary. Nucleic acids and cellular membranes are negatively charged, which results in electorostatic repulsion when in proximity to negatively charged groups. Several studies have shown the possibility of binding functionalized CNTs to nucleic acids via electrostatic interactions [265], π-π interactions [266].

6.3.6.1   CNTs as a Contrast Agent in Imaging and Identification of Cancer Cells

Methotrexate (an anticancer agent), given along with fluoroscein-functionalized nanotubes showed better visibility in the body due to the fluorescence produced by fluoroscein on the surface of the CNTs [267].

6.3.6.2   CNT-Based Nanobiosensors

A number of papers deal with CNTs applied as biosensors, some of which have shown their use for detecting specific fragments of DNA related to cancer production, and identifying the genes and antibodies associated with human autoimmune diseases [268–270]. Nanotube-based bionanosensors can be used to make a difference between metastatic and non-metastatic cancer in the case of, e.g., prostate cancer, since they can detect change in hormone levels [271].

6.3.6.3   Bone Issues and CNTs

Zhang *et al.* have concluded that CNTs with polysaccharide-functionalized surfaces (positively charged groups) significantly improve cell growth. Several groups [272–274] have found that CNT functionalized with negatively charged groups (hydroxyapatite, with which implants such as artificial bones and dental implants are often coated) allowed better cell proliferation, therefore better connection between the tissue and an implant. Hyaluronan (HA), or sodium hyaluronate (hyaluronic acid–HY), is a polysaccharide with high molecular weight that plays a crucial role in tissue repair during wound healing and inflammation by stimulating migration, adhesion, proliferation and cell differentiation, which all together lead to improved bone formation [275, 276]. In order to overcome HY instability, Mendes *et al.* [277] have investigated the effects of single-wall carbon nanotubes and its functionalization with sodium hyaluronate in terms of bone repair, and demonstrated that SWNT with HY can improve the physical properties of the HY without losing its biological effects. Mao *et al.* [278] have tested the uptake and intracellular distribution of collagen SWNTs and suggested that SWNTs functionalized by collagen should be suitable for application in biomedicine and biotechnology.

6.3.6.4   *CNTs and Photothermal Therapy of Cancer*

Unlike biological systems, CNTs strongly absorb light within near infrared (NIR) spectral window, which allows them to be applied for selective cancer cell destruction by acting as NIR heating devices that can cause cell death due to excessive local heating [279–286]. Cherukuri *et al.* [287] demonstrated that ingested CNTs could be imaged with high contrast. They also applied this technique to visualize CNTs in living organisms [287]. *In vivo* imaging was achieved in *Drosophila* (fruit fly) larvae, which were raised on food containing 10 ppm of disaggregated SWNTs [288]. Lacerda *et al.* [289] succeeded in complete tumor elimination in a large number of photothermally-treated mice without any toxic side effects. They used the amino functionalized SWNTs, which have UV-VIS luminescence properties that made it easy to monitor their intracellular trafficking with no need for special NIR lasers [289]. N-doped MWNTs were shown to produce photoablative killing of model kidney cancer cells when NIR light was applied. Moreover, the length of nanotubes was found to be a major determinant of the nanotube's ability to transfer heat and kill the tumor [290]. Gannon *et al.* [291] have succedeed in heating up nanotubes by using a radiofrequency field (13.56 MHz), which allows deeper tissue penetration than NIR irradiation. Beside *in vitro* studies, they have also completed *in vivo* experiments on rabbits treated with SWNTs, and demonstrated the killing of hepatic tumor cells. Al-Faraj *et al.* [292] and Vittorio *et al.* [293] used magnetic resonance imaging (MRI) for visualizing CNTs in cells or living organisms based on present iron oxide impurities. Hong *et al.*

[294] have intravenously applied Na^{125}I-filled glyco-SWCNTs in mice and tracked them *in vivo* by using single-photon emission computed tomography. They achieved specific tissue accumulation (here lung) coupled with high *in vivo* stability that prevented leakage or excretion of radionuclide to high-affinity organs (thyroid/stomach), and resulted in ultrasensitive imaging [294].

Being a nanosized material, nanotubes are used for preparing nanoscale surgical instruments that improve surgery since they are easier and more sophisticated [295]. Both SWNTs and MWNTs were found to be good candidates for several biomedical surgical devices [296]. Baughman *et al.* [297] showed that CNTs are able to behave like an artificial muscle since they can convert electrical into mechanical energy.

6.3.7 Toxicity

In opposition to the versatile benefits of chemical functionalization of CNTs stand the toxicity and responses of CNTs and their derivatives in *in vivo* systems and environment. Ma *et al.* [298] have shown that SWNTs alter cytochrome c electron transfer and modulate mitochondrial function. The *in vitro* and *in vivo* studies of Poland *et al.* [299] showed detrimental pulmotoxic effects of long CNTs on cell cultures and tissues, which leave them inappropriate for medical application. Chou *et al.* have also reported that SWNTs can induce pulmonary injury on mouse model [300] by alveolar macrophage activation, induction of a large number of proinflammatory genes, recruitment of leukocytes, and severe pulmonary granuloma formation.

Rotoli *et al.* [301] demonstrated an alteration of the permeability of human airway epithelial cells after treatment with MWNTs. Muller *et al.* [302, 303] and Fenoglio *et al.* [304] investigated the respiratory toxicity of MWNTs and demonstrated that they cause protein exudation and interstitial granulomas, and induced inflammatory and fibrotic reactions. Poland *et al.* [299] conducted a pilot study which proved CNTs show asbestos-like pathogenicity.

Monteiro-Riviere *et al.* [305] examined the dermal toxicity of MWNTs and noticed that after exposition of epidermal keratinocytes to MWNTs, particles were uptaken, and cell viability parameters decreased up to 20% within 24-hour exposure (in comparison with control cell cultures). Besides this, the expression of interleukin-8 was increased up to 6 times.

Nygaard *et al.* [306] examined the immune responses induced by CNTs (both SWNTs and MWNTs) and proved allergic immune responses. Watanabe *et al.* [307], Chen *et al.* [308] and Radomski *et al.* [309] have also investigated immune responses to CNTs. According to Bottini *et al.* [310], the strong tendency of CNTs to agglomerate could be the reason why CNTs impair cell viability. They also found that in terms of toxicity on human T-cells, MWNTs when compared to oxidized MWNTs, were more toxic and induced massive loss of cell viability through apoptosis. Another consequence of MWNTs strong tendency to agglomerate is, as Qu *et al.* [311] have demonstrated, injury to the lungs and heart due to delayed clearance of carboxylated MWNTs in mice.

Lanone *et al.* [312] have proved that CNTs toxicity originates from the nanomaterial size and surface area, composition and shape. Residual impurities from the synthesis of CNTs can promote their toxicity [313–315] founded that the increase in reactive oxygen species (ROS) production is a concequence of the residual metal particles in nanotubes, since ultra-purified form of CNTs had no biological adverse effects. Murray *et al.* confirmed these findings by showing that in comparison with SWCNTs, SWNTs containing 30% iron increased inflammatory responses as they produced a greater amount of ROS [316, 315].

The length of CNTs is one of the most important factors that influence bioclearance and bioretention times. However, a decrease in size causes the nanomaterial surface to become more reactive towards itself (aggregation) and to its surrounding environment (biological components) [317, 318].

Dutta *et al.* [319] have concluded that bovine or human serum albumin adsorbed onto the CNT surface results in inflammatory responses upon uptake by macrophage cells.

6.3.8 The Fate of CNTs

Several studies [265, 236] suggested that CNTs loaded with drug molecules could pass into cells and enter the cell nucleus, which enables the achievement of both cellular- and nuclear-targeted delivery. Mu *et al.* [320] suggested that the uptake by cells could be done through either energy-dependent endocytotic processes (clusters), or direct membrane permeation (single nanotubes). Raffa *et al.* [321] have demonstrated that MWNTs are more easily internalized through an energy-independent pathway. Pogodin *et al.* [322] suggested that the coatings of CNTs by biomolecules commonly present in cell culture supernatants enhance the possibility of transduction through cell membranes. Heister *et al.* have demonstrated that functionalization, which reduces non-specific bindings of biomolecules to CNTs, is crucial for controlling the interactions between nanotubes and cells [323].

Singh *et al.* [324] have demonstrated that after entering the lungs, CNTs distribute in the central nervous system, peripheral nervous system, lymph, and blood, after which they undergo distribution in the heart, spleen, kidney, bone marrow, and liver.

Cherukuri *et al.* [287] dispersed SWCNTs with a surfactant, Pluronic F108, and found that Pluronic-F108-dispersed SWCNTs accumulated mostly in the liver, while the pulmonary uptake was undetectable. Tween-80-dispersed MWCNTs accumulated in the liver and spleen with a short blood circulation time [325]. Serum-dispersed MWCNTs accumulated in the lungs and liver [326].

CNTs can be excreted via the biliary or renal pathway [327, 328]. Short and well-functionalized CNTs were shown to be eliminated by the kidneys in quite a short blood circulation time (approx. 30 min) [329], which is inapropriate for the treatment of chronic diseases. PEGylated CNTs increase their blood half-life and decrease RES uptake, but cause an accumulation of nanotubes in the dermis of the skin [330].

Despite all the obstacles and adverse effects of the application of CNTs, drug delivery undoubtedly stands as one of their most promising bioapplications.

6.4   Genotoxicity of Carbon Nanomaterials

Increased development of nanotechnology in the past decades has raised questions concerning the toxicity of nanomaterials (NM) and their influence not only on the human population, but on the environment as well [331]. Therefore, investigating the influence of NM on DNA plays an important role in health risk assessment. It is well-known that genotoxic substances induce DNA damage in somatic cells, which can initiate and promote cancerogenesis, while DNA damage in reproductive cells could affect future offspring as well. Genotoxicity of materials can also promote ageing. Therefore, it is of great importance to investigate DNA damage, by applying particular genotoxicity tests after exposure of biological material to genotoxic substances. Examination of DNA strand disturbance, point mutations, presence of DNA adducts (8-hydroxy-2'-deoxigu-anosine) and change in gene expression, makes it possible to determine whether NM possesses genotoxic properties or not. Changes in cells that might occur at lower concentrations of nanoparticles (NP) have not been thoroughly examined yet. Also alarming is the fact that the genotoxic potential of nanoparticles could be unpredictable because of their tiny size, large surface and variety of physico-chemical characteristics. Finally, the importance of investigating genotoxicity is not only for health risk assessment, but also for evaluation of the mutagenicity or carcinogenicity of examined nanoparticles—an important part of preclinical testing of pharmaceutics. Concerning the genotoxicity of nanomaterials, primary and secondary genotoxicity should be taken into account [332].

Primary genotoxicity stands for the genetic damage caused directly by NP themselves. Namely, when NP reach the nucleus they can react with DNA or proteins that bind to DNA, which both cause DNA damage. Furthermore, if NP accumulate in cytoplasm, during mytosis they can come in close contact with DNA. Primary genotoxic effects can also occur since NP can cause ROS formation, which is in close connection with characteristics of particles such as surface properties, or presence of metal ions and process of lipid peroxidation [333, 334].

Secondary genotoxicity occurs when NP interact with cell proteins involved in vital processes in the cell, after which different cellular answers such as oxidative stress, inflammation and aberant signaling response are induced [332]. Genotoxicity of NP can be conditioned by some of the following: particle size and shape, NP functionalization, charge, water solubility, chemical reactivity, interaction with primary biomolecules and particle uptake by cells in physiological conditions, presence of impurities and aggregation affinity. The most prominent tests for genotoxicity assessment of NM are a micronuclei test, Comet assay [335], as well as chromosome aberration test and Ames test on *Salmonela tifimurium*.

6.4.1   Genotoxicity of Graphene in *In Vitro* and *In Vivo* Models

Given the fact that the investigation of the genotoxic potential of some materials is conducted at their subcytotoxic concentrations, data concerning the genotoxicity of graphene should be researched in studies that deal with low concentrations of graphene.

The toxicity of pristine graphene is not so important, since prior to being applied for biomedical purposes they should be functionalized. Literature data show that stable functionalized graphene-based nanomaterials are less toxic than their nonfunctionalized counterparts [336]. Namely, graphenes (highly dispersed and non-coated) exhibit high *in vitro* cellular toxicity. Unlike them, GO proved not to be toxic, whatsoever, and promotes growth and proliferation of *E. coli* cell lines and human adenocarcinoma cells HT-29 [179]. Toxicity results for graphene and its derivatives vary from biocompatibility to toxicity. Taking into account the interaction of graphene with proteins in physiological medium, it was revealed that unlike GO, which adsorbs serum proteins in a non-specific manner, PEGylated GO has not only decreased protein binding, but has the affinity for only several proteins, e.g., immune related factors [337].

Acellular models: Besides in *in vivo* and *in vitro* models, the influence of DNA is also being investigated in acellular models. These investigations are based on the examination of isolated DNA and present one of the easiest ways to determine whether some NP cause DNA damage or not. These studies can help in describing the mechanisms believed to take part in genotoxicity of particles. Study results show that graphene can efficiently bind to ss-DNA, but not to ds-DNA, and that it can protect olygonucleotides from enzymatic cleavage [338, 173]. One of the most attractive topics for investigation is functionalization of graphene with DNA/RNA [339, 216, 217, 340], since the major challenge in gene therapy is development of gene vectors that can protect DNA from degradation and enable cellular uptake of DNA with great efficacy. DNA, negatively charged, exhibit repulsive interactions towards graphene, however, DNA bases can interact with GO surface via hydrophobic π-π interactions. In ds-DNA bases are well protected inside the double helix and outer phosphate groups have low affinity for binding since they are negatively charged [341]. According to a study by Ren *et al.* [339], GO can intercalate between base pairs inside ds-DNA in presence of Cu^{2+} ions, which results in DNA cleavage. This opens a new frontier in terms of potential application of graphene as nanocarrier of antineoplastic drugs that express their activity via DNA intercalation.

In vitro **models:** Several studies pointed out that GO induces concentration-dependent cytotoxicity in cell, decreasing cell adhesion, inducing cell apoptosis, and entering into various cellular compartments [342, 178, 187]. According to Wang *et al.* [174], in *in vitro* medium GO adheres to the surface of human cells enabling signal transduction. Signal is being transduced to nucleus, thus causing down-regulation of adhesion-associated genes and corresponding adhesive proteins, which results in cell detachment. Meanwhile, GO enter the cytoplasm via endocytosis and locate themselves in lysosomes, mitochondria and nucleus, which can a disturbance in cell metabolism, as well as in the transcription and translation of genes, and cause apoptosis and cell death [174]. Dependence of carbon nanoparticles genotoxicity on their physicochemical properties is well-known. Genotoxicity of rGO is determined by sheet dimensions that influence cell (nucleus) passing through. Namely, it was investigated how the size and concentration of rGO influence cyto- and genotoxicity to human mesenchymal stem cells (hMSCs). Genotoxicity of rGO was analyzed by Comet assay and chromosome aberrations. Concentration- and size-dependent DNA fragmentation was detected after 1 hour of treatment of mesenchymal stem cells with rGO sheets. DNA damage was detected even at concentrations of rGO (dimensions up to 11±4nm and 91±37nm) lower than 0.01 µg/ml. Larger nanoparticles (418±56nm and 3.8±04 µm) caused no damage to DNA even at concentrations of 100µg/ml. The authors suggested oxidative stress and physical injury of the membrane as the major mechanisms of rGO cytotoxicity, nevertheless, none of the proposed mechanisms can completely explain the cell destruction observed at lower concentrations ≤ 1.0 µg/ml [343]. The powerful impact on the biological/toxicological response of graphene and GO, apart from particle size, has particulate state and oxygen content/surface charge. By comparing genotoxicity of pristine GO and non-reduced GO, the correlation between NP dimensions and DNA damage was established by investigating three cell lines: human lung carcinoma cells A549, human colorectal carcinoma and mammalian cells of Vero nephrone [344]. In addition, graphene toxicity depends on whether aggregation occurred or not. It was revealed that GO normally forms stable water suspension, aggregates in salt and other biological solutions [146]. Examination of graphene and GO cytotoxicity on human skin fibroblasts showed more toxic effects of the "compacted graphene sheets" compared to "less densely packed graphene oxide," which confirms the impact of aggregation and the impact of the type of interaction with the cells [345].

In vivo **models:** Graphene-based materials such as graphite nanoplatelets proved not to cause genotoxic effects on *Caenorhabditis elegans* [191].

6.4.2 Genotoxicity of SWNT and MWNT

Although CNTs have been widely investigated, data concerning their genotoxicity are yet to be completed. Besides, comparing the biological response of SWNT and MWNT is still quite a demanding task [346]. Carbon nanotubes proved to enter different types of cells such as human epithelial lung cells A549, keratynocytes and rat and mouse macrophages [347]. Commercially used SWNT, unlike purified SWNT, can pass through the cell membrane of rats' macrophages and cause the increase in intracellular ROS production and decrease in mitochondrial membrane potential. However, it has not been clearly determined whether that effect was caused by structural characteristics of CNT, or the presence of impurities [315], since it is well-known that metal residues, especially iron, which are used during processing of CNT, are the first responsible for induction of oxidative stress [314], therefore for genotoxic effect of CNT. Apart from the impact of residual impurities, one of the key factors for genotoxic effect is the length of CNT themselves. In addition, high adsorption capacity of CNT can result in them being coated with proteins in physiological mediums [348], which depends on NP size, concentration, aggregation, charge, functional groups and surface [349]. Protein coating of CNT can influence their effective size and charge, which can consequently change the biological response of cells [350], their cellular uptake, and finally the genotoxic potential of nanoparticles. Adsorption of nutrients and growth factors from medium can lead to DNA damage as well as to bad interpretation of cell viability and apoptosis [351].

In vitro **models:** The majority of *in vitro* studies state that SWNT and MWNT induce DNA damage, which is probably connected to the intracellular increase in ROS production [347, 352, 315, 314]. The SWNTs can cause a broad spectrum of detrimental effects, including activation of molecular signaling in connection to oxidative stress [352]. Migliore *et al.* [347] have investigated the genotoxicity of SWNT on murine macrophages RAW 264.7 and came to the conclusion that concentrations of SWNT higher than 0.1µg/ml result in higher frequency of micronuclei, while Comet assay showed DNA damage at SWNT concentrations from 1 to 100µg/

ml. Kagan *et al.* [314] revealed dose- and time-dependent production of intracellular ROS and change in mitochondrial membrane potential in rat macrophage NR8383 and human lung cells A549. H. Yang *et al.* [353] have deduced that SWNT long 8nm might penetrate into cell nucleus through nucleopores and then mehanically destruct the DNA double helix. In this study, SWNT investigated on primary mouse embryo fibroblast showed particle shape-dependent genotoxicity [353]. The MWNTs were also proven to induce DNA damage, i.e., purine bases oxidation, at concentrations of 1–10μg/ml, while pyrimidine bases were significantly damaged only at the highest applied concentration [347]. A MWNTs concentration of 5mg/ml cause the increased expression of proteins that repair DNA, induce apoptosis and activate the tumor suppressor protein p53, 2 hours after the exposure of mouse embryonic stem cells [354]. The MWNTs can also express clastogenic and aneugenic effects on human breast carcinoma cells MCF-7 since they induce chromosome interruptions at 5, 10 and 50 μg/ml [355]. Lindberg *et al.* [356] have proved time-dependent genotoxicity by micronuclei test on human bronchial epithelial cells BEAS 2B. Patlolla *et al.* [357] confirmed that MWNT can induce massive damage of DNA and apoptosis, probably due to direct interaction between MWNT and DNA, which was determined by Comet assay and DNA ladder technique on cell line of normal human dermal fibroblasts (NHDF) at concentrations of 40, 200 and 400 μg/ml. Li *et al.* [358] have suggested that CNTs are effective in interaction with biomolecules of DNA dimensions.

In vivo **models:** Concerning the *in vivo* effects of CNTs, a majority of studies emphasize their positive genotoxic effects [359, 360]. Chronic exposure of mice to SWNT for 7, 28 and 60 days, as well as once every other week for eight weeks (20 mg/mouse) results in mitochondrial DNA damage [359]. The SWNTs also cause inflammation in mice lungs [361], and higher level of 8-hydroxy guanine in rat lungs and liver [360], which were, according to the authors, a probable consequence of ROS generation rather than respiratory system inhibition. Donaldson *et al.* [362] confirmed the role of CNT length in *in vivo* genotoxicity [362, 299]. Future prospects concerning CNT should deal with more detailed physico-chemical characterization of CNTs and carefully planned and precised experiment conditions, which will enlighten severity of CNTs genotoxicity and precisely define under which conditions it is exhibited.

6.4.3   Genotoxicity of Polyhydroxylated Fullerene Derivatives

So far, polyxydroxyl derivatives of fullerene, fullerenols $C_{60}(OH)_n$, have not shown significant toxicity in conducted experiments, which is mostly explained by their ability to scavenge ROS and protect cells from cellular death. Physico-chemical properties of fullerenol size, surface functionalization, solubility and aggregation/agglomeration, that define their biological activity, have proved to influence potential toxicity [363].

Acellular models: Acellular model experiments showed that $C_{60}(OH)_{24}$ binds polynucleotid chain of native DNA, i.e., base pairs [364], which was believed to be impossible in the presence of a chain of protein systems such as histones that are responsible for packing DNA into chromosomes. Recent studies have suggested that interaction between fullerenol and DNA changes DNA structure by which DNA protection of the restriction endonuclease cleavage at a higher temperature was guaranteed. Namely, at high temperatures, fullerenol can bind λDNK, after which restriction endonuclease activities Dnase I and *Hind*III were resticted, thus stabilizing DNA [365].

In vitro models: Numerous studies suggested that the genotoxic potential of fullerene decreases with the increase in its solubility [363, 366–368]. In addition, several studies confirmed the absence of $C_{60}(OH)_{24}$ genotoxicity towards DNA [369, 370, 78]. Mrđanović *et al.* [370] proved that fullerenol, $C_{60}(OH)_{24}$, leads to dose-dependent decrease in micronuclei incident and chromosome aberrations at subcytotoxic concentrations during 3-hour and 24-hour treatment of hamster ovary cell line CHO-K1 [370], and periferal blood primary lymphocyte culture throughout G0, G1/S, G1/S/G2/M phases [369]. Furthermore, fullerenol effectively protected previously mentioned cell lines from chromosome aberrations caused by Mytomycine C. The DNA protective effect of fullerenol is attributed to its ROS scavenging potential that reduces the levels of intracellulary and extracellulary produced ROS [371]. Bogdanović *et al.* [372] confirmed these results by presenting antioxidative properties of $C_{60}(OH)_{24}$ on CHO-K1 within 3 and 24 h. Kojić *et al.* [78] conducted a micronuclei test and sister chromatide exchange on human breast cancer cell lines MCF7 and MDA-MB-231 and confirmed that $C_{60}(OH)_{24}$ did not express any genotoxic potential. On the other hand, chronic exposure of hamster ovary cell line CHO-K1, human cervical epithelial carcinoma (HeLa) and human embryonic kidney (HEK293) cells to very low fullerenol concentrations (10, 100 or 1000 pg/ml) within 80 days leads to increased frequency of micronuclei incidence, not for genotoxic effects of fullerenol, but for genomic instability that causes transformation of cell lines [86]. The $C_{60}(OH)_{24}$ induces gene expression for enzymes that are involved in cell protection, proving its antigenotoxicity [373, 374]. The $C_{60}(OH)_{24}$ shows its protective role via induction of expression of gene Nrf, which is an important regulator of antioxidative response and some enzymes involved in detoxication [373]. Some recent studies have revealed that fullerenol nanoparticles significantly supress doxorubicin-induced inhibition of manganese-dependent SOD, as well as glutathione synthesis in the redox cycle [375], which confirms the antigenotoxic potential of fullerenol in conditions of induced oxidative stress, based on its antioxidative properties. The fullerene derivative of malonic acid, $C_{60}(C(COOH)_2)_2$, also showed its protective side by inducing the increase in expression of Hsp 7 that has a role in stress protection of cells, thus stabilizing lysosomes, which results in TNF-α-induced apoptosis inhibition [376].

In vivo **models:** Genotoxicity analysis for fullerene and polyhydroxylated fullerene derivatives on *in vivo* models gave different results which, however, all pointed out the importance of fullerene concentration and solvent choice. Namely, C_{60} at lower doses from 0.064 to 0.64 mg/kg bw in saline or corn oil, after oral application, induce damaged DNA followed via 8-hydroxydeoxyguanosine in the liver and lung, but not in the colon. The author explains this by possible NP agglomeration, but also by intensive turnover, which could dilute any possible transient DNA damage, whereas liver and lung cells have a low proliferation rate, allowing damage to accumulate. It is also proven that solvent corn oil exhibited higher genotoxicity than C_{60} itself [360]. Intratracheal instillation of C_{60} using male C57BL/6J or *gpt* delta transgenic mice was followed by Comet assay that proved DNA damage and mutations. In mice lungs, increased frequencies of *gpt* and Spi- mutant were observed. Analysis of mutation spectra showed that transversions were predominant [368]. Contrary to the findings of Totsuka *et al.* [368], Ema *et al.* [377] published results of Comet assays according to which C_{60} nanoparticles had no potential for DNA damage at single doses of 0.5 or 2.5 mg/kg or repeated doses of 0.1 or 0.5 mg/kg during five weeks of treatment, nor even in the development of inflammation processes. Other studies confirmed these findings. Namely, Shinohara *et al.* [367] conducted an experiment on mice where they orally applied C_{60} at concentrations of 22, 45, 88 mg/kg, and observed no DNA damage, which was determined by analysis of micronuclei in bone marrow cells. Unlike fullerene, low concentrations of its polyhydroxylated derivatives possess an anti-inflammatory effect when applied to mice trachea; while at higher concentrations express pro-inflammatory effects in lung [378]. One of the possible mechanisms by which water-soluble fullerenes can decrease inflammatory response is via ROS elimination [18, 379]. Therefore, it is to be concluded that the anti-inflammatory response of fullerenol is based on its antioxidative characteristics [77, 41]. Furthermore, by comparing anti-inflammatory effects of fullerenol and the commercial drugs amifostine and indomethacin on caragenin-induced inflammation of mice paws, fullerenol nanoparticles significantly reduced edema. The authors suggested this was a consequence of their inhibitory influence on infiltration of polymorphonuclear leukocytes, and oxidative property as free radical scavenger [77]. Ryan *et al.* [41] showed that polyhydroxyl C_{60} in nanogram concentrations not only soothes inflammation, but also inhibits inflammatory response. Previously confirmed anti-inflammatory properties of fullerenol nanoparticles, $C_{60}(OH)_{24}$, are witness to the lack of its potential for being an inducer of secondary cytotoxicity *in vivo*.

6.4.4   Conclusion

Nanomaterials possess a broad spectrum of characteristics that may allow their direct interaction with DNA. Differences between toxicity of nanomaterials and their derivatives in *in vitro* and *in vivo* models are believed to depend on the biodistribution, toxicokinetics and respiratory antioxidative defense of investigated models themselves [380]. However, problems such as nanomaterials hydrophobicity, aggregation, and size/shape/surface reactivity differences, have to be overcome since they can result in completely different toxicity/genotoxicity effects already observed in similar biological systems [335].

Future research based on type of NM-DNA interaction will help in genotoxicity assessment, thus thoroughly explaining carbon nanomaterials' health benefits and positive properties that these materials undoubtedly possess.

6.5 Ecotoxicological Effects of Carbon Nanomaterials

For its stable structure of closed cage, fullerene nC_{60} is believed not to be susceptible to microbiological degradation. However, when it is, via abiotic processes, being transformed into a better water-soluble form [381], carbon cage becomes more susceptible to biological degradation. For example, fullerenol $C_{60}(OH)_{19-27}$ can be oxidized by white rot basidiomycete fungi to CO_2 [382].

Agreggation of fullerene nanoparticles (nC_{60}) is of the utmost importance for its fate and transport in the environment [383], its production of reactive oxygen species [384], toxicity agains microorganisms [13, 384], as well as for its sorption capacity of nonpolar organic contaminants (e.g., atrazine and naphtalene) [385, 386] (Figure 6.5). Widely present natural organic materials such as

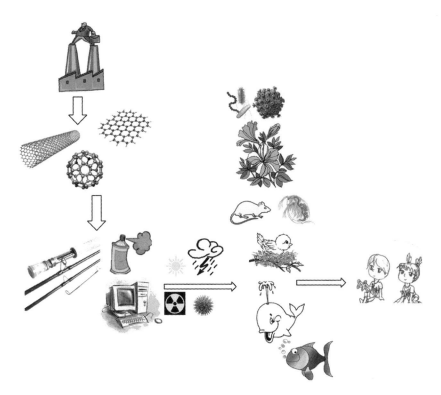

Figure 6.5 Carbon nanoparticles transporter for nonpolar toxicat mater in the environment.

humic and fulvic acids can significantly influence nC_{60} agreggation in water and stabilize the suspension, thus increasing the persistance of nC_{60} agreggates in aquatic ecosystems [387].

Hyung et al. [388] have shown that natural organic materials in aquatic ecosystems strongly interact with carbon nanotubes, greatly supporting their stability and transport in that environment. Powerful adsorptive interactions were proven to exist between CNTs (SWNT and MWNT) and nytroaromatic compounds in aqueous medium, which were quite weak between CNTs and non-polar aromatic compounds and weakest between CNTs and non-polar aliphatic molecules [389].

Phenols and anilines, relevant environmental organic contaminants, are predominantly found in industrial waste waters as products of some pesticide degradation [390]. In aqueous phase they can exist in the form of molecules or ions. Given the fact that they could be highly hazardous to organisms even at low concentrations, the majority of national environmental protection agencies list these compounds as pollutants of high priority, while many of them are classified as hazardous pollutants for their potentially toxic effects to human health. Being greatly water soluble, they are easy to transport in the environment. It was revealed that CNTs adsorption of phenolic derivatives depends on the number of hydroxyl groups, where metha-functionalized derivatives are being adsorbed more than ortho- and para-functionalized ones [391].

In order to overcome the hydrophobic character of carbon nanomaterials and achieve their better water solubility, dispersive agents are being used. Thus, for example, preparation of C_{60} suspension in water demands previously dissolved particle in a non-polar organic solvent, e.g., tetrahydrofurane [392, 393]. Tetrahydrofurane oxidative products showed toxic effects on zebrafish Danio rerio larva [394], Daphnia magna [395], fathead minnow [396], and largemouth bass [397]. Tao et al. [398] have observed bioaccumulation of stable nC_{60} in aqueous solution in Daphnia magna, under different environmental factors. Results showed time-dependent increase in organisms loading with nC_{60}, which additionally grows with the increase in nC_{60} size and concentration. As for the environmental impact, bioaccumulation increases with an increase in water hardness, lower pH and higher temperatures. Nanoparticles of nC_{60} also have an influence on increased accumulation of pesticides in agricultural crop species [399]. In this way, not only the toxicity of cocontaminants is being increased, but also cocontaminants adsorbed on nC_{60}

can reach aquatic ecosystems by water runoff and thus endanger those ecosystems. Furthermore, surfactants are greatly responsible for the toxicity of SWNTs suspended in various surfactants [400] such as sodium dodecyl sulfate, sodium dodecylbenzene sulfonate, gum arabic and Triton X-100 [401, 402].

It is well-known that surface derivatization and light exposure impact toxicity of different nanoparticles, among which are fullerenes as well [403, 404]. It is widely believed that surface coatings have the potential to make significant changes in the toxicity, solubility, reactivity, bioavailability and catalytic properties of the material underneath, thus soothing their impact on the environment and in health. Unfortunately, these coatings are not permanently present after nanoparticles are released in the environment, since prolonged light and oxygen exposure can oxidize either their surface or superficially bonded ions, and the coating may preferentially partition into the local environment, and microbs may utilize the coating in chemical reaction activity.

Microbiological symbiosis is essential for the survival of every ecosystem. Microbs in aquatic ecosystems have different roles such as being primary producers and recycling nutritional material, and are the most important members of the pollutant degradation chain. Carbon-based nanomaterials (CBNs) interact with microbial communities in natural and engineered aquatic systems. Exposure pathways include deposition of used products containing CBNs [405], direct application of CBNs in water treatment and application in the environment remediation [406], and accidental release of nanomaterials into the environment. The dispersion and ecotoxicity of CBNs will be influenced by the physico-chemical properties of the manufactured nanomaterial and parameters including pH, ionic strength, and dissolved organic matter concentration. In waters and ecosystems of waste waters, physico-chemical properties and biological activity of CBNs would be modified by adsorption of natural organic molecules, such as polysaccharides and biomacromolecules. The recent studies suggest that natural organic substances reduce deposition and bacteria attachment to nanomaterial surface, without soothing SWNTs toxicity in already attached cells. Although some bacteria are resistant to the toxic effects of CBNs, prolonged exposure, however, affects viability and integrity of membrane [407]. In addition, it was shown that individually dispersed SWNTs were more toxic than SWNT agregagates against bacteria (gram-negative *Escherichia coli, Pseudomonas aeruginosa,* and gram-positive *Staphylococcus aureus, Bacillus subtilis*). Bacteria mortality is in tight junction with their mechanical properties. Namely, "softer" cells are more susceptible to SWNT goring [245].

The toxicity of CNTs depends on their unique properties: physical, chemical, mechanical, presence of inpurities (especially catalytic metals Fe, Y, Mo, Ni and Cu) that are not being removed during CNTs preparation and purifying, surface charge, length, shape, agglomeration and number of layers (walls) [408, 409]. Nevertheless, ways in which animals are being exposed to CNTs include inhalation [410], vein injection, and dermal or oral exposure, all of which can mainly influence the *in vivo* behavior and fate of CNTs. Mechanisms on which CNTs toxicity are based on involve oxidative stress, inflammatory responses, malignant transformations, DNA damage and mutations, formation of granuloma and interstitial fibrosis [411].

Numerous studies conducted experiments in which CNTs were detected in different aquatic organisms such as *Daphnia magna,* zebrafish, rainbow trout, copepods, and which at the same time pointed out CNTs potential toxic activity [412–415]. Easy cellular uptake of CNTs and their detection in aquatic organisms implied their potential bioaccumulation. If organisms uptake carbon nanoparticles, these materials can be easily distributed to the food chain and be found in organisms at the higher trophic level in significant quantity. Besides the risk caused by pure CNTs, these materials can strongly adsorb non-polar organic (PAHs) and inorganic ionized toxic metals (Pb $^{2+}$ Cu $^{2+}$ and Cd $^{2+}$), and thus can affect the intake of such pollutants into the environment [416, 417]. Released PAHs in water present a huge risk for the environment. The cylindrical outer surface of both SWNTs and MWNTs are a pyrene, phenantrene and naphtalen-friendly area. Desorption hysteresis of PAHs was detected in fullerenes, but not in CNT, which is explained by deformational rearrangment that is a consequence of the formation of closed interstitial space in spherical aggregates of fullerene. However, cylindrical CNTs are not able to form such structures for their length, therefore, do not show significant hysteresis. The high adsorptive capacity and reversible adsorption of PAHs onto CNTs implies potential release of PAHs in case they are inhaled by animals or humans, carrying a high risk for the environment and health [418].

Graphene was shown to significantly inhibit the growth and bio-mass increase in certain plants: cabbage, tomato, red spinach and lettuce. Toxic effects proved to be dose-dependent, where phyto-toxic mechanism involves necrosis caused by oxydative stress [419].

References

1. H. Kroto, J. Health, R. O'Brien, R. Curl, R. Smalley, *Nature*, Vol. 318, p. 162, 1985.
2. W. Krätschmer, L.D. Lamb, K. Fostiropoulos, D. Huffman, *Nature*, Vol. 347, p. 354, 1990.
3. Y. Feldman, A. Zak, R. Popovitz-Biro, R. Tenne, *Solid State Sci*, Vol. 2, p. 663, 2000.
4. A. Hirsch, M. Brettreich, Fullerenes-Chemistry and Reactions, *Fullerenes: Chemistry, Physics and Technology*, WILEY-VCH Verlag GmbH & Co. KG, Weinheim, New York, 2004.
5. T. Da Ros, *Carbon Materials: Chemistry and Physics, Medicinal Chemistry and Pharmacological Potential of Fullerenes and Carbon Nanotubes*, p. 1, Springer Science, 2008.
6. B. Vileno, P.R. Marcoux, M. Lekka, A. Sienkiewicz, T. Fehér, and L. Forró, *Advanced Functional Materials*, Vol. 16, p. 120, 2006.
7. S. Bosi, T. Da Ros, G. Spalluto, and M. Prato, *Eur J Med Chem*, Vol. 38, p. 913, 2003.
8. K. Kokubo, K. Matsubayashi, H. Tategaki, H. Takada, and T. Oshima, *ACS Nano*, Vol. 2, p. 327, 2008.
9. K. Semenov, D. Letenko, N. Charykov, V. Nikitin, M. Matuzenko, V. Keskinov, V. Postnov, and A.A. Kopyrin, *Russian Journal of Applied Chemistry*, Vol. 83, p. 2076, 2010.
10. S. Bosi, L. Feruglio, T. Da Ros, G. Spalluto, B. Gregoretti, M. Terdoslavich, G. Decorti, S. Passamonti, S. Moro, and M. Prato, *Journal of Medicinal Chemistry*, Vol. 47, p. 6711, 2004.
11. M. Makha, A. Purich, C.L. Raston, and A.N. Sobolev, *European Journal of **Inorganic** Chemistry*, p. 507, 2006.
12. S. Deguchi, S.A. Mukai, M. Tsudome, and K. Horikoshi, *Advanced Materials*, Vol. 18, p. 729, 2006.
13. D. Lyon, L. Adams, J. Falkner, and P. Alvarez, *Environmental Science & Technology*, Vol. 40, p. 4360, 2006.
14. J.A. Brant, J. Labile, J. Bottero, and M. Wiesner, *Langmuir*, Vol. 22, p. 3878, 2006.
15. B. Zhao, P. Bilski, Y-Y. He, L. Feng, and C. Chignell, *Photochemistry and Photobiology*, Vol. 84, p. 1215, 2008.

16. C. Sayes, A. Gobin, K. Ausman. J. Mendez, J. West, and V. Colvin, *Biomaterials*, Vol. 26, p. 7587, 2005.
17. C. Sayes, J. Fortner, D. Lyon, A.M. Boyd, K. Ausman, B. Sitharaman, L. Wilson, J. Hughes, J. West, V. Colvin, and Y. Tap, *Nano Lett*, Vol. 4, p. 1881, 2004.
18. A. Isaković, Z. Marković, B. Todorović-Marković, N. Nikolić, S. Vranješ-Djurić, M. Mirković, M. Dramićanin, L. Harhaji, N. Raičević, Z. Nikolić, and V. Trajković, *Toxicology Sci*, Vol. 91, p. 173, 2006.
19. N. Gharbi, M. Pressac, M. Hadchouel, H. Szwarc, S.R. Wilson, and F. Moussa, *Nano Letters*, Vol. 5, p. 2578, 2005.
20. J. Fortner, D. Lyon, C. Sayes, A. Boyd, J. Falkner, E. Hotze, L. Alemany, Y. Tao, W. Guo, K. Ausman, V. Colvin, and J. Hughes, *Environ Sci Technol*, Vol. 39, p. 4307, 2005.
21. G. Djordjević, G. Bogdanović, and S. Dobrić, *Journal of BUON*, Vol. 11, p. 391, 2006.
22. G.D. Nielsen, M. Roursgaard, K. Jensen, S. Poulsen, S. Larsen, *Basic Clinical Pharmacology and Toxicology*, Vol. 103, p. 197, 2008.
23. R. Injac, N. Radic, B. Govedarica, A. Djordjevic, and B. Strukelj, *Afr J Biotechnol*, Vol. 25, p. 4940, 2008.
24. R. Partha, and J.L. Conyers, *International Journal of Nanomedicine*, Vol. 4, p. 261, 2009.
25. F. Cataldo, T. daRos, *Medicinal Chemistry and Pharmacological Potential of Fullerenes and Carbon Nanotubes*, Springer, 2008.
26. H. Ma, H. Liang, and X.J. Liang, *Science China Chemistry*, Vol. 53, p. 2233, 2010.
27. P. Anilkumar, F. Lu, L. Cao, P.G. Luo, J.-H. Liu, S. Sahu, K.N. Tackett, Y. Wang, and Y.-P. Sun, *Current Medicinal Chemistry*, Vol. 18, p. 2045, 2011.
28. R.V. Bensasson, M.N. Berberan-Santos, M. Brettreich, J. Frederiksen, H. Göttinger, A. Hirsch, E.J. Land, S. Leach, D.J. McGarvey, H. Schönberger, and C. Schröderd, *Physical Chemistry Chemical Physics*, Vol. 3, p. 4679, 2001.
29. L. Xiao, H. Takada, K. Maeda, M. Haramoto, and N. Miwa, *Biomedicine and Pharmacotherapy*, Vol. 59, p. 351, 2005.
30. J. Yin, F. Lao, P. Fu, W. Wamer, Y. Zhao, P. Wang, Y. Qiu, B. Sun, G. Xing, J. Dong, X. Liang, and C. Chen, *Biomaterials*, Vol. 30, p. 611, 2009.
31. F. Lao, W. Li, D. Han, Y. Qu, Y. Liu, Y.L. Zhao, and C.Y. Chen, *Nanotechnology*, Vol. 20, p. 225103, 2009.
32. N. Tsao, T. Luh, C. Chou, T. Chang, J. Wu, C. Liu, and H. Lei, *J Antimicrob Chemother*, Vol. 49, p. 641, 2002.
33. A.A. Corona-Morales, A. Castell, A. Escobar, R. Drucker-Colin, and L Zhang, *Journal of Neuroscience Research*, Vol. 71, p. 121, 2003.
34. M. Bisaglia, B. Natalini, R. Pelicciari, E. Straface, W. Malorni, D. Monti, C. Franceschi, and G. Schettini, *J Neurochem*, Vol. 74, p. 1197, 2000.

35. T.-Y Lin, S. Fang, S. Lin, C. Chou, T. Luh, and L. Ho, *Neurosci Res*, Vol. 43, p. 317, 2002.

36. B. Sitharaman, S. Asokan, I. Rusakova, M. Wong, L. Wilson, *Nano Letters*, Vol. 4, p. 1759, 2004.

37. N. Tsao, T. Luh, C. Chou, J. Wu, Y. Lin, H. Lei, *Antimicrob Agents Chemother*, Vol. 45, p. 1788, 2001.

38. J. Cui, Y. Yang, Y. Cheng, F. Gao, C. Liu, C. Zhou, Y. Cheng, B. Li, and J. Cai, *Free Radic Res*, Vol. 47, p. 301, 2013.

39. L. Huang, Y. Xuan, Y. Koide, T. Zhiyentayev, M. Tanaka, and M. Hamblin, *Lasers Surg Med*, Vol. 44, p. 490, 2012.

40. B. Belgorodsky, L. Fadeev, V. Ittah, H. Benyamini, S. Zelner, D. Huppert, A.B. Kotlyar, and M. Gozin, *Bioconjugate Chem*, Vol. 16, p. 1058, 2005.

41. J. Ryan, H. Bateman, A. Stover, G. Gomez, S. Norton, W. Zhao, L. Schwartz, R. Lenk, and C. Kepley, *Journal of Immunology*, Vol. 179, p. 665, 2007.

42. Y.L. Lai, P. Murugan, K.C. Hwang, *Life Sci.*, Vol. 72, p. 1271, 2003.

43. Z. Hu, C. Zhang, P. Tang, C. Li, Y. Yao, S. Sun, L. Zhang, and Y. Huang, *Cell Biol Int*, 36, p. 677, 2012.

44. S. Nakamura, and T. Mashino, *Journal of Physics: Conferec Series*, Vol. 159, p. 209, 2003.

45. M. Patel, U. Harikrishnan, N. Valand, N. Modi, and S. Menon, *Arch. Pharm. Chem. Life Sci.*, Vol. 346, p. 210, 2013

46. A. Kumar, G. Patel, S. Menon, *Chem Biol Drug Des*, Vol. 73, p. 553, 2009.

47. F. Cheng, X. Yang, C. Fan, and H. Zhu, *Tetrahedron*, Vol. 57, p. 7331, 2001.

48. F. Rancan, S. Rosan, F. Boehm, A. Cantrell, M. Brellreich, H. Schoenberger, A. Hirsch, and F. Moussa, *J Photochem Photobiol B: Biol*, Vol. 67, p. 157, 2002.

49. T. Zakharian, A. Seryshev, B. Sitharaman, B. Gilbert, V. Knight, and L. Wilson, *J Am Chem Soc*, Vol. 127, p. 12508, 2005.

50. H. Isobe, W. Nakanishi, N. Tomita, S. Jinno, H. Okayama, and E. Nakamura, *Mol Pharmaceutics*, Vol. 3, p. 124, 2005.

51. J. Ashcroft, D. Tsyboulski, K. Hartman, T. Zakharian, J. Marks, B. Weisman, M. Rosenblum, and L. Wilson, *Chem Commun (Camb)*, Vol. 28, p. 3004, 2006.

52. B.C. Braden, F.A. Goldbaum, B.X. Chen, A.N. Kirschner, S.R. Wilson, and B.F. Erlanger, *Proc Natl Acad Sci USA*, Vol. 97, p. 12193, 2000.

53. S. Foley, C. Crowley, M. Smaihi, C. Bonfils, B. Erlanger, P. Seta, and C. Larroque, *Biochemical and Biophysical Research Communications*, Vol., 294, p. 116, 2002.

54. J.E. Kim and M. Lee, *Biochem. Biophys. Res. Commu.*, Vol. 303, p. 576, 2003.

55. J. Shi, H. Zhang, L. Wang, L. Li, H. Wang, Z. Wang, Z. Li, C. Chen, L. Hou, C. Zhang, and Z. Zhang, *Biomaterials*, Vol. 34, p. 251, 2013.
56. J. Li, A. Takeuchi, M. Ozawa, X.H. Li, K. Saigo, and K. Kitazawa, *J. Chem. Soc., Chem. Communications*, Vol. 23, p. 1784, 1993.
57. L.Y. Chiang, R.B. Upasani, J.W. Swirczewshi, and S. Soled, *Journal of The American Chemical Society*, Vol. 115, p. 5453, 1993.
58. N. Schneider, A. Daiwish, H. Kroto, R. Taylor, and D. Walton, Journal of the Chemical Society, Chemical Communications, p. 463, 1994.
59. K. Kokubo, S. Shirakawa, N. Kobayashi, H. Aoshima, and T. Oshima, *Nano Research*, Vol. 4, p. 204, 2011.
60. L. Yao, F. Kang, Q. Peng, and X. Yang, *Chinese Journal of Chemical Engineering*, Vol. 18, p. 876, 2010.
61. A. Djordjević, M. Vojinović-Miloradov, N. Petranović, A. Devečerski, D. Lazar, and B. Ribar, *Fullerenes Sciences & Technology*, Vol. 6, p. 689, 1998.
62. S. Mirkov, A. Djordjević, N. Andrić, S. Andrić, T. Kostić, G. Bogdanović, M. Vojinović-Miloradov, and R. Kovacević, *Nitric Oxide: Biol and Chem*, Vol. 11, p. 201, 2004.
63. J.A. Brant, J. Labile, C.O. Robichaund, and M. Wiesner, *J Colloid Interf Sci*, Vol. 314, p. 281, 2007.
64. B. Zhao, Z. He, P. Bilski, C. Chignell, *Chem Res Toxicol*, Vol. 21, p. 1056, 2008.
65. M. Brettreich, A. Hirsch, *Tetrahedron Lett*, Vol. 39, p. 2731, 1998.
66. A. Djordjević, J. Canadanovic-Brunet, M. Vojinović -Miloradov, and G. Bogdanović, *Oxid Commun*, Vol. 4, p. 806, 2005.
67. R. Injac, M. Boškovic, M. Perse, E. Koprivec-Furlan, A. Cerar, A. Djordjević, and B. Štrukelj, B., *Pharm Rep*, Vol. 60, p. 742, 2008a.
68. R. Injac, A. Djordjević, and B. Štrukelj, *Hem Ind*, Vol. 62, p. 197, 2008.
69. R. Injac, N. Kocevar, and B. Štrukelj, *Farm Vest*, Vol. 59, p. 257, 2008.
70. R. Injac, M. Perse, M. Boškovic, V. Djordjevic-Milic, A. Djordjević, A. Hvala, A. Cerar, and B. Štrukelj, *Tech Cancer Res*, Vol. T 7, p. 15, 2008.
71. R. Injac, M. Perse, N. Obermajer, V. Djordjevic-Milic, M. Prijatelj, A. Djordjević, A. Cerar, and B. Štrukelj, *Biomaterials*, Vol. 29, p. 3451, 2008.
72. R. Injac and B. Štrukelj, *Technol Cancer Res*, Vol. T 7, p. 497, 2008.
73. R. Injac, M. Perse, M. Cerne, N. Potocnik, N. Radić, B. Govedarica, A. Djordjević, A. Cerar, and B. Štrukelj, *Biomaterials*, Vol. 30, p. 1184, 2009.
74. R. Injac, N. Radić, B. Govedarica, M. Perse, A. Cerar, A. Djordjević, and B. Štrukelj, *Pharm Rep*, Vol. 61, p. 335, 2009.
75. V. Djordjević-Milić, A. Djordjević, S. Dobrić, R. Injac, D. Vučković, K. Stankov, V. Dragojević-Simić, and L. Suvajdžić, *Mater Sci Forum*, Vol. 518, p. 525, 2006.

76. V. Djordjević-Milić, K. Stankov, R. Injac, A. Djordjević, B. Srdjenović, B. Govedarica, N. Radić, V. Dragojević-Simić, and B. Štrukelj, *Toxicol Mech Method*, Vol. 19, p. 24, 2009.
77. V. Dragojević-Simić, V. Jacević, S. Dobrić, A. Djordjević, D. Bokonjić, M. Bajcetić, and R. Injac, *Dig J Nanomater Bios*, Vol. 6, p. 819, 2011.
78. V. Kojić, D. Jakimov, G. Bogdanović, and A. Djordjević, *Mater Sci Forum*, Vol. 494, p. 543, 2005.
79. Y. Chen, K. Hwang, C. Yen, and Y. Lai, *Am J Physiol Regul Integr Comp Physiol*, Vol. 287, p. 21, 2004.
80. G. Bogdanović, V. Kojić, A. Djordjević, J. čanadanović-Brunet, M. Vojinović-Miloradov, and V.V. Baltić, *Toxicol In Vitro*, Vol. 18, p. 629, 2004.
81. L. Lu, Y. Lee, H. Chen, L. Chiang, and H. Huang, *Br J Pharmacol*, Vol. 123, p. 1097, 1998.
82. V. Bogdanović, K. Stankov, I. Icević, D. Žikić, A. Nikolić, S. Šolajić, A. Djordjević, and G. Bogdanović, *J Radiat Res*, Vol. 49, p. 321, 2008.
83. H.U. Bergmeyer, *Methods of Enzymatic Analysis*, Weinheim, Basel, 1983.
84. H.P. Misra, and I. Fridovich, *J Biol Chem*, Vol. 247, p.6960, 1972.
85. E. Beutler, *Red Cell Metabolism: A Manual of Biochemical Methods*, Grune & Stratton, New York, 1984.
86. Y. Niwa, and N. Iwai, *Environ Health Prev Med*, Vol. 11, p. 292, 2006.
87. Q. Zhao, Y. Li, J. Xu, R. Liu, and W. Li, *Int J Radiat Biol*, Vol. 81, p. 169, 2005.
88. B. Daroczi, G. Kari, M.F. McAleer, J.C. Wolf, U. Rodeck, and A.P. Dicker, *Clin Cancer Res*, Vol. 12, p. 7086, 2006.
89. T-H. Ueng, J-J. Kang, H-W. Wang, Y-W. Cheng, and L.Y. Chiang, *Toxicol Lett*, Vol. 93, p. 29, 1997.
90. H. Yamawaki, and N. Iwai, *Am J Physiol – Cell Physiol*, Vol. 290, p. 1495, 2006.
91. Y. Su, J. Xu, P. Shen, J. Li, L. Wang, Q. Li, W. Li, G. Xu, C. Fan, and Q. Huang, *Toxicol*, Vol. 269, p. 155, 2010.
92. M.P. Gelderman, O. Simakova, J.D. Clogston, A. Patri, S. Siddiqui, A. Vostal, and J. Simak, *Internat J Nanomed*, Vol. 3, p. 59, 2008.
93. R.H. Burdon, V. Gill, (1993) *Free Radical Res Commun*, Vol. 19, p. 203, 1993.
94. Y. Wei, and H. Lee, *Exp Biol Med*, Vol. 227, p. 671, 2002.
95. N. Tagmatarchis, and H. Shinohara, *Mini Rev Med Chem*, Vol. 1, p. 339, 2001.
96. H.S. Lai, W.J. Chen, and L. Chiang, *World J Surg*, Vol. 24, p. 450, 2000.
97. H.S. Lai, Y. Chen, W.J. Chen, J. Chang, and L.Y. Chiang, *Transplant Proc*, Vol. 32, p. 1272, 2000.

98. C. Zimmermann, K. Winnefeld, S. Streck, M. Roskos, R. Haberl, *Eur Neurol*, Vol. 51, p. 157, 2004.

99. D.N. Johnson-Lyles, K. Peifley, S. Lockett, B. Neun, M. Hansen, J. Clogston, S. Stern, and S. McNeil, Toxicol Appl Pharmacol, Vol. 248, p. 249, 2010.

100. K Stankov, I. Icević, V .Kojić, L. Rutonjski, A. Djordjević, G. Bogdanović, *Journal of Nanoscience and Nanotechnology*, Vol.13, p. 105, 2013.

101. J. Mrdjanovic, V. Solajić, V. Bogdanović, A. Djordjević, G. Bogdanović, R. Injac, and Z. Rakočević, *Dig J Nanomater Bios*, Vol. 7, p. 673, 2012.

102. Y. Zha, B. Yang, M. Tang, Q-C. Guo, J-T. Chen, L-P. Wen, and M. Wang, *Int J Nanomed*, Vol. 7, p. 3099, 2012.

103. A.L. Buege, D.S. Aust, *Methods in Enzimology*, Academic Press, New York, 1978.

104. S. Trajković, S. Dobrić, A. Djordjević, V. Dragojević-Simić, and Z. Milovanović, *Mater Sci Forum*, Vol. 494, p. 549, 2005.

105. S. Trajković, S. Dobrić, V. Jacević, V. Dragojević-Simić, Z. Milovanović, and A. Djordjević, *Coll Surf B*, Vol. 58, p. 39, 2007.

106. I. Icević, S. Vukmirovic, B. Srdjenovic, J. Sudji, A. Djordjević, R. Injac, and V. Vasović, *Hem Ind*, Vol. 65, p. 329, 2011.

107. T. Maksim, D. Djokić, D. Janković, A. Djordjević, O. Nesković, *J Optoelectron Adv M*, Vol. 9, p. 2571, 2007.

108. J. Zhu, Z. Ji, J. Wang, R. Sun, X. Zhang, Y. Gao, H. Sun, Y. Liu, Z. Wang, A. Li, J. Ma, T. Wang, G. Jia, and Y. Gu, *Small*, Vol. 4, p. 1168, 2008.

109. F. Jiao, Y. Liu, Y. Qu, W. Li, G. Zhou, C. Ge, Y. Li, B. Sun, and C. Chen, *Carbon*, Vol. 48, p. 2231, 2010.

110. J-Y. Xu, K. Han, S-X. Li, J-S. Cheng, G-T. Xu, W-X. Li, and Q-N. Li, *J Appl Toxicol*, Vol. 29, p. 578, 2009.

111. P. Chaudhuri, A. Paraskar, S. Soni, A. Raghunath, M., and S. Sengupta, *ACS Nano*, Vol. 3, p. 2505, 2009.

112. X.-J. Liang, C. Chen, Y. Zhao, J. Lee, and P. Wang, *Curr Drug Metabol*, Vol. 9, p. 697, 2008.

113. J. Wang, C. Chen, B. Li, H. Yu, Y. Zhao, J. Sun, Y. Li, G. Xing, H. Yuan, J. Tang, Z. Chen, H. Meng, Y. Gao, C. Ye, Z. Chai, C. Zhu, B. Ma, X. Fang, and L. Wan, *Biochem Pharmacol*, Vol. 71, p. 872, 2006.

114. T. Watanabe, H. Ichikawa, and Y. Fukumori, *Eur J Pharm Biopharm*, Vol. 54, p. 119, 2002.

115. M. Mikawa, H. Kato, M. Okumura, M. Narazaki, Y. Kanazawa, N. Miwa, and H. Shinohara, *Bioconjug Chem*, Vol. 12, p. 510, 2001.

116. H. Sies, *Free Radic Biol Med*, Vol. 27, p. 916, 1999.

117. I. Vapa, V. Milic-Torres, A. Djordjević, V. Vasović, B. Srdjenovic, V. Dragojević Simić, and J. Popović, *Eur J Drug Metab Pharmacokinet*, Vol. 37, p. 301, 2012.

118. Slavić, A. Djordjević, R. Radojicić, S. Milovanović, Z. Orescanin-Dusić, Z. Rakocević, M.B. Spasić, D. Blagojević, *Journal of Nanoparticle Research*, in press.

119. K. Novoselov, A. Geim, S. Morozov, D. Jiang, Y. Zhang, S. Dubonos, I. Grigorieva, and A. Firsov, *Science*, Vol. 306, p. 666, 2004.

120. X.S. Li, W.W. Cai, J.H. An, S. Kim, J. Nah, D.X. Yang, R. Piner, A. Velamakanni, I. Jung, E. Tutuc, S.K. Banerjee, L. Colombo, and R.S. Ruoff, *Science*, Vol. 324 , p. 1312, 2009.

121. X.L. Li, X.R. Wang, L. Zhang, S.W. Lee, H.J. Dai, *Science,* Vol. 319, p. 1229, 2008.

122. R.R. Nair, P. Blake, A.N. Grigorenko, K.S. Novoselov, T.J. Booth, T. Stauber, N.M.R. Peres, and A.K. Geim, *Science,* Vol. 320, p. 1308, 2008.

123. H. Chen, T.N. Cong, W. Yang, C. Tan, Y. Li, and Y. Ding, *Prog. Nat. Sci.,* Vol. 19, p. 291, 2009.

124. X. Huang, Z. Yin, S. Wu, X. Qi, Q. He, Q. Zhang, Q. Yan, F. Boey, and H. Zhang, *Small*, Vol. 7, p. 1876, 2011.

125. M. Nagatsu, T. Yoshida, M. Mesko, A. Ogino, T. Matsuda, T. Tanaka, H. Tatsuoka, and K. Murakami, *Carbon*, Vol. 44, p. 3336, 2006.

126. M. Acik, C. Mattevi, C. Gong, G. Lee, K. Cho, M. Chhowalla, and Y.J. Chabal, *ACS Nano*, Vol. 4, p. 5861, 2010.

127. X.L. Li, G.Y. Zhang, X.D. Bai, X.M. Sun, X.R. Wang, E. Wang, H.J. Dai, *Nat. Nanotech.,* Vol. 3, p. 538, 2008b.

128. S. Stankovich, R. Piner, S. Nguyen, and R. Ruoff, *Carbon*, Vol. 44, p. 3342, 2006.

129. S. Niyogi, E. Bekyarova, M. Itkis, M. Hamon, and R. Haddon, *J Am Chem Soc*, Vol. 128, p. 7720, 2006.

130. R. Muszynski, B. Seger, and P. Kamat, *J Phys Chem C*, Vol. 112, p. 5263, 2008.

131. A.A. Balandin, S. Ghosh, W. Bao, I. Calizo, D. Teweldebrhan, F. Miao, and C.N. Lau, *Nano Lett.,* Vol. 8, p. 902, 2008.

132. K.A. Ritter, and J. W. Lyding, *Nat Mater*, Vol. 8, p. 235, 2009.

133. D.V. Kosynkin, A.L. Higginbotham, A. Sinitskii, J.R. Lomeda, A. Dimiev, B.K. Price, J.M. Tour, *Nature*, Vol. 458, p. 872, 2009.

134. M. Kim, N.S. Safron, E. Han, M.S. Arnold, and P. Gopalan, *Nano Lett.,* Vol. 10 , p. 1125, 2010.

135. S. Park, J.H. An, I.W. Jung, R.D. Piner, S.J. An, X.S. Li, A. Velamakanni, and R.S. Ruoff, *Nano Lett*, Vol. 9, p. 1593, 2009.

136. W.S. Hummers and R.E. Offeman, *J Am Chem Soc*, Vol 80, p. 1339, 1958.

137. D.C. Marcano, D.V. Kosynkin, J.M. Berlin, A. Sinitskii, Z. Sun, A. Slesarev, L.B. Alemany, W. Lu, and J.M. Tour, *ACS Nano*, Vol. 4, p. 4806, 2010.

138. J. Shen, Y. Hu, M. Shi, X. Lu, C. Qin, C. Li, and M. Ye, *Chem Mater*, Vol. 21, p. 3514, 2009.

139. J.Q. Liu, Z.Y. Yin, X.H. Cao, F. Zhao, A.P. Lin, L.H. Xie, Q.L. Fan, F. Boey, H. Zhang, and W. Huang, *ACS Nano*, Vol. 4, p. 39879, 2010.

140. X. Fan, W. Peng, Y. Li, X. Li, S. Wang, G. Zhang, and F. Zhang, *Adv Mater,* Vol. 20, p. 4490, 2008.

141. G.K. Ramesha, and S. Sampath, *J Phys Chem C*, Vol. 113, p. 7985, 2009.

142. X. Huang, X.Z. Zhou, S.X. Wu, Y.Y. Wei, X.Y. Qi, J. Zhang, F. Boey, H. Zhang, *Small*, Vol. 6, p. 513, 2010.

143. T.V. Cuong, V.H. Pham, T.T. Quang, J.S. Chung, E.W. Shin, J.S. Kim, and E.J. Kim, *Mater Lett*, Vol. 64, p. 765, 2010.

144. G. Wang, J. Yang, J. Park, X. Gou, B. Wang, H. Liu, and J. Yao, *J Phys Chem C*, Vol. 112, p. 8192, 2008.

145. C.-Y. Su, Y. Xu, W. Zhang, J. Zhao, A. Liu, X. Tang, C.-H. Tsai, Y. Huang, and L.-J. Li, *ACS Nano*, Vol. 4, p. 5285, 2010.

146. D. Li, M.B. Muller, S. Gilje, R.B. Kaner, and G.G. Wallace, *Nat Nanotech*, Vol. 3, p. 101, 2008.

147. S. Dubin, S. Gilje, K. Wang, V.C. Tung, K. Cha, A.S. Hall, J. Farrar, R. Varshneya, Y. Yang, and R.B. Kaner, *ACS Nano*, Vol. 4, p. 3845, 2010.

148. K.A. Mkhoyan, A.W. Contryman, J. Silcox, D. A. Stewart, G. Eda, C. Mattevi, S. Miller, M. Chhowalla, *Nano Lett*, Vol. 9, p. 1058, 2009.

149. C. Gomez-Navarro, J.C. Meyer, R.S. Sundaram, A. Chuvilin, S. Kurasch, M. Burghard, K. Kern, and U. Kaiser, *Nano Lett*, Vol. 10, p. 1144, 2010.

150. Y. Si, and E.T. Samulski, *Nano Lett*, Vol. 8, p. 1679, 2008.

151. Y. Liang, D. Wu, X. Feng, and K. Müllen, *Adv Mater*, Vol. 21, p. 1679, 2009.

152. M. Quintana, K. Spyrou, M. Grzelczak, W.R. Browne, P. Rudolf, and M. Prato, *ACS Nano*, Vol. 4, p. 3527, 2010.

153. S. Singh, M. Singh, P. Kulkarni, V. Sonkar, J. Gracio, and D. Dash, *ACS Nano*, Vol. 6, p. 2731, 2012.

154. H. Wen, C. Dong, H. Dong, A. Shen, W. Xia, X. Cai, Y. Song, X. Li, Y. Li, and D. Shi, *Small*, Vol. 8, p. 760, 2012.

155. C. Peng, W. Hu, Y. Zhou, C. Fan, and Q. Huang, *Small*, Vol. 6, p. 1686, 2010.

156. L. Jin, K. Yang, K. Yao, S. Zhang, H. Tao, S.-T. Lee, Z. Liu, and R. Peng, *ACS Nano*, Vol. 6, p. 4864, 2012.

157. G. Gollavelli and Y.-C. Ling, *Biomaterials*, Vol. 33, p. 2532, 2012.

158. C. Shan, H. Yang, D. Han, Q. Zhang, A. Ivaska, and L. Niu, *Langmuir*, Vol. 25, p. 12030, 2009.

159. L. Zhang, J. Xia, Q. Zhao, L. Liu, and Z. Zhang, *Small*, Vol. 6, p. 537, 2010.

160. S.A. Zhang, K. Yang, L.Z. Feng, and Z. Liu, *Carbon*, Vol. 49, p. 4040, 2011.

161. K.-H. Liao, Y.-S. Lin, C. W. Macosko, and C.L. Haynes, *ACS Appl. Mater. Interfaces*, Vol. 3, p. 2607, 2011.

162. H. Bao, Y. Pan, Y. Ping, N.G. Sahoo, T. Wu, L. Li, J. Li, and L.H. Gan, *Small*, Vol. 7, p. 1569, 2011.

163. L. Zhang, Z. Lu, Q. Zhao, J. Huang, H. Shen, and Z. Zhang, *Small*, Vol. 7, p. 460, 2011.

164. J.T. Robinson, S.M. Tabakman, Y.Y. Liang, H.L. Wang, H.S. Casalongue, D. Vinh, and H.J. Dai, *J Am Chem Soc*, Vol. 133, p. 6825, 2011.

165. J. Liu, W. Yang, L. Tao, D. Li, C. Boyer, and T.P. Davis, *J Polym Sci, Part A: Polym Chem*, Vol. 48, p. 425, 2010.

166. W. Hu, C. Peng, M. Lv, X. Li, Y. Zhang, N. Chen, C. Fan, and Q. Huang, *ACS Nano*, Vol. 5, p. 3693, 2011.

167. M.C. Duch, G.R.S. Budinger, Y.T. Liang, S. Soberanes, D. Urich, S.E. Chiarella, L.A. Campochiaro, A. Gonzalez, N.S. Chandel, M.C. Hersam, and G.M. Mutlu, *Nano Lett*, Vol. 11, p. 5201, 2011.

168. W. Yang, Q. Ren, Y-N. Wu, V.K. Morris, A.A. Rey, F. Braet, A.H. Kwan, and M. Sunde, *Biopolymers*, Vol. 99, 2012

169. J. Liu, Y. Li, Y. Li, J. Li, and Z. Deng, *J Mater Chem*, Vol. 20, p. 10944, 2010.

170. S. Park, N. Mohanty, J.W. Suk, A. Nagaraja, J. An, R.D. Piner, W. Cai, D.R. Dreyer, V. Berry, and R.S. Ruoff, *Adv Mater*, Vol. 22, p. 1736, 2010.

171. J. Shen, M. Shi, N. Li, B. Yan, H. Ma, Y. Hu, and M. Ye, *Nano Res*, Vol. 3, p. 339, 2010.

172. X. Ma, H. Tao, K. Yang, L. Feng, L. Cheng, X. Shi, Y. Li, L. Guo, and Z. Liu, *Nano Res*, Vol. 5, p.199, 2012.

173. V.C. Sanchez, A. Jackhak, R. Hurt, A. Kane, *Chem Res Toxicol*, Vol. 25, p. 15, 2012.

174. K. Wang, J. Ruan, H. Song, J. Zhang, Y. Wo., S. Guo, and D. Cui, *Nanoscale Res Lett*, Vol. 6, p. 1, 2011.

175. Y. Chang, S.-T. Yang, J.-H. Liu, E. Dong, Y. Wang, A. Cao, Y. Liu, and H. Wang, *Toxicol Lett*, Vol. 200, p. 201, 2011.

176. Y. Li, Y. Liu, Y. Fu, T. Wei, L. Le Guyader, G. Gao, R.-S. Liu, Y.Z. Chang, and C. Chen, *Biomaterials*, Vol. 33, p. 402, 2012.

177. A. Schinwald, F.A. Murphy, A. Jones, W. MacNee, and K. Donaldson, *ACS Nano*, Vol. 6, p. 736, 2012.

178. Y. Zhang, S. Ali, E. Dervishi, Y. Xu, Z. Li, D. Casciano, and A. Biris, *ACS Nano*, Vol. 4, p. 3181, 2010.

179. O. Ruiz, K. Fernando, B. Wang, N. Brown, P. Luo, N. McNamara, M. Vangsness, Y. Sun, and C. Bunker, *ACS Nano*, Vol. 5, p. 8100, 2011.

180. N. Li, X. Zhang, Q. Song, R. Su, Z. Qi, T. Kong, L. Liu, G. Jin, M. Tang, and G. Cheng, *Biomaterials*, Vol. 32, p. 9374, 2011.

181. S. Agarwal, X. Zhou, F. Ye, Q. He, G.C. Chen, J. Soo, F. Boey, H. Zhang, and P. Chen, *Langmuir*, Vol. 26, p. 2244, 2010.

182. M. Wojtoniszak, X. Chen, R. Kalenczuk, A. Wajda, J. Łapczuk, M. Kurzewski, M. Drozdzik, P. Chu, and E. Borowiak-Palen, *Colloids Surf B*, Vol. 89, p. 79, 2012.

183. O. Akhavan, and E. Ghaderi, *ACS Nano*, Vol. 10, p. 5731, 2010.

184. W. Hu, C. Peng, W. Luo, M. Lv, X. Li, D. Li, Q. Huang, and C. Fan, *ACS Nano*, Vol. 7, p. 4317, 2010.

185. S. Liu, T.H. Zeng, M. Hofmann, E. Burcombe, J. Wei, R. Jiang, J. Kong, and Y. Chen, *ACS Nano*, Vol. 5, p. 6971, 2011.

186. O. Akhavan, and E. Ghaderi, *Carbon*, Vol. 50, p. 1853, 2012.

187. A. Sasidharan, L.S. Panchakarla, P. Chandran, D. Menon, S. Nair, C. Rao, and M. Koyakutty, *Nanoscale*, Vol. 3, p. 2461, 2011.

188. L. Feng, S. Zhang, and Z. Liu, *Nanoscale*, Vol. 3, p. 1252, 2011.

189. Z. Markovic, L. Harhaji-Trajkovic, B. Todorovic-Markovic, D. Kepic, K. Arsikin, S. Jovanovic, A. Pantovic, M. Dramicanin, and V. Trajkovic, *Biomaterials*, Vol. 32, p.1121, 2011.

190. Z. Liu, J. Robinson, X. Sun, H. Dai, *J Am Chem Soc*, Vol. 130, p. 10876, 2008.

191. E. Zanni, G. De Bellis, MP. Bracciale, A. Broggi, M.L. Santarelli, M.S. Sarto, C. Palleschi, and D. Uccelletti, *Nano Lett*, Vol. 12, p. 2740, 2012.

192. X. Zhan, J.L. Yin, C. Peng, W. Hu, Z. Zhu, W. Li, C. Fan, and Q. Huang, *Carbon*, Vol. 49, p. 986, 2011.

193. X. Sun, Z. Liu, K. Welsher, J. T. Robinson, A. Goodwin, and S. Zaric, *Nano Res*, Vol. 1, p. 203, 2008.

194. K. Yang, J. Wan, S. Zhang, Y. Zhang, S-T. Lee, and Z. Liu, *ACS Nano*, Vol. 5, p. 516, 2011.

195. K. Yang, H. Gong, X. Shi, J. Wan, Y. Zhang, and Z. Liu, *Biomaterials*, Vol 34, p. 2787, 2013.

196. K. Yang, S. Zhang, G. Zhang, X. Sun, ST. Lee, and Z. Liu, *Nano Lett*, Vol. 10, p.3318, 2010.

197. Y. Wang, J-T. Chen, and X-P. Yan, *Anal Chem*, Vol. 85, p. 2529, 2013.

198. K. Yang, J. Wan, S. Zhang, B. Tian, Y. Zhang, and Z. Liu, *Biomaterials*, Vol. 33, p. 2206, 2012.

199. K. Yang, L. Hu, X. Ma, S. Ye, L. Cheng, X. Shi, C. Li, Y. Li, and Z. Liu, *Adv Mater*, Vol. 24, p. 1868, 2012.

200. H. Hong, K. Yang, Y. Zhang, J.W. Engle, L. Feng, Y. Yang, T.R. Nayak, S. Goel, J. Bean, C.P. Theuer, T. E. Barnhart, Z. Liu, and W. Cai, *ACS Nano*, Vol. 6, p. 2361, 2012.

201. H. Hong, Y. Zhang, J.W. Engle, T.R. Nayak, C. P. Theuer, R.J. Nickles, T.E. Barnhart, and W. Cai, *Biomaterials*, Vol. 33, p. 4147, 2012.

202. T.N. Narayanan, B.K. Gupta, S.A. Vithayathil, R.R.burto, S.A. Mani, J. Taha-Tijerina, B. Xie, B.A. Kaipparettu, S.V. Torti, and P.M. Ajayan, *Adv Mater*, Vol. 24, p. 2992, 2012.

203. D. Ma, J. Lin, Y. Chen, W. Xue, and L.-M. Zhang, *Carbon*, Vol. 50, p. 3001, 2012.

204. H. Hu, J. Yu, Y. Li, J. Zhao, and H. Dong, *J Biomed Mater Res, Part A*, Vol. 100A, p. 141, 2012.

205. K. Liu, J.-J. Zhang, F.-F. Cheng, T.-T. Zheng, C. Wang, and J.-J. Zhu, *J Mater Chem*, Vol. 21, p. 12034, 2011.

206. Y. Pan, H. Bao, N. G. Sahoo, T. Wu, and L. Li, *Adv Funct Mater*, Vol. 21, p. 2754, 2011.

207. N.G. Sahoo, H. Bao, Y. Pan, M. Pal, M. Kakran, H.K.F. Cheng, L. Li, and L. P. Tan, *Chem Commun*, Vol. 47, p. 5235, 2011.

208. M. Kakran, N.G. Sahoo, H. Bao, Y. Pan, and L. Li, *Curr Med Chem*, Vol. 18, p. 4503, 2011.

209. Y.-J. Lu, H.-W. Yang, S.-C. Hung, C.-Y. Huang, S.-M. Li, C.-C. M. Ma, P.-Y. Chen, H.-C. Tsai, K.-C. Wei, and J.-P. Chen, *Int J Nanomed*, Vol. 7, p. 1737, 2012.

210. X.T. Zheng, and C.M. Li, *Mol Pharmaceutics*, Vol. 9, p. 615, 2012.

211. X. Yang, X. Zhang, Y. Ma, Y. Huang, Y. Wang, and Y. Chen, *J Mater Chem*, Vol. 19, p. 2710, 2009.

212. X. Yang, Y. Wang, X. Huang, Y. Ma, Y. Huang, R. Yang, H. Duan, and Y. Chen, *J Mater Chem*, Vol. 21, p. 3448, 2011.

213. Z. Hu, Y. Huang, S. Sun, W. Guan, Y. Yao, P. Tang, and C. Li, *Carbon*, Vol. 50, p. 994, 2012.

214. H. Kim, R. Namgung, K. Singha, I.-K. Oh, and W.J. Kim, *Bioconjugate Chem*, Vol. 22, p. 2558, 2011.

215. X. Yang, G. Niu, X. Cao, Y. Wen, R. Xiang, H. Duan, and Y. Chen, *J Mater Chem*, Vol. 22, p. 6649, 2012.

216. C.-H. Lu, C.-L. Zhu, J. Li, J.-J. Liu, X. Chen, and H.-H. Yang, *Chem Commun*, Vol. 46, p. 3116, 2010.

217. Y. Wang, Z. Li, D. Hu, C.-T. Lin, J. Li, and Y. Lin, *J Am Chem Soc*, Vol. 132, p. 9274, 2010.

218. Z. Yin, Q. He, X. Huang, J. Zhang, S. Wu, P. Chen, G. Lu, P. Chen, Q. Zhang, Q. Yan, and H. Zhang, *Nanoscale*, Vol. 4, p. 293, 2012.

219. V.C. Sanchez, A. Jackhak, R.H. Hurt, A.B. Kane, *Chemical Research in Toxicology*, Vol. 25, p. 15, 2012.

220. J. Yuan, H. Gao, C. Ching, *Toxicol Lett*, Vol. 207, p. 213, 2011.

221. L.V. Radushkevich, and V.M. Lukyanovich, Journal of Physical Chemistry (in Russian), Vol. 26, p. 88, 1952.

222. S. Iijima, *Nature*, Vol. 354, p. 56, 1991.

223. S. Iijima, and T. Ichihashi, *Nature*, Vol. 363, p. 603, 1993.

224. M.S. Dresselhaus, G. Dresselhaus, and A. Jorio, *Annual Review of Material Research*, Vol. 34, p. 247, 2004.

225. M.S. Dresselhaus, G. Dresselhaus, J.C. Charlier, and E. Hernández, *Phil Trans A Math Phys Eng Sci*, Vol. 362, p. 2065, 2004.

226. R. Saito, M. Fujita, G. Dresselhaus, and M.S. Dresselhaus, *Appl. Phys. Lett.*, Vol. 60, p. 2204, 1992.

227. P. Nithiyasri, K. Balaji, P. Brindha, and M. Parthasarathy, *Nanotechnology*, Vol. 23, p. 465603, 2012.

228. M.J. Treacy, T.W. Ebbesen, and J.M. Gibson, *Nature*, Vol. 381, p. 678, 1996.
229. E. Joselevich, *Chem Phys*, Vol. 5, p. 619, 2004.
230. S. Iijima, M. Yudasaka, R. Yamada, S. Bandow, K. Suenaga, F. Kokai, and K. Takahashi, *Chem Phys Lett*, Vol. 309, p. 165, 1999.
231. D.S. Bethune, C.H. Kiang, M. DeVries, G. Gorman, R. Savoy, J. Vazquez, and R. Beyers, *Nature*, Vol. 363, p. 605, 1993.
232. Guo, P. Nikolaev, A. Thess, D. Colbert, and R. Smalley, Chem Phys Lett, Vol. 243, p. 49, 1995.
233. G. Che, B.B. Lakshmi, C.R. Martin, and E.R. Fisher, *Chem Mater*, Vol. 10, p. 260, 1998.
234. N. Tagmatarchis, and M. Prato, *J Mater Chem*, Vol. 14, p. 437, 2004.
235. V. Georgakilas, N. Tagmatarchis, D. Pantarotto, A. Bianco, J.P. Briand, and M. Prato, *Chem Commun*, Vol. 24, p. 3050, 2002.
236. D. Pantarotto, J.P. Briand, M. Prato, and A. Bianco, *Chem Commun*, Vol.1, p. 16, 2004.
237. H. Peng, L.B. Alemany, J.L. Margrave, and V.N. Khabashesku, *J Am Chem Soc*, Vol. 125, p. 15174, 2003.
238. Y. Zhang, J. Li, Y. Shen, M. Wang, and J. Li, *J Phys Chem B*, Vol. 108, p. 15343, 2004.
239. K. Shiba, M. Yudasaka, and S. Iijima, *Nippon Rinsho*, Vol. 64, p. 239, 2006.
240. N. Sinha, and J.T.W. Yeow, *IEEE Trans Nanobioscience*, Vol. 4, p. 180, 2005.
241. A. Bianco, K. Kostarelos, C.D. Partidos, and M. Prato, *Chem Commun*, Vol. 5, p. 571, 2005.
242. C.A. Dyke, and J.M. Tour, *Chem Eur J*, Vol. 10, p. 812, 2004.
243. D. Tasis, N. Tagmatarchis, V. Georgakilas, and M. Prato, *Chem Eur J*, Vol. 9, p. 4000, 2003.
244. E. Heister, V. Neves, C. Tilmaciu, K. Lipert, V.S. Beltran, H. Coley, S. Ravi P. Silva, and J. McFadden, *Carbon*, Vol. 47, p. 2152, 2009.
245. Z. Liu, A.C. Fan, K. Rakhra, S. Sherlock, A. Goodwin, X. Chen, Q. Yang, D.W. Felsher, and H. Dai, *Angew. Chem. Int. Ed.*, Vol. 48, p. 7668, 2009.
246. Z. Liu, X.M. Sun, N. Nakayama-Ratchford, and H. Dai, *ACS Nano*, Vol. 1, p. 50, 2007.
247. X.K. Zhang, L.J. Meng, Q.G. Lu, Z.F. Fei, P.J. Dyson, *Biomaterials*, Vol. 30, p. 6041, 2009.
248. R.B. Li, R.A. Wu, L.A. Zhao, Z.Y. Hu, S.J. Guo, X.L. Pan, and H.F. Zou, *Carbon*, Vol. 49, p. 1797, 2011.
249. H. Huang, Q. Yuan, J.S. Shah, and R.D.K. Misra, *Adv. Drug Del. Rev.*, Vol. 63, p. 1332, 2011.
250. M. Arlt, D. Haase, S. Hampel, S. Oswald, A. Bachmatiuk, R. Klingeler, R. Schulze, M. Ritschel, A. Leonhardt, S. Fuessel, B. Buchner, K. Kraemer, and M.P. Wirth, *Nanotechnology*, Vol. 21, p. 335101, 2010.

251. R.B. Li, R. Wu, L. Zhao, M.H. Wu, L. Yang, and H.F. Zou, *ACS Nano*, Vol. 4, p. 1399, 2010.

252. P. Luksirikul, B. Ballesteros, G. Tobias, M.G. Moloney, and M.L.H. Green, *Carbon*, Vol. 48, p. 1912, 2010.

253. A.A. Bhirde, V. Patel, J. Gavard, G.F. Zhang, A.A. Sousa, A. Masedunskas, R.D. Leapman, R. Weigert, J.S. Gutkind, and J.F. Rusling, *ACS Nano*, Vol. 3, p. 307, 2009.

254. P. Chaudhuri, S. Soni, and S. Sengupta, *Nanotechnology*, Vol. 21, p. 025102, 2010.

255. K. Knop, R. Hoogenboom, D. Fischer, and U.S. Schubert, *Angewandte Chemie-International Edition*, Vol. 49, p. 6288, 2010.

256. Z. Liu, K. Chen, C. Davis, S. Sherlock, Q.Z. Cao, X.Y. Chen, and H.J. Dai, *Cancer Res*, Vol. 68, p. 6652, 2008a.

257. S. Liu, L. Wei, L. Hao, N. Fang, M. W. Chang, R. Xu, Y. Yang and Y. Chen, *ACS Nano*, Vol. 3, p. 3891, 2009.

258. Z. Tian, M. Yin, H. Ma, L. Zhu, H. Shen, and N. Jia, N. *J Nanosci Nanotechnol*, Vol. 11, p. 953, 2011.

259. S.S. Mohapatra, and A. Kumar, Method of drug delivery by carbon nanotube-chitosan nanocomplexes, US20080214494, 2008.

260. F. Yang, D.L. Fu, J. Long, and Q.X. Ni, *Med Hypotheses*, Vol. 70, p. 765, 2008.

261. W.M. Pardridge, *Drug Discov Today*, Vol. 12, p. 54, 207.

262. J. Rautioa, and P.J. Chikhale, *Curr Pharm Des*, Vol. 10, p. 1341, 2004.

263. M. Benincasa, S. Pacor, W. Wu, M. Prato, A. Bianco, and R. Gennaro, *ACS Nano*, Vol. 25, p. 199, 2011.

264. M.R. McDevitt, D. Chattopadhyay, B.J. Kappel, J.S. Jaggi, S.R. Schiffman, C. Antczak, J.T. Njardarson, R. Brentjens, and D.A. Scheinberg, *J Nucl Med*, Vol. 48, p. 1180, 2007.

265. D. Pantarotto, R. Singh, D. McCarthy, M. Erhardt, J.P. Briand, M. Prato, K. Kostarelos, and A. Bianco, *Chem Int Ed Engl*, Vol. 43, p. 5242, 2004.

266. M. Zheng, A. Jagota, E.D. Semke, B.A. Diner, R.S. Mclean, S.R. Lustig, R.E. Richardson, and N.G. Tassi, *Nat Mater*, Vol 2, p. 338, 2003.

267. M. Prato, K. Kostarelos, and A. Bianco, *Acc Chem Res*, Vol. 41, p. 60, 2008.

268. H. Wang, W. Zhou, D.L. Ho, K.I. Winey, J.E. Fischer, C.J. Glinka, and E.K. Hobbie, *Nano Letters*, Vol. 4, p. 1789, 2004.

269. Y. Xu, Y. Jiang, H. Cai, P.G. He, and Y. Fang, *Anal Chim Acta*, Vol. 516, p. 19, 2004.

270. P. He, and L. Dai, *Chem Commun*, Vol. 3, p. 348, 2004.

271. Y. Yun, Z. Dong , V.N. Shanov, and M.J. Schulz, *Nanotechnology*, Vol. 18, p. 465, 2007.

272. X.K. Zhang, L.J. Meng, and Q. Lu, *ACS Nano*, Vol. 3, p. 3200, 2009.

273. B. Zhao, H. Hu, S.K. Mandal, and R.C. Haddon, *Chem. Mat.*, Vol. 17, p. 3235, 2005.

274. X. Ji, W. Lou, Q. Wang, J. Ma, H. Xu, Q. Bai, C. Liu, and J. Liu, *International Journal of Molecular Sciences*, Vol. 13, p. 5242, 2012.
275. M. David-Raoudi, F. Tranchepain, B. Deschrevel, J.-C. Vincent, P. Bogdanowicz, K. Boumediene, J.-P. Pujol, *Wound Repair and Regeneration*, Vol. 16, p. 274, 2008.
276. G. Pasquinelli, C. Orrico, L. Foroni, F. Bonafè, M. Carboni, C. Guarnieri, S. Raimondo, C. Penna, S. Geuna, P. Pagliaro, A. Freyrie, A. Stella A, C.M. Caldarera, and C. Muscari, *Journal of Anatomy*, Vol. 213, p. 520, 2008.
277. R.M. Mendes, G.A.B. Silva, M.V. Caliari, E.E. Silva, L.O. Ladeira, and J. Anderson, *Life Sciences*, Vol. 87, p. 215, 2010.
278. H. Mao, N. Kawazoe, and G. Chen, *Biomaterials*, Vol. 34, p. 2472, 2013.
279. N.W.S. Kam, M. O'Connell, J.A. Wisdom, H.J. Dai, *Proc Natl Acad Sci USA*, Vol. 102, p. 11600, 2005.
280. P. Chakravarty, R. Marches, N.S. Zimmerman, A.D.E. Swafford, P. Bajaj, I.H. Musselman, I. P. Pantano, R.K. Draper, and E.S. Vitetta, *Proc Natl Acad Sci.USA, 2008*, Vol. 105, p. 8697, 2008.
281. R. Marches, P. Chakravarty, I.H. Musselman, P. Bajaj, R.N. Azad, P. Pantano, R.K. Draper, E.S. Vitetta, *Int J Cancer*, Vol. 125, p. 2970, 2009.
282. C.H. Wang, Y.J. Huang, C.W. Chang, W.M. Hsu, and C.A. Peng, *Nanotechnology*, Vol. 20, p. 315101, 2009.
283. Y. Xiao, X.G. Gao, O. Taratula, S. Treado, A. Urbas, R.D. Holbrook, R.E. Cavicchi, C.T. Avedisian, S. Mitra, R. Savla, P.D. Wagner, S. Srivastava, and H.X. He, *BMC Cancer*, Vol. 9, p. 351, 2009.
284. J.W. Fisher, S. Sarkar, C.F. Buchanan, C.S. Szot, J. Whitney, H.C. Hatcher, S.V. Torti, C.G. Rylander, and M.N. Rylander, *Cancer Res.*, Vol. 70, p. 9855, 2010.
285. J.T. Robinson, G. Hong, Y. Liang, B. Zhang, O.K. Yaghi, H. Dai, *J. Am. Chem. Soc.*, Vol. 134, p. 10664, 2012.
286. S. Diao, G. Hong, J.T. Robinson, L. Jiao, A.L. Antaris, J.Z. Wu, C.L. Choi, H. Dai, *J. Am. Chem. Soc.*, Vol. 134, p. 16971, 2012.
287. P. Cherukuri, C.J. Gannon, T.K. Leeuw, H.K. Schmidt, R.E. Smalley, S.A. Curley, and R.B. Weisman, *Proc Natl Acad Sci USA*, Vol. 103, p. 18882, 2006.
288. T.K. Leeuw, R.M. Reith, R.A. Simonette, M.E. Harden, P. Cherukuri, D.A. Tsyboulski, K.M. Beckingham, and R.B. Weisman, *Nano Lett*, Vol. 7, p. 2650, 2007.
289. L. Lacerda, G. Pastorin, D. Gathercole, J. Buddle, M. Prato, A. Bianco, and K. Kostarelos, *Adv Mater*, Vol. 19, p. 1480, 2007.
290. S.V. Torti, F. Byrne, O. Whelan, N. Levi, B. Ucer, M. Schmid, F.M. Torti, S. Akman, J. Liu, P.M. Ajayan, O. Nalamasu, and D.L. Carroll, *Int J Nanomedicine*, Vol. 2, p. 707, 2007.
291. C.J. Gannon, P. Cherukuri, B.I. Yakobson, L.Cognet, J.S. Kanzius, C. Kittrell, R.B.Weisman, M. Pasquali, H.K. Schmidt, R.E. Smalley, and S.A. Curley, *Cancer*, Vol. 110, p. 2654, 2007.

292. Al-Faraj, F. Fauvelle, N. Luciani, G. Lacroix, M. Levy, Y. Cremillieux, and E. Canet-Soulas, *Int J Nanomedicine*, Vol. 6, p. 351, 2011.

293. O. Vittorio, S.L. Duce, A. Pietrabissa, and A. Cuschieri, *Nanotechnology*, Vol. 22, p. 095706, 2011.

294. S.Y. Hong, G. Tobias, K.T. Al-Jamal, B. Ballesteros, H. Ali-Boucetta, S. Lozano-Perez, P.D. Nellist, R.B. Sim, C. Finucane, S.J. Mather, M.L.H. Green, K. Kostarelos, and B.G. Davis, *Nat Mater*, Vol. 9, p. 485, 2010.

295. A.G. Mamalis, *J Mat Process Technol*, Vol. 181, p. 52, 2007.

296. M. Gao, L. Dai, R.H. Baughman, G.M. Spinks, and G.G. Wallace, *Proc SPIE*, Vol. 3987, p. 18, 2000.

297. R.H. Baughman, C.X. Cui, A.A. Zakhidov, Z. Iqbal, J.N. Barisci, G.M. Spinks, G.G. Wallace, A. Mazzoldi, D. De Rossi, A.G. Rinzler, O. Jaschinski, S. Roth, and M. Kertesz, *Science*, Vol. 284, p. 1340, 1999.

298. X. Ma, L.-H. Zhang, Y. Wu, G. Zou, X. Wu, P.C. Wang, W.G. Wamer, J.-J. Yin, K. Zheng, and X.-J. Liang, *ACS Nano*, Vol. 6, p. 10486, 2012.

299. C.A. Poland, R. Duffin, I. Kinloch, A. Maynard, W.A.H. Wallace, A. Seaton, V. Stone, S. Brown, W. MacNee, and K. Donaldson, *Nat Nanotechnol*, Vol. 3, p. 423, 2008.

300. C.C. Chou, H.Y. Hsiao, Q.S. Hong, C.H. Chen, Y.W. Peng, H.W. Chen, and P.C. Yang, *Nano Lett*, Vol. 8, p. 437, 2008.

301. B.M. Rotoli, O. Bussolati, M.G. Bianchi, A. Barilli, C. Balasubramanian, S. Bellucci, and E. Bergamaschi, *Toxicol Lett*, Vol. 178, p. 95, 2008.

302. J. Muller, F. Huaux, N. Moreau, P. Misson, J.F. Heilier, M. Delos, M. Arras, A. Fonseca, J.B. Nagy, and D. Lison, *Toxicol Appl Pharmacol*, Vol. 207, p. 221, 2005.

303. J. Muller, F. Huaux, A. Fonseca, J.B. Nagy, N. Moreau, M. Delos, E. Raymundo-Piñero, F. Béguin, M. Kirsch-Volders, I. Fenoglio, B. Fubini, and D. Lison, *Chem Res Toxicol*, Vol. 21, p. 1698, 2008.

304. I. Fenoglio, G. Greco, M. Tomatis, J. Muller, E. Pinero, F. Beguin, A. Fonseca, J.B. Nagy, D. Lison, and B. Fubini, *Chem Res Toxicol*, Vol, 21, p.1690, 2008.

305. N.A. Monteiro-Riviere, R.J. Nemanich, A.O. Inman, Y.Y. Wang, and J.E. Riviere, *Toxicol Lett*, Vol. 155, p. 377, 2005.

306. U.C. Nygaard, J.S. Hansen, M. Samuelsen, T. Alberg, C.D. Marioara, and M. Løvik, *Toxicol. Sci. Vol*, 109, p. 113, 2009.

307. H. Watanabe, K. Numata, T. Ito, K. Takagi, and A. Matsukawa, *Shock*, Vol. 22, p. 460, 2004.

308. X. Chen, J.J. Oppenheim, and O.M. Howard, *J. Leukoc. Biol*, Vol. 78, p. 114, 2005.

309. P. Radomski, D. Jurasz, M. Alonso Escolano, M. Drews, T. Morandi, M.W. Malinski, and W. Radomski, *Br J Pharmacol*, Vol. 146, p. 882, 2005.

310. M. Bottini, S. Bruckner, K. Nika, N. Bottini, S. Bellucci, A. Magrini, A. Bergamaschi, and T. Mustelin, *Toxicol Lett*, Vol. 160, p. 121, 2006.

311. G. Qu, Y. Bai, Y. Zhang, Q. Jia, W. Zhang, and B. Yan, *Carbon*, Vol. 47, p. 2060, 2009.
312. S. Lanone, and J. Boczkowski, *Curr Mol Med*, Vol. 6, p. 651, 2006.
313. J.M. Wörle-Knirsch, K. Pulskamp, and H.F. Krug, *Nano Lett*, Vol. 6, p. 1261, 2006.
314. V.E. Kagan, Y.Y. Tyurina, V.A. Tyurin, N.V. Konduru, A.I. Potapovich, A.N. Osipov, E.R. Kisin, D. Schwegler-Berry, R. Mercer, V. Castranova, A.A. Shvedova, *Toxicol Lett*, Vol. 165, p. 88, 2006.
315. K. Pulskamp, S. Diabaté, and H.F. Krug, *Toxicol Lett*, Vol. 168, p. 58, 2007.
316. A.R. Murray, E. Kisin, S.S. Leonard, S.H. Young, C. Kommineni, V.E. Kagan, V. Castranova, and A.A. Shvedova, *Toxicology*, Vol. 257, p. 161, 2009.
317. K. Soto, K.M. Garza, and L.E. Murr, *Acta Biomater*, Vol. 3, p. 351, 2007.
318. P. Wick, P. Manser, L.K. Limbach, U. Dettlaff-Weglikowska, F. Krumeich, S. Roth, W.J. Stark, A. Bruinink, *Toxicol Lett*, Vol. 168, p. 121, 2007.
319. D. Dutta, S.K. Sundaram, J.G. Teeguarden, B.J. Riley, L.S. Fifield, J.M. Jacobs, S.R. Addleman, G.A. Kaysen, B.M. Moudgil, and T.J. Weber, *Toxicol Sci*, Vol. 100, p. 303, 2007.
320. Q.X. Mu, D.L. Broughton, and B. Yan, *Nano Lett*, Vol. 9, p. 4370, 2009.
321. V. Raffa, G. Ciofani, S. Nitodas, T. Karachalios, D. D'Alessandro, M. Masini, and A. Cuschieri, *Carbon*, Vol. 46, p. 1600, 2008.
322. S. Pogodin, N.K.H. Slater, and V.A. Baulin, *ACS Nano*, Vol. 5, p. 1141, 2011.
323. E. Heister, C. Lamprecht, V. Neves, C. Tilmaciu, L. Datas, E. Flahaut, B. Soula, P. Hinterdorfer, H.M. Coley, H. S.R.P. Silva, S. J. McFadden, *ACS Nano*, Vol. 4, p. 2615, 2010.
324. R. Singh, D. Pantarotto, L. Lacerda, G. Pastorin, C. Klumpp, M. Prato, A. Bianco, and K. Kostarelos, *Proc Natl Acad Sci USA*, Vol. 103, p. 3357, 2006.
325. X. Deng, S. Yang, H. Nie, H. Wang, and Y. Liu, *Nanotechnology*, Vol. 19, p. 075101, 2008.
326. L. Lacerda, H. Ali-Boucetta, M.A. Herrero, G. Pastorin, A. Bianco, M. Prato, and K. Kostarelos, *Nanomedicine*, Vol. 3, p. 149, 2008.
327. A. Ruggiero, C.H. Villa, E. Bander, D.A. Rey, M. Bergkvist, C.A. Batt, K. Manova-Todorova, W.M. Deen, D.A.Scheinberg, and M.R. McDevitt, *Proc Natl Acad Sci USA*, Vol. 107, p. 12369, 2010.
328. L. Lacerda, M.A. Herrero, K. Venner, A. Bianco, M. Prato, and K. Kostarelos, *Small*, Vol. 4, p. 1130, 2008.
329. L. Lacerda, A. Soundararajan, R. Singh, G. Pastorin, K.T. Al-Jamal, J. Turton, P. Frederik, M.A. Herrero, S.L.A. Bao, D. Emfietzoglou, S. Mather, W.T. Phillips, M. Prato, A. Bianco, B. Goins, and K. Kostarelos, *Adv Mater*, Vol. 20, p. 225, 2008.

330. X.W. Liu, H.Q. Tao, K. Yang, S.A. Zhang, S.T. Lee, and Z.A. Liu, *Biomaterials*, Vol. 32, p. 144, 2011.

331. S. Singh, and H.S. Nalwa, *J Nanosci Nanotechnol*, Vol. 7, p. 3048, 2007.

332. V. Stone, H. Johnston, and R.P.F. Schins, *Critical Reviews in Toxicology*, Vol. 39, p. 613, 2009.

333. H. Greim, P. Borm, R. Schins, K. Donaldson, K. Driscoll, A. Hartwig, E. Kuempel, G. Oberdorster, and G. Speit, Report of the workshop held in Munich, Germany, 26–27 October 2000. *Inhal Toxicol*, Vol. 13, p. 737, 2001.

334. D.B. Kell, *BMC Medical Genomics*, Vol. 2, p. 1, 2009.

335. R. Landsiedel, M.D. Kapp, M. Schulz, K. Wiench, and F. Oesch, *Mutat Res*, Vol. 681, p. 241, 2009.

336. Y. Zhang, T.R. Nayak, H. Hongb, and W. Cai, *Nanoscale*, Vol. 4, p. 3833, 2012.

337. Tan, L. Feng, J. Zhang, K. Yang, S. Zhang, Z. Liu, and R. Peng, *ACS Appl Mater Interfaces*, 2013.

338. Z. Qian, Q. Yun, H. Fei, Z. Ling, W. Shuyao, L. Ying, L. Jinghong, and S. Xi-Ming, *Chem Eur J*, Vol. 16, p. 8133, 2010.

339. H. Ren, C. Wang, J. Zhang, X. Zhou, D. Xu, J. Zheng, and S. Guo, *ACS Nano*, Vol. 4, p. 7169, 2010.

340. Y. Xu, Q. Wu, Y. Sun, H. Bai, and G. Shi, *ACS Nano*, Vol. 4, p. 7358, 2010.

341. M. Wu, R. Kempaiah, P.J. Huang, V. Maheshwari, and J. Liu, *Langmuir*, Vol. 27, p. 2731, 2011.

342. W. Kan, J. Ruan, H. Song, J. Zhang, Y. Wo, S. Guo, and D. Cui, *Nanoscale Res Lett*, Vol. 6, p. 8, 2011.

343. O. Akhavan, E. Ghaderi, and A. Akhavan, *Biomaterials*, Vol. 33, p. 8017, 2012.

344. L. Ottaviano, NanoTP - 3rd Annual Scientific Meeting Berlin, Germany Poster Presentations, 2012.

345. L. Ken-Hsuan, Y.S. Lin, C.W. Macosko, and C.L. Haynes, *ACS Appl. Mater. Interfaces*, Vol. 3, p. 2607, 2011.

346. L. Gonzalez, D. Lison, and M. Kirsch-Volders, *Nanotoxicology*, Vol. 2, p. 252, 2008.

347. L. Migliore, D. Saracino, A. Bonelli, R. Colognato, M.R.D'Errico, A. Magrini, A. Bergamaschi, and E. Bergamaschi, *Environmental and Molecular Mutagenesis*, Vol. 51, p. 294, 2010.

348. R.S. Kane, and A.D. Stroock, *Biotechnol Prog*, Vol. 23, p. 316, 2007.

349. T. Cedervall, I. Lynch, S. Lindman, T. Berggard, E. Thulin, H. Nilsson, K.A. Dawson, and S. Linse, *Proc Natl Acad Sci U.S.A.*, Vol. 104, p. 2050, 2007.

350. C. Schulze, A. Kroll, C.M. Lehr, U.F. Schäfer, K. Becker, J. Schnekenburger, C. Schulze Isfort, R. Landsiedel, and W. Wohleben, *Nanotoxicology*, Vol. 2, p. 51, 2008.

351. L. Guo, A. Von Dem Bussche, M. Buechner, A. Yan, A.B. Kane, and R.H. Hurt, *Small*, Vol. 4, p. 721, 2008.
352. M. Pacurari, X.J. Yin, J. Zhao, M. Ding, S.S. Leonard, D. Schwegler-Berry, B.S. Ducatman, D. Sbarra, M.D. Hoover, V. Castranova, and V. Vallyathan, *Environ Health Perspect*, Vol. 116, p. 1211. 2008.
353. H. Yang, C. Liu, D. Yang, H. Zhang, and Z. Xi, *J Appl Toxicol*, Vol. 29, p. 69, 2009.
354. L. Zhu, D.W. Chang, L. Dai, and Y. Hong, *Nano Lett*, Vol. 7, p. 3592, 2007.
355. J. Muller, I. Decordier, P.H. Hoet, N. Lombaert, L. Thomassen, F. Huaux, D. Lison, and M. Kirsch-Volders, *Carcinogenesis*, Vol. 29, p. 427, 2008.
356. H.K. Lindberg, G.C. Falck, S. Suhonen, M. Vippola, E. Vanhala, J. Catalán, K. Savolainen, and H. Norppa, *Toxicol Lett*, Vol. 186, p. 166, 2009.
357. Patlolla, B. Knighten, and P. Tchounwou, Multi-Walled Carbon Nanotubes Induce Cytotoxicity, *Ethn Dis*, Vol. 20 p.65, 2010.
358. S. Li, P. He, J. Dong, Z. Guo, and L. Dai, *J Am Chem Soc*, Vol. 127, p. 14, 2005.
359. Z. Li, T. Hulderman, R. Salmen, R. Chapman, S.S. Leonard, S.H. Young, A. Shvedova, M.I. Luster, and P.P. Simeonova, *Environ Health Perspect*, Vol. 115, p. 377, 2007.
360. J.K. Folkmann, L. Risom, N.R. Jacobsen, H. Wallin, S. Loft, and P. Møller, *Environ Health Perspect*, Vol. 117, p. 703, 2009.
361. A.A. Shvedova, E.R. Kisin, R. Mercer, A.R. Murray, V.J. Johnson, A.I. Potapovich, Y.Y. Tyurina, O. Gorelik, S. Arepalli, D. Schwegler-Berry, AF Hubbs, J. Antonini, D.E. Evans, B.K. Ku, D. Ramsey, A. Maynard, V.E. Kagan, V. Castranova, and P. Baron, *Am J Physiol Lung Cell Mol Physiol*, Vol. 289, p. L698, 2005.
362. K. Donaldson, F.A. Murphy, R. Duffin, C.A. Poland, *Particle and Fibre Toxicology*, Vol. 7, p. 1, 2010.
363. K. Aschberger, H. Johnston, V. Stone, R. Aitken, L C.ang Tran, S. Hankin, S. Peters, and F. Christensen, *Regulatory Toxicology and Pharmacology*, Vol. 58, p. 455, 2010.
364. M. Pinteala, A. Dascalu, and C. Ungurenasu, *International Journal of Nanomedicine*, Vol. 4, p. 193, 2009.
365. H. An, and B. Jin, *Environ Sci Technol*, Vol. 1, p. 6608, 2 011
366. H.J. Johnston, G.R. Hutchison, F.M. Christensen, K. Aschberger, and V. Stone, *Toxicological Sciences*, Vol. 114, p. 162, 2010
367. N. Shinohara, K. Matsumoto, S. Endoh, J. Maru, and J. Nakanishi, *Toxicol Lett*, Vol. 191, p. 289, 2009.
368. Y. Totsuka, T. Higuchi, T. Imai, A. Nishikawa, T. Nohmi, T. Kato, S. Masuda, N. Kinae, K. Hiyoshi, S. Ogo, M. Kawanishi, T. Yagi,

T. Ichinose, N. Fukumori, M. Watanabe, T. Sugimura and K. Wakabayashi, *Particle and Fibre Toxicology*, Vol. 6, p. 23, 2009.

369. J. Mrđanović, S. Šolajić, V. Bogdanović, A. Djordjevic, G. Bogdanović, R. Injac, Z. Rakočević, Digest Journal of Nanomaterials and Biostructures, Vol. 7, p. 673, 2012.

370. J. Mrđanović, S. Šolajić, V. Bogdanović, K.V. Stankov, G. Bogdanović, and A. Djordjević, *Mutation Research-Genetic Toxicology and Environmental Mutagenesis*, Vol. 680, p. 25, 2009.

371. F. Marano, S. Hussain, F. Rodrigues-Lima, A. Baeza-Squiban, and S. Boland, *Arch Toxicol*, Vol. 85, p. 733, 2010.

372. V. Bogdanović, M. Slavić, J. Mrđanović, S. Šolajić, and A. đorđević, *Hemijska industrija*, Vol. 63, p. 143, 2009.

373. X. Cai, H. Jia, Z. Liu, B. Hou, C. Luo, Z. Feng, W. Li, and J. Liu, *Journal of Neuroscience Research,* Vol. 86, p. 3622, 2008.

374. J. Gao, Z.R. Zhu, H.Q. Ding, Z. Qian, L. Zhu, and Y. Ke, *Neurochem Int*, Vol. 50, p. 379, 2007.

375. K. Stankov, G. Bogdanović, J. Katanic, V. Bogdanović, S. Stankov, K. Katic Bajin, and A. Djordjević, *FEBS Journal*, Vol. 278, p. 74, 2011.

376. W. Li, L. Zhao, T. Wei, Y. Zhao, and C. Chen, *Biomaterials*, Vol. 32, p. 4030, 2011.

377. M. Ema, J. Tanaka, N. Kobayashi, M. Naya, S. Endoh, J. Maru, M. Hosoi, M. Nagai, M. Nakajima, M. Hayashi, and J. Nakanishi, *Regulatory Toxicology and Pharmacology*, Vol. 62, Issue 3, p. 419, 2012.

378. M. Roursgaard, S.S. Poulsen, C.L. Kepley, M. Hammer, G.D. Nielsen, and S.T. Larsen, *Basic Clin Pharmacol Toxicol*, Vol. 103, p. 386, 2008.

379. L.L. Dugan, J.K. Gabrielsen, S.P. Yu, T.S. Lin, and D.W. Choi, *Neurobiol Dis,* Vol. 3, p. 129, 1996.

380. C.M. Sayes, A.A. Marchione, K.L. Reed, and D.B. Warheit, *Nano Lett*, Vol. 7, p. 2399, 2007.

381. J. Lee, M. Cho, J.D. Fortner, J.B. Hughes, and J.H Kim, *Environ Sci Technol*, Vol. 43, p. 4878, 2009.

382. K.M. Schreiner, T.R. Filley, R.A. Blanchette, B.B Bowen, R. D. Bolskar, W.C. Hockaday, C.A. Masiello and J.W. Raebiger, *Environ Sci Technol*, Vol. 43, p. 3162, 2009.

383. A.R. Petosa, D.P. Jaisi, I.R. Quevedo, M. Elimelech, and N. Tufenkji, *Environ Sci Technol*, Vol. 44, p.6532, 2010.

384. S.R. Chae, A.R. Badireddy, J.F. Budarz, S.H. Lin, Y. Xiao, M. Therezien, and M.R. Wiesner, *Acs Nano*, Vol. 4, p. 5011, 2010.

385. K. Gai, B. Shi, X. Yan, and D. Wang, *Environ Sci Technol*, Vol. 45, p. 5959, 2011.

386. A. Baun, S.N. Sørensen, R.F. Rasmussen, N.B. Hartmann, and C.B. Koch, *Aquatic Toxicology*, Vol. 86, p. 379, 2008.
387. W. Zhang, U. Rattanaudompol, H. Li, and D. Bouchard, *Water Research*, Vol. 47, p. 1793, 2013.
388. H. Hyung, J.D. Fortner, J.B. Hughes, and J.H. Kim, *Environ Sci Technol*, Vol. 41, p.179, 2007.
389. W. Chen, L. Duan, and D. Zhu, *Environ Sci Technol*, Vol. 41, p. 8295, 2007.
390. K. Laszlo, E. Tombacz, and C. Novak, *Colloids and Surfaces A: Physicochemical and Engineering Aspects*, Vol. 306, p. 95, 2007.
391. Q. Liao, J. Sun, and L. Gao, *Colloids and Surfaces A: Physicochemical and Engineering Aspects*, Vol. 312, p. 160–165, 2008.
392. S. Deguchi, R. G. Alargova, and K. Tsujii, *Langmuir*, Vol. 17, p. 6013, 2001.
393. J. Gao, S. Youn, A. Hovsepyan, V.L. Llaneza, Y. Wang, G. Bitton, and J.C. Bonzongo, *Environ Sci Technol*, Vol. 43, p. 3322, 2009.
394. T.B. Henry, F.M. Menn, J.T. Fleming, J. Wilgus, R.N. Compton, and G.S. Sayler, Environ Health Perspect., Vol. 115, p. 1059, 2007.
395. S.B. Lovern, and R. Klaper, *Environ Toxicol Chem*, Vol. 25, p. 1132, 2006.
396. S. Zhu, E.Oberdörster, and M.L. Haasch, *Mar Environ Res.*, Vol. 62, p.5, 2006.
397. E. Oberdörster, *Environ Health Perspect*, Vol. 112, p. 1058, 2004.
398. X. Tao, J.D. Fortner, B. Zhang, Y. He, Y. Chen, and J.B. Hughes, *Chemosphere*, Vol. 77, p. 1482, 2009.
399. R. De La Torre-Roche, J. Hawthorne, Y. Deng, B. Xing, W. Cai, L.A. Newman, C. Wang, X. Ma, and J.C. White, *Environ Sci Technol*, Vol. 46, p. 9315, 2012.
400. L. Dong, K.L. Joseph, C.M. Witkowski, and M.M. Craig, *Nanotechnology*, Vol. 25, p. 255702, 2008.
401. J. Wang, G. Liu, and M. Rasul, *J Am Chem Soc*, Vol. 126, p. 3010, 2004.
402. J. Gao, V. Llaneza, S. Youn, C.A. Silvera-Batista, K.J. Ziegler, and J.C. Bonzongo, *Environ Toxicol Chem*, Vol. 31, p. 210, 2012.
403. A. Sakai, Y. Yamakoshi, and N. Miyata. *Fullerene Sci Technol*, Vol. 7, p. 743, 1999.
404. N. Nakajima, C. Nishi, F.M. Li, and Y. Ikada, *Fullerene Sci Technol*, Vol. 4, p.1, 1996.
405. N.C. Mueller, and B. Nowack, *Environ Sci Technol*, Vol. 42, p. 4447, 2008.
406. S. Brady-Estévez, S. Kang, and M. Elimelech, *Small*, Vol. 4, p. 481, 2008.
407. S. Kang, M.S. Mauter, and M. Elimelech, *Environ Sci Technol.*, Vol. 43, p. 2648, 2009.
408. Y.L. Zhao, G.M Xing, and Z.F. Chai, *Nature Nanotechnology*, Vol. 3, p. 191, 2008.

409. Z. Chen, H. Meng, G.M. Xing, C.Y. Chen, and Y.L. Zhao, *Int J Nanotechnology*, Vol. 4, p. 179, 2007.
410. J.G. Li, W.X. Li, J.Y. Xu, X.Q. Cai, R.L. Liu, Y.J. Li, Q.F. Zhao, and Q.N. Li, *Environ Toxicol*, Vol. 22, p. 415, 2007.
411. Y. Liu, Y. Zhao, B. Sun, and C. Chen, *Acc Chem Res*, Vol. 46, p. 702, 2013.
412. A.P. Roberts, A.S. Mount, B. Seda, J. Souther, R. Qiao, S. Lin, P.C. Ke, A.M. Rao, and S.J. Klaine, *Environ Sci Technol*, Vol. 41, p. 3025, 2007.
413. J. Cheng, E. Flahaut, and S.H. Cheng, *Environ Toxicol Chem*, Vol. 26, p. 708, 2007.
414. C.J. Smith, B.J. Shaw, and R.D. Handy, *Aquat Toxicol*, Vol. 82, p. 94, 2007.
415. R.C. Templeton, P.L. Ferguson, K.M. Washburn, W.A. Scrivens, and G.T. Chandler. *Environ Sci Technol*, Vol. 40, p. 7387, 2006.
416. Y. Li, J. Ding, Z. Luan, Z. Di, Y. Zhu, C. Xu, D. Wu, and B. Wei, *Carbon*, Vol. 41, p. 2787, 2003.
417. K. Yang, L. Zhu, and B Xing. *Environ Sci Technol*, Vol. 40, 1855, 2006.
418. K. Yang, and B. Xing, *Environmental Pollution*, Vol. 145, p. 529, 2007.
419. P. Begum, R. Ikhtiari, and B. Fugetsu, *Carbon*, Vol. 49, p. 3907, 2011.

Part 2

COMPOSITE MATERIALS

7

Advanced Optical Materials Modified with Carbon Nano-Objects

Natalia V. Kamanina

*Head of the Lab at Vavilov State Optical Institute, St. Petersburg, Russia
Professor of Saint-Petersburg Electrotechical University (LETI), St. Petersburg,
Russia Professor of Saint-Petersburg Technical University (IFMO),
St. Petersburg, Russia*

Abstract

Using different types of modern effective nanoparticles such as fullerenes, carbon nanotubes, shungites, graphene oxides, quantum dots, etc., the properties of organic and inorganic systems have been considered after applying nano-objects to modify the bulk and interface features of these systems. Some evidence of the nanostructures influence on the photorefractive, photoconductive, and dynamic characteristics of the organic conjugated materials with initial donor-acceptor interaction has been shown. A correlation between photorefractive and photoconductive characteristics has been found, and the mechanisms responsible for this correlation are discussed. Moreover, the spectral and mechanical properties of the inorganic compounds have been shown under the conditions of the nano-objects modification.

Keywords: Materials properties modification, fullerenes, carbon nanotubes, shungites, graphene oxides, quantum dots, laser-matter interaction

7.1   Introduction

In the last decade there has been a systematic search for new optical materials and methods to optimize the properties of optoelectronic systems and telecommunication schemes, as well as laser,

Corresponding author: nvkamanina@mail.ru

Ashutosh Tiwari and S.K. Shukla (eds.) Advanced Carbon Materials and Technology, (273–316)
2014 © Scrivener Publishing LLC

display, solar energy, gas storage and biomedicine techniques. It can be shown that simple manufacturing, design, ecology points of view, etc., indicate the good advantages of nanostructured materials with improved photorefractive parameters among other organic and inorganic systems. Indeed, it should be stated that changes in photorefractive properties are correlated with the changes of the spectral, photoconductive and dynamic ones. The change in non-linear refraction and cubic nonlinearity reveals the modification of barrier free electron pathway and dipole polarizability. From one side it is connected with the change of the dipole moment and the charge carrier mobility, while from the other side, it is regarded to be the change of absorption cross section. Thus, these features show the unique place of photorefractive characteristics among the other ones in order to characterize the spectral, photoconductive, photorefractive and dynamic properties of the optical materials.

It is well known that promising nano-objects such as fullerenes, carbon nanotubes (CNTs), quantum dots (QDs), shungites, and graphenes are found to permit different areas of applications of these nano-objects [1–9]. The general view of some of the nano-objects indicated above is shown in Figure 7.1. All of these structures have many C–C bonds.

It is very interesting that the unique structure configuration responsible for the high mechanical features was predicted in the 18th century by Leonard Euler (1707–1783). A Swiss-born mathematician, physicist and astronomer, He was a member of the Russian Academy of Sciences from the period of 1726–1741 and from 1766 on. In his mathematical calculations he established that the relationship between the numbers of vertices (V), edges (E) and faces (S) for convex polyhedron should satisfy the condition of Equation 7.1 [10]:

$$V - E + F = 2 \qquad (7.1)$$

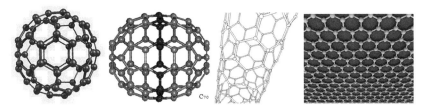

Figure 7.1 The structures of fullerene C_{60}, C_{70}, carbon nanotubes and graphenes (from left to right).

Figure 7.2 K.S. Melnikov's building in Moscow (from ref. [11]), created in 1929 (left), and R.B. Fuller's Biosphere (from ref. [12]) created in 1959 for the American Exhibition in Moscow (right).

In this case the maximum stable construction can be designed. For example, for fullerene C_{60} this equation is: 60–90+32=2. Thus, the chemical stability of the C_{60} molecule could be easily explained. It should be remarked that many architects have used the configuration containing the hexagons and pentagons in their architectural constructions. For example, the buildings of K.S. Melnikov and B. Fuller have applied Euler's geometry (see Figure 7.2).

It should be mentioned that the main reasons to use the fullerenes, shungites, and quantum dots are connected with their high value of electron affinity energy and unique energy levels. For example, the electron affinity energy of shungite structure is ~2 eV, the one for fullerenes is ~2.65 eV, and the one for quantum dots is 3.8–4.2 eV, which is larger than that of most dyes and organic molecule intramolecular acceptor fragment. It can stimulate the efficient charge transfer complex (CTC) formation in the nano-objects-doped organic conjugated materials.

Moreover, the energetic diagram permits the use of modern nanosensitizers at the visible and near-infrared spectral range. For example, for a molecule of fullerene C_{60} the energy scheme is shown in Figure 7.3. It provokes the effective light absorption due to an increase in the absorption cross section in the exited states [1–4]. The times of the singlet-triplet transition are shown in Table 7.1. This diagram is responsible for the effective attenuation of the laser beam in the fullerene-doped structures. Absorption increases

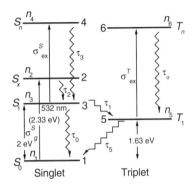

Figure 7.3 The energy levels of fullerenes C_{60}.

Table 7.1 Value of the transition and relaxation times of energy levels of fullerene C_{60}.

τ	Transition	Time
τ_0	$S_1 \rightarrow S_0$	650 picoseconds
τ_2	$S_x \rightarrow S_1$	~1 picosecond
τ_3	$S_n \rightarrow S_1$	~1 femtosecond
τ_6	$T_n \rightarrow T_1$	~1 femtosecond
τ_1	$S_1 \rightarrow T_1$	1.2 nanoseconds
τ_5	T_1	40±4 microseconds

with increasing incident energy because of the increase in the population of excited states of fullerenes [3, 4]. For example, for the nanosecond pulse laser region with the duration of laser pulses t_p ~ 10–20 ns, whereas the fullerene singlet–triplet interaction time is τ_1~1.2 ns, i.e., the inequality $t_p > t_{S1 \rightarrow T1}$ is satisfied and the fullerene molecules are accumulated in the excited triplet state. In this case, optical limitation of laser radiation occurs via the $T_n \rightarrow T_1$ channel.

Moreover, due to large surface energy the fullerenes and carbon nanotubes are provoked to organize the homogenous thin-film systems based on most organic conjugated materials, for example, polyimide compounds. The views of polyimide materials with fullerene skeleton are shown in Figure 7.4(a,b). The films were investigated with a scanning electron microscope (SEM) HU-11B

(a) (b)

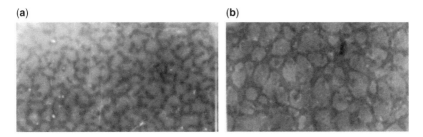

Figure 7.4 (a) SEM image of fullerene structures with different configuration, forming chains of quasi-pentagonal and hexagonal shapes. (b) SEM image of the polyimide film formed on the fullerene skeleton. The dimension is: 1 cm is equal to 200 nm.

at the accelerating voltage of 75 kV. The multiplication was 12500$^{\times}$. It should be noted that the dimension of fullerene molecules is close to 6.5–7Å; the diameter of carbon nanotubes can be placed in the range of 2–10 nm, but the length of nanotubes can be longer than 100 nm.

The basic features of carbon nanotubes and graphenes are attributed to their high conductivity and strong hardness of their C–C bonds, as well as their complicated and unique mechanisms of charge carrier moving. It should be remarked that Young's modulus of carbon nanotubes is close to 0.32–1.47 TPa [5, 6].

These peculiarities of carbon nano-objects and their possible optoelectronics, solar energy, gas storage, medicine, display and biology applications connected with dramatic improvement of photorefractive, spectral, photoconductive and dynamic parameters will be under consideration in this chapter. In comparison with other effective nano-objects, the main accent will be put namely on carbon nanotubes (CNTs) and their unique features for modifying the bulk and surface properties of optical materials.

7.2   Photorefractive Features of the Organic Materials with Carbon Nanoparticles

In the present paragraph the emphasis is placed on improvement of the photorefractive characteristics of conjugated organic materials doped by CNTs in comparison with the fullerenes, quantum dots, etc. The possible mechanism to increase the laser-induced change in the refractive index, nonlinear refractive index and cubic

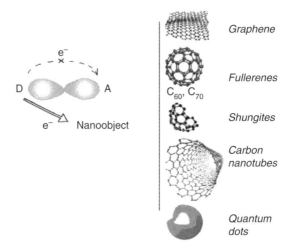

Figure 7.5 Schematic diagram of possible charge transfer pathways in organic molecule–nano-objects.

nonlinearity has been explained in the papers [8, 9, 13–15]. The dominant effect has been banded with intermolecular CTC formation. It is connected with the increased dipole moment of the nano-objects-doped systems in comparison with the pure ones, due to the increase of barrier-free electron pathway and the large value of electron affinity energy of the sensitized materials. The possible scheme of charge transfer is schematically depicted in Figure 7.5. Regarding CNTs, it was necessary to take into account the variety of charge transfer pathways, including those along and across a CNT, between CNTs, inside a multiwall CNT, between organic molecules and CNTs, and between the donor and acceptor moieties of an organic matrix molecule. These features of CNTs with added charge moving are shown in Figure 7.6(a). Indeed, the electron pathway in CNTs will depend on the placement of the CNTs' fullerene top at the different distance from donor part of the organic molecules doped with CNTs. The possible scheme shown in Figure 7.6(b) is when the intermolecular charge transfer process can dominate the intramolecular one.

It should be noted that the variations in the angle of nano-object orientation relative to the intramolecular donor can significantly change the pathway of charge carrier transfer, which will lead to changes in the electric field gradient, dipole moment (proportional to the product of charge and distance), and mobility of charge carriers. In addition, the barrier-free charge transfer will be influenced

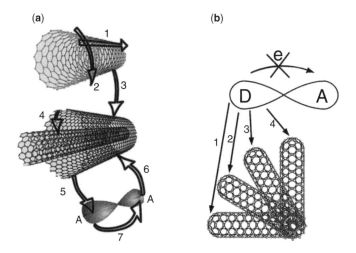

Figure 7.6 (a) Schematic diagram of possible charge transfer pathways in the organic molecule–CNTs using different ways for electron. (b) Schematic diagram of possible charge transfer pathways depending on the arrangement of introduced intermolecular acceptor relative to the intramolecular donor.

by competition between the diffusion and drift of carriers during the creation of diffraction patterns with various periods and, hence, different charge localization at the grating nodes and antinodes. Indeed, in the case of a nanocomposite irradiated at small spatial frequencies (large periods of recorded grating), a drift mechanism of the carrier spreading in the electric field of an intense radiation field will most probably predominate, while at large spatial frequencies (short periods of recorded grating) the dominating process is diffusion. As a result, the lower values of photoinduced changes in the refractive index of nanocomposites were observed at high spatial frequencies.

The systems doped with nano-objects can be studied using a four-wave mixing scheme analogous to that described previously [16] to study the dynamic parameters of the spatial light modulators (SLMs) based on the electro-optical liquid crystal (LC) media. The general view of the experimental scheme is shown in Figure 7.7. By monitoring the diffraction response manifested in this laser scheme, it is possible to study the dynamics of a high frequency Kerr effect to estimate the variation of the diffraction efficiency and the photoinduced change in the refractive index of a sample, and to calculate the nonlinear refraction and nonlinear third order optical susceptibility (cubic nonlinearity). An increase in the latter

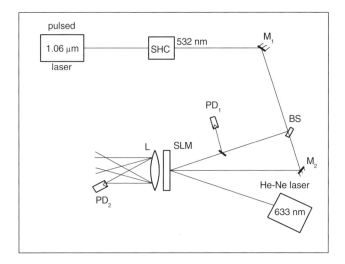

Figure 7.7 SHC – second harmonic converter; M_1 and M_2 – rotating mirrors; BS – beam-splitting mirror; PD_1 and PD_2 – photodetectors; LC–SLM (or thin films); L – lens.

parameter characterizes a change in the specific (per unit volume) local polarizability and, hence, in the macroscopic polarization of the entire system.

First let us provide some background. We can consider the high frequency Kerr effect in a simpler isotropic classical optical system, for example, as quartz, and then extend this consideration to the local volume of the nanostructured materials with dimensions significantly less than the laser wavelength-irradiated materials.

In nonlinear media, the lowest nontrivial nonlinearity is the cubic one. The matter equation of that medium is Equation 7.2 according to the paper [17],

$$P = \chi^{(1)}E + \chi^{(3)}E^3, \tag{7.2}$$

where P is nonlinear polarization of the systems, E is field intensity of the light beam, and $\chi^{(1)}$ and $\chi^{(3)}$ are linear and nonlinear optical susceptibilities, respectively.

In that approximation, the refractive index n is defined by Equation 7.3.

$$D = E + 4\pi P = \varepsilon E = n^2 E, \tag{7.3}$$

that yields

$$n = \sqrt{1 + 4\pi P / E}$$ (7.4)

With Equation 7.2 and neglecting the nonlinear term, one can obtain:

$$n = n_0 + \frac{2\pi}{n_0} \chi^{(3)} E^2,$$ (7.5)

where

$$n_0 = \sqrt{1 + 4\pi \chi^{(1)}},$$ (7.6)

We will use the equation for the light intensity in the form as $I = cE^2/8\pi$. Therefore,

$$n = n_0 + n_2 I,$$ (7.7)

where

$$n_2 = \frac{16\pi^2}{n_0 c} \chi^{(3)},$$ (7.8)

n_0 is the linear refractive index and c is the light velocity.

It follows from Equation 7.7 that the refractive index depends on the light intensity in the media with the cubic nonlinearity. This effect causes self-interaction of the light waves resulting in self-focusing of a light beam, phase self-modulation of pulses, etc. Thus, n_2 is an adequate characteristic of the cubic nonlinearity. The mechanism of an anisotropic molecule turn can result in the Equation 7.7 nonlinearity under the effect of the intense polarized lightwave. The process is quite slow in comparison to the electron polarizability of the medium. Because this mechanism provides the birefringence induced by the dc field (the Kerr effect), the dependence of the refractive index on the light intensity is the high-frequency Kerr effect, and the Equation 7.7 nonlinearity is the Kerr nonlinearity.

Let us return to the organic conjugated structures modified with nano-objects in order to use the consideration mentioned above for the systems with registered high cubic nonlinearity. As has been remarked, the main idea consists in the creation of an additional field gradient due to a more efficient charge transfer between the organic intramolecular donor fragment (inside the monomer, polymer, or liquid crystal structures) and the intermolecular acceptor (nano-objects based on the fullerenes, shungites, quantum dots, CNTs, etc.). By way of example, we consider a small local volume of our medium, substantially smaller than the incident wavelength. Indeed, for a system of dimensions smaller than the optical wavelength (532 nm in our experiment; for comparison, fullerene molecules are 0.65 ± 0.7 nm in size), the most important optical characteristic is the induced dipole, whose dependence on the applied local field can be expressed through dipole polarizabilities $\alpha^{(n)}$ [18]. These are in turn related by the proportional dependence to the nonlinear sensitivities $\chi^{(n)}$ and are inversely proportional to the considered unit cell volume υ in Equation 7.9.

$$\chi^{(n)} = \frac{a^{(n)}}{\upsilon} \qquad (7.9)$$

Based on the experimentally observed laser-induced change in the refractive index, we can consider the nano-objects-doped (fullerene-, nanotubes-, quantum dots-doped, etc.) organic π-conjugated systems as the materials with higher cubic optical susceptibility. The experiments were performed under the Raman–Nath diffraction conditions (where $\Lambda^{-1} \geq d$; Λ is the spatial frequency, Λ^{-1} indicates the grating period, and d is the thickness of the structure irradiated with laser beam) for thin gratings with spatial frequencies of 100 and 150 cm^{-1} recorded at an energy density varied within 0.05–0.8 J×cm^{-2}.

Model organic compounds based on polyimide (PI) and 2-cyclooctylamine-5-nitropyridine (COANP) have been studied. The structures of molecules mentioned above are shown in Figure 7.8(a,b). Triphenylamine fragment of polyimide [19] and NH-group of pyridine [20] compound are the donor fragments of these complicated organic conjugated molecules.

The thicknesses of thin-film samples were within 2–4 μm. The view of the thin films studied is shown in Figure 7.9. The organic matrices were sensitized by doping with commercially available fullerenes C$_{60}$ and C$_{70}$ (purchased from Alfa Aesar Company,

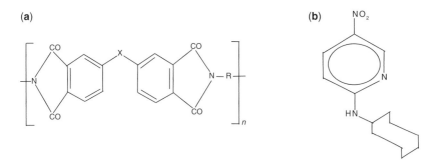

Figure 7.8 The structures of photosensitive polyimide (a) with electrodonor part R and pyridine (b) molecules with HN-group as donor fragment.

Figure 7.9 Photos of the polyimide (nk10) and pyridine (nk62) nanostructured films in comparison with the pure one (nk8).

Karlsruhe, Germany), and CNTs (received from various Russian institutions, among which were Vladimir State University, Vladimir, Russia and Boreskov Institute of Catalysis, SB RAS, Novosibirsk, Russia). The concentration of dopants was varied within 0.1–5 wt% for fullerenes and 0.1 wt% or below this value for CNTs.

Indeed, before revealing the nonlinear optical features, all systems have been treated via spectrometry and mass-spectrometry analysis [20]. Figure 7.10 shows the change in the mass spectra in, for example, the model polyimide and pyridine compounds after the nanosensibilization. Mass spectrometry data point out the formation of fullerene-triphenylamine charge transfer complex (CTC) formation for the polyimide–C_{70} system and of fullerene–HN group CTC for the COANP–C_{70} system. Figure 7.10 shows the rate of release of C_{70} from the sensitized polyimide and COANP upon heating. Curve 1 corresponds to a COANP sample with 5 wt% of C_{70}, and curve 2 belongs to polyimide with 0.5 wt% of C_{70}. Curve 1 exhibits two peaks. The first one, at a temperature of about 400°C, relates to the release of fragments with the mass of a fullerene

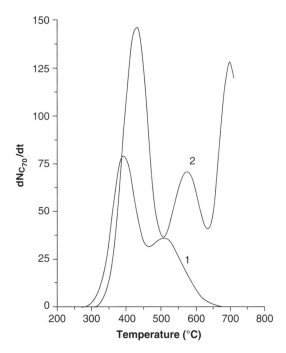

Figure 7.10 The rate of release of C_{70} molecules on heating of (1) COANP with 5 wt% of C_{70}, and (2) polyimide with 0.5 wt% of C_{70}.

molecule. The second peak is shifted to the temperature range around 520°C and apparently corresponds to the decomposition of HN group–fullerene complexes. The maximum release of the C_{70} molecules from the polyimide–C_{70} system (curve 2) is observed in three temperature ranges. The first peak, similarly to the COANP–C_{70} system, is observed near 400°C. The second peak located near 560°C is probably caused by the decomposition of triphenylamine–fullerene CTCs. It should be noted that the melting temperature these polyimides lie within is ~700–1000°C; therefore, the third peak at a temperature above 700°C corresponds to the total decomposition of the polyimide film itself.

The spectra shifts are shown in Figures 7.11 and 7.12. Comparison of curves 2 and 3 in Figure 7.11 for pure and fullerene-sensitized polyimide shows that fullerene causes a bathochromic shift of the absorption spectrum and an increase in the absorption intensity in the visible region. This facilitates the efficient excitation transfer between the fullerene and the polyimide and the increase in the number of excited states, which results, for example, in increasing

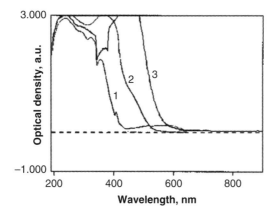

Figure 7.11 (1) Absorption spectrum of pure C_{60}; absorption spectra of 0.3% solutions of pure (2) C_{60} polyimide, and sensitized with 0.5 wt% of C_{60} polyimide (3) in chloroform. The ordinate is the optical density D.

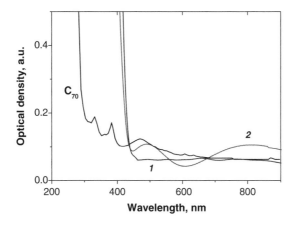

Figure 7.12 Absorption spectra of thin films of pure COANP (1), COANP–C_{70} system (2), and fullerene C_{70}.

of the photoconductivity. Upon addition of fullerene to the COANP solution, its absorption spectrum considerably changes. An increase in the optical density of the solution in the visible region and of additional absorption peaks near 500 and 800 nm are observed. The films change their color from light-yellow (pure COANP) to light-brown (COANP–C_{70}), which is also consistent with the formation of donor–acceptor complexes between the donor fragments of the COANP molecules (HN groups) and fullerene molecules. Remember that the fullerenes have large electron affinity energy (~2.65 eV) and can act as the strong acceptors for many organic molecules, including polyimide and COANP. It should be noted that the acceptor fragment of the COANP molecule is a NO_2 group bound to the donor fragment by a benzene ring. An individual NO_2 molecule or NO_2 radical typically has an electron affinity of 2.3 eV, but when bound to the benzene ring, the electron affinity of the NO_2 group becomes as low as 0.54 eV, which is lower than that of fullerene by a factor of more than four. Because of this, the fullerene molecule dominates over the NO_2 acceptor group in the COANP molecule, changing the intramolecular donor–acceptor interaction. The formation of the of COANP–C_{70} complexes is additionally confirmed by an increase in conductivity of fullerene-containing system by an order of magnitude.

The main nonlinear optical results of this study are summarized in refs. [21–24] in Table 7.2 (for the nanostructured COANP) and in Table 7.3 (for nanostructured polyimides). An analysis of data presented in Tables 7.2 and 7.3 for various organic systems shows that the introduction of nano-objects as active acceptors of electrons significantly influences the charge transfer under conditions where the intermolecular interaction predominates over the intramolecular donor–acceptor contacts. Indeed, the electron affinity of the acceptor fragments (which is close to 1.1–1.4 eV in polyimide-based composites and 0.4–0.54 eV in pyridine-based systems) is 2.5–5 times less than that of fullerenes (2.65–2.7 eV). Redistribution of the electron density during the recording of gratings in nanostructured materials changes the refractive index by at least an order of magnitude as compared to that in the initial matrix. This results in the formation of a clear interference pattern with a distribution of diffraction orders shown in Figure 7.13. The diffusion of carriers from the bright to dark region during the laser recording of the interference pattern proceeds in three (rather than two) dimensions, which is manifested by a difference in the distribution of diffraction orders along the horizontal and vertical axes. Thus, the grating displacement takes place in a three-dimensional (3D) medium formed as a result of the nanostructurization (rather than in a 2D medium).

The light-induced refractive index change Δn_i in the thin nano-objects-doped films could be estimated from the experimental data of increase in diffraction efficiency using the Equation 7.10 [25].

$$\eta = I_1 / I_0 = \left(\pi \, \Delta n_i d / 2\lambda \right)^2$$

(7.10)

Table 7.2 Laser-induced change in the refractive index in the systems based on COANP.

Structures	Nanoobjects contents, wt%	Wave-length, nm	Energy density, $J \cdot m^{-2}$	Laser pulse width, ns	Change in the refractive index, Δn
Pure COANP	0	532	0.9	10-20	10^{-5}
COANP+dye [21]	0.1	676	$2.2\ W \cdot m^{-2}$		2×10^{-5}
COANP+C_{60}	5	532	0.9	10-20	6.21×10^{-3}
COANP+C_{70}	5	532	0.9	10-20	6.89×10^{-3}
Polymer-dispersed liquid crystal (PDLC) based on COANP+C_{70}	5	532	17.5×10^{-3}	10-20	1.4×10^{-3}

Table 7.3 Laser-induced change of the refractive index of the sensitized polyimides.

Structure studied	Nanoobjects content, wt.%	Wave-length, nm	Energy density, $J \times cm^{-2}$	Spatial frequency, mm^{-1}	Laser pulse width, ns	Laser-induced change in the refractive index, Δn_i	References
Pure polyimide	0	532	0.6	90	20	10^{-4}-10^{-5}	(14)
Polyimide+malachite green	0.2	532	0.5-0.6	90-100	10-20	2.87×10^{-4}	(8)
Polyimide+QDs CdSe(ZnS)	0.003	532	0.2-0.3	90-100		2.0×10^{-3}	(22)
Polyimide+shungite	0.2	532	0.063-0.1	150	10	3.8-5.3×10^{-3}	(23)
Polyimide+C_{60}	0.2	532	0.5-0.6	90	10-20	4.2×10^{-3}	(14)
Polyimide+C_{70}	0.2	532	0.6	90	10-20	4.68×10^{-3}	(14)
Polyimide+nanotubes	0.1	532	0.5-0.8	90	10-20	5.7×10^{-3}	(14)
Polyimide+nanotubes	0.05	532	0.3	150	10	4.5×10^{-3}	(24)
Polyimide+nanotubes	0.07	532	0.3	150	10	5.0×10^{-3}	(24)
Polyimide+nanotubes	0.1	532	0.3	150	10	5.5×10^{-3}	(24)

Structure studied	Nanoobjects content, wt.%	Wavelength, nm	Energy density, $J \times cm^{-2}$	Spatial frequency, mm^{-1}	Laser pulse width, ns	Laser-induced change in the refractive index, Δn_i	References
Polyimide+double-walled CNT powder	0.1	532	0.063-0.1	100	10	9.4×10^{-3}	(15)
Polyimide+double-walled CNT powder	0.1	532	0.063-0.1	150	10	7.0×10^{-3}	(15)
Polyimide+carbon nanofibers (type MIG)	0.1	532	0.6	90-100	10	11.7×10^{-3}	(23)
Polyimide+carbon nanofibers (type MIG)	0.1	532	0.3-0.6	150	10	11.2×10^{-3}	(15)
Polyimide+carbon nanofibers (type 65BR)	0.1	532	0.1-0.3	90-100	10	12.0×10^{-3}	(23)
Polyimide+carbon nanofibers (type 65BR)	0.1	532	0.1	90	10	15.2×10^{-3}	(15)

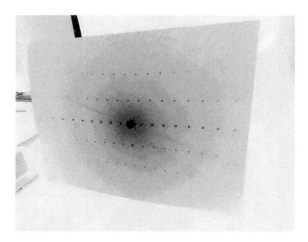

Figure 7.13 The visualization of the diffraction response in the organics doped with nano-objects.

Where η is the diffraction efficiency, I_1 is the intensity of the first diffraction order, I_0 is the incident laser beam, d is the film thickness, and λ is the laser wavelength.

It should be noted that the thermal part of delta n in the materials studied is close to a value of 10^{-5}. Thus, the increase in the diffraction efficiency in the current experiments and, hence, in the light-induced refractive index change could be explained by the high frequency Kerr photorefractive effect (see Equation 7.9) stimulated by intermolecular charge transfer processes in these compounds.

In addition, the data in Table 7.3 show that the introduced CNTs produce almost the same change in the refractive properties as do fullerenes, at a much lower percentage content of the CNTs as compared to that of C_{60} and C_{70}, and for the irradiation at higher spatial frequencies (150 mm^{-1} for CNTs versus 90–100 mm^{-1} for fullerenes). This implies that the possibility of various charge transfer mechanisms in the systems with CNTs is quite acceptable and may correspond to that depicted in Figure 7.6(a).

Using the obtained results, we have calculated the nonlinear refraction n_2 and nonlinear third order optical susceptibility (cubic nonlinearity) $\chi^{(3)}$ for all systems using the method described above. It was found that these parameters fall within $n_2 = 10^{-10}$–10^{-9} cm$^2 \times$W^{-1} and $\chi^{(3)} = 10^{-10}$–10^{-9} cm$^3 \times$erg^{-1}.

It should be remarked that all results (see Table 7.3) have been obtained in the reversible mode close to the threshold before the nonreversible one. The view of the nonreversible grating recorded

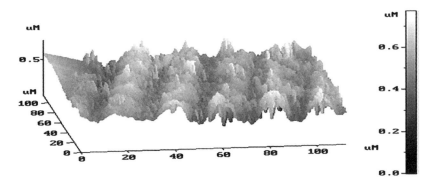

Figure 7.14 The visualization of the grating recorded at the energy density extended energy threshold.

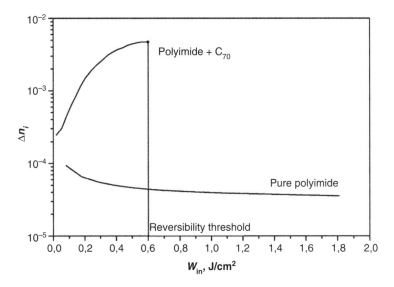

Figure 7.15 The threshold (vertical line) between the reversible and nonreversible process when thin holographic grating has been recorded on polyimide structures.

and established via atomic force microscopy method is shown in Figure 7.14. This case corresponded to that where the laser energy density increased drastically and extended the threshold level. The thermal-induced grating can be registered. Some evidence of the threshold obtained for polyimide-based materials is shown in Figure 7.15 (vertical line). Some calculation results for n_2 and $\chi^{(3)}$ are shown in Table 7.4 [26–39] with comparison with other optical

Table 7.4 Nonlinear refraction coefficient and cubic nonlinearity of conjugated materials.

Structure	n_2, cm^2 W^{-1}	$\chi^{(3)}$, cm^3 erg^{-1} (esu)	References
CS_2	3×10^{-14}	10^{-12}	(17)
SiO_2	3×10^{-16}	10^{-14}	(17)
C_{60} film		0.7×10^{-11}	(26)
C_{60} film		8.7×10^{-11}	(27)
C_{60} film		2×10^{-10}	(28)
C_{70} film		1.2×10^{-11}	(29)
C_{70} film		2.6×10^{-11}	(27)
Cu - phthalocyanine		$2.1\pm0.2\times10^{-12}$	(30)
Pb - phthalocyanine		2×10^{-11}	(31)
α-TiO- phthalocyanine		1.59×10^{-10}	(32)
bis- phthalocyanine	-2.87×10^{-9}	$2-5\times10^{-9}$	(33)
Polyimide - C_{70}	0.78×10^{-10}	2.64×10^{-9}	(34,35)
Polyimide - C_{70}	-1.2×10^{-9}	1.9×10^{-10}	(36)
Polyimide+nanotubes (0.05 wt.% CNTs)	0.15×10^{-10}	0.51×10^{-9}	(9)
Polyimide+nanotubes (0.07 wt.% CNTs)	0.17×10^{-10}	0.6×10^{-9}	(9)
Polyimide+nanotubes (0.1 wt.% CNTs)	0.18×10^{-10}	0.62×10^{-9}	(9)
COANP- C_{60}	$0,69\times10^{-10}$	$2,14\times10^{-9}$	(37)
COANP - C_{70}	$0,77\times10^{-10}$	$2,4\times10^{-9}$	(37,38)
Polymer-dispersed LC based on COANP- C_{70}	1.6×10^{-9}	4.86×10^{-8}	(39)
Polymer-dispersed LC based on PANI-C_{60} (1.0 wt.% C_{60})	0.04×10^{-10}		(9)

Structure	n_2, cm^2 W^{-1}	$\chi^{(3)}$, cm^3 erg^{-1} (esu)	References
Polymer-dispersed LC based on PANI-C$_{60}$ (6.0 wt.% C$_{60}$)	0.1×10^{-10}		(9)
Si	10^{-10}	10^{-8}	(17)
Liquid crystal	10^{-4}	10^{-3}	(17)

and nonlinear optical materials, including, for example, polyaniline (PANI) compound, which is traditionally applied in the solar energy and gas storage systems.

One can see from the data in Table 7.4 that the nonlinear optical parameters of the nano-objects-doped conjugated structures are larger than those obtained for traditional nonlinear systems that permit the application of these materials as effective holographic recording element, spatial light modulator, switchers, and nonlinear absorber in the visible and near-infrared spectral ranges. Moreover, the data testify that the nonlinear characteristics of the materials studied are close to the Si-based structures and also to some semiconductor materials (see Figure 7.16) that provoke the organic conjugated structures with nano-objects to be used in organic solar energy technique.

It should be mentioned that the observed refractive features can be considered as an additional mechanism responsible for the optical limiting effects. Indeed, the energy losses due to refraction provoke the attenuation of the laser beam transferred through nonlinear optical media in order to protect the human eye and technical devices from high laser irradiation. Moreover, it should be remarked that photorefractive characteristics are correlated with the photoconductive one. The data shown in Table 7.5 support this correlation. Next the charge carrier mobility has been estimated using the Child–Langmuir current-voltage relationship analogous to that shown in paper [15]. We have calculated the absolute values of the charge carrier mobility, μ, and then compared it with the one for pure polyimide films. It should be noted that the estimation has been made at a thin-film thickness of $d = 2$ micrometer, a dielectric

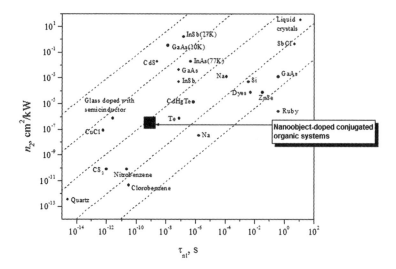

Figure 7.16 A visual depiction of the studied materials placed between other optoelectronics structures shown in ref. [17].

Table 7.5 Volt-current characteristics of model conjugated polyimide materials doped with different types of nano-objects.

Bias voltage, V	Current, A, under light irradiation			
	Pure PI	PI+0.2wt.%C$_{70}$	PI+0.1 wt.% CNTs	PI+0.003wt.% QD
0	2.77×10^{-12}	4.33×10^{-12}		
1			Close to 10^{-4} - 10^{-3}	
5	2.98×10^{-11}	5.0×10^{-11}		
10	6.96×10^{-11}	1.1×10^{-10}		
15	1.01×10^{-10}	2.3×10^{-10}		
20	1.44×10^{-10}	3.7×10^{-10}		
30	2.5×10^{-10}	8.0×10^{-10}		Close to 10^{-9} - 10^{-8}
40	3.8×10^{-10}	1.4×10^{-9}		

constant of $\varepsilon \cong 3.3$, and an upper electrode contact area with a diameter of 2 mm. The results of the calculations of the changes in the carrier mobility at the broad range of the bias voltage, V, testify to the dramatic increase of charge carrier mobility that is more than the one previously observed for the pure polyimide organic systems. Two orders of magnitude difference of charge carrier mobility for the pure polyimide and for the nano-objects-doped polyimide have been found. It probably also supports the prognosis for the perspective of use of the thin-film organic nanostructured materials instead of bulk inorganic compounds in different civil areas of applications. Indeed, it can be possible under the condition when the electro-optical, the nonlinear optical, and the photoconductive characteristics are the same.

7.3   Homeotropic Alignment of the Nematic Liquid Crystals Using Carbon Nanotubes

It is well known that electro-optical nematic liquid crystal (NLC) structure is a good model in order to consider the realistic technical cells as laser radiation switching devices, electrically and optically addressed spatial light modulators, and analogs of display elements [40–45]. Mostly liquid crystal (LC) cells, namely from the cyanobiphenyl group (see Figure 7.17[a,b]), operate in S- and T- configurations that realize a planar orientation of the LC mesophase on the aligning substrate surface.

Figure 7.18 shows various types of alignments of LC molecules on a substrate surface, including planar (molecules are aligned parallel to the substrate surface, $\theta = 0$, see Figure 7.18[a]), homeotropic (molecules are perpendicular to the substrate surface, $\theta = 90°$, see Figure 7.18[b]), and tilted (LC director is tilted at a certain angle, $0 < \theta < 90°$, see Figure 7.18[c]) orientations of LC molecules.

The planar and the tilted orientations can be achieved using some oxides and polymer alignment coatings, such as cerium oxide (CeO), silicon oxides (SiO, SiO_2), germanium oxide (GeO), poly(vinyl alcohol), and polyimide nonphotosensitive coatings. As is well known, rubbing of the glass substrates or irradiation of them by holographic methods also leads to the planar orientation of LC molecules [40–46]. Homeotropic alignment is frequently obtained using surfactants, such as lecithin, fused quartz, etc. Sometimes irradiation of polyimide by UV light provokes the homeotropic

(a)   (b)

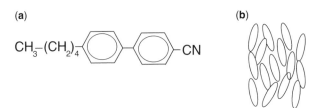

Figure 7.17 Structure (a) and ordering (b) of nematic liquid crystal molecules.

(a)   (b)   (c)

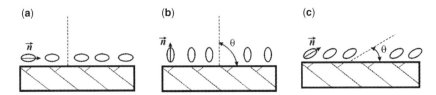

Figure 7.18 Orientation of LC molecules on a substrate surface in cases of (a) planar, (b) homeotropic, and (c) tilted orientations.

alignment. A new alternative method for obtaining a surface nanorelief that favors the homeotropic alignment of an LC mesophase is offered by nanoimprinting technology [47]. Realization of this method, while making possible the formation of a surface relief with a good optical quality, requires the use of toxic substances, in particular, acids. It is a disadvantage of the nanoimprinting method.

The importance of determining the type of LC alignment is related to the fact that the anchoring energy of LC molecules on a substrate and the alignment conditions significantly influence all physical properties of the LC mesophase and the electro-optical characteristics of related devices, such as optically-addressed LC SLMs. For example, the free surface energy density F_s in traditional approximation based on Equation 7.11, is related to the surface anchoring energy W_s and the LC direct tilt angle θ as follows [44]:

$$F_s = \frac{1}{2} W_s \sin^2 \theta$$

(7.11)

Therefore, various types of the alignment of LC molecules on a substrate correspond to different values of anchoring energy and different conditions for choosing a compromise between viscoelastic and dielectric forces applied to the LC mesophase.

Figure 7.18(b) shows the homeotropic alignment of LC molecules on a solid substrate. In crossed polarizers, light will not be transmitted through the cell for the vertical orientation of molecules relative to the analyzer in the absence of a bias voltage; application of the voltage leads to rotation of the LC dipoles by 90°, after which the light is transmitted to form a bright spot on a screen.

Because the solution of some problems requires obtaining namely the initial black field, the problem of finding a new method to orient LC molecules in the homeotropic mode is important. Recently the perspective method for the homeotropic alignment of LC molecules has been proposed [48]. It is based on a contactless technique of relief formation on the surface of a glass (quartz) substrate using the deposition of carbon nanotubes (CNTs) and their additional orientation in an electric field. The procedure can be briefly described as follows. The glass or quartz substrates were covered with ITO contact and then with CNTs using laser deposition technique. Additionally, CNTs were oriented at the electric field close to 100–250 V×cm^{-1}. To decrease the roughness of the relief the surface electromagnetic wave (SEW) treatment was used. The SEW source was a quasi-CW gap CO_2 laser generating p-polarized radiation with a wavelength of 10.6 micrometers and a power of 30 W. The skin layer thickness was ~0.05 micrometers. The reliefs obtained before and after SEW treatment of the CNT layers are shown in Figures 7.19(a,b).

The homeotropic alignment of the LC molecules was studied using two sandwich-type cells with a nematic LC mesophase confined between two glass plates. The general view of the LC cells is shown in Figure 7.20. The classical view of the construction is shown in Figure 7.21.

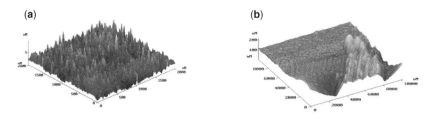

Figure 7.19 New orientation reliefs obtained after laser deposition oriented CNTs: before (a) and after (b) SEW treatment.

Figure 7.20 General pictures of the LC cells or LC modulator.

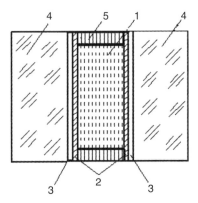

Figure 7.21 The structural scheme of LC cell: 1 – LC layer, 2 – orienting layers, 3 – conducting layers, 4 – glass or quartz substrates, 5 – spacers.

The reference cell represented a classical nematic LC structure, in which the alignment surfaces were prepared by rubbing of non-photosensitive polyimide layers. In the experimental cell, both alignment surfaces were prepared using CNTs as mentioned above. Experiments with the second cell confirmed that a homeotropic alignment of the LC molecules was achieved. The results of these experiments are presented in the Table 7.6.

The spectral data in a 400–860 wavelength range were obtained using an SF-26 spectrophotometer. Reference samples with planar alignments and homeotropically-aligned samples prepared using the new nanotechnology of substrate surface processing were mounted in a holder and the transmission of both cells was simultaneously measured at every wavelength in the indicated range. It should be noted that the two cells had the same thickness of 10 μm and contained the same nematic LC composition belonging to the class of cyanobiphenyls. The experimental data were reproduced in several sets of cells.

Table 7.6 Optical transmission of LC cells prepared using different methods of alignment.

Wavelength, nm	Transmission, % reference cell (planar)	Transmission, % experimental cell (homeotropic)	Wavelength, nm	Transmission, % reference cell (planar)	Transmission, % experimental cell (homeotropic)
400	0	0	620	24.1	0.4
410	2.8	0	630	23.8	0.4
420	11.1	0	640	23.3	0.4
430	19	0	650	22.5	0.3
440	23.2	0	660	22.1	0.3
450	25.6	0.5	670	21.7	0.3
460	26.7	0.5	680	21.4	0.3
470	27.3	0.5	690	21.0	0.3
480	27.6	0.5	700	20.6	0.3
490	27.5	0.5	710	20.2	0.3
500	27.4	0.5	720	20.0	0.3
510	27.2	0.5	730	19.3	0.3

(continued)

Table 7.6 (*Cont.*)

Wavelength, nm	Transmission, %		Wavelength, nm	Transmission, %	
	reference cell (planar)	experimental cell (homeotropic)		reference cell (planar)	experimental cell (homeotropic)
520	27.0	0.5	740	19.0	0.3
530	26.8	0.5	750	18.6	0.3
540	26.6	0.5	760	18.2	0.3
550	26.3	0.4	770	17.9	0.3
560	26.1	0.4	780	17.4	0.3
570	25.8	0.4	790	17.1	0.3
580	25.6	0.4	800	16.7	0.3
590	25.2	0.4	820	16.1	0.3

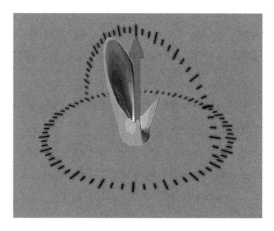

Figure 7.22 The model view of the LC molecules orientation along the direction of some degrees from the vertical axes.

Thus, the new features of CNTs have been demonstrated in order to obtain the initial black field of LC cells using a homeotropic orientation of LC molecules on the CNTs relief. It should be remarked that the same spectral peculiarities can be obtained when the tilt of the LC director from vertical axes will be in the range of 5–15 degrees. The model picture to show the LC orientation under the condition of the CNTs relief SEW treatment is shown in Figure 7.22.

The results of this investigation can be used both to develop optical elements for displays with vertical orientation of NLC molecules (for example, in MVA-display technology) and to use as laser switchers.

7.4   Thin Film Polarization Elements and Their Nanostructurization via CNTs

The functioning of various optoelectronic devices implies the use of polarization elements. As it is known, the operation of polarization elements is based on the transverse orientation of fields in electromagnetic waves. Polarization devices transmit one component of the natural light, which is parallel to the polarizer axis, and retard the other (orthogonal) component. There are two main approaches to creating thin film polarizers. The first method employs metal stripes deposited onto a polymer base. The metal layer reflects the incident light, while the polymer film transmits and partly absorbs the light, so that only the light of a certain polarization is transmitted. Another method is based on the creation of polymer-dispersed compositions, e.g., iodinated poly(vinyl alcohol) (PVA) films, which transmit the parallel component of the incident light and absorb the orthogonal component. Thus, the principle of operation of the iodinated PVA film polarizer is based on the dichroism of light absorption in PVA-iodine complexes.

This paragraph considers the possibility of improving the optical and mechanical properties of iodine-PVA thin film polarizers by the application of modern nano-objects carbon nanotubes. The polarizers' structures comprising iodinated PVA films with a thickness of 60–80 μm, which were coated from both sides by ~0.05μ-thick layers of single-walled CNTs have been studied. The polarizers contained polarization films with either parallel or mutually perpendicular (crossed) orientations, depending on the need to obtain the initial bright or dark field. It was found that the modification (nanostructuring) of the surface of polarization films by CNTs led to some increase in the optical transmission for the parallel component of incident light (see upper curves of Figure 7.23), while

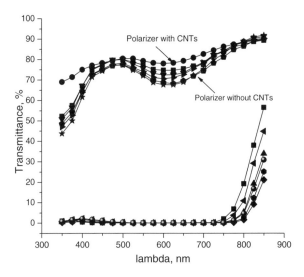

Figure 7.23 Dependence of percentage transmission versus wavelength of parallel (upper curves) and orthogonal (lower curves) polarized light for iodinated PVA film polarizers without and with CNTs-modified nanostructured surfaces.

retaining minimum transmission for the orthogonal component (see lower curves of Figure 7.23).

It should be mentioned that CNTs were laser deposited onto the surface of polarization films in vacuum using p-polarized radiation of a quasi-CW CO_2 laser. During the deposition, the CNTs were oriented by applying an electric field with a strength of 50–150 V×cm^{-1}. For the first time, the method of laser deposition of CNTs onto iodine-PVA structures has been described in paper [49]. The optical transmission measurements were performed using an SF-26 spectrophotometer operating in a 200–1200 nm wavelength range. The results of spectral measurements were checked using calibrated optical filters. The error of transmission measurements has not exceeded 0.2%.

The nanostructuring of the surface of polarization films by CNTs led to a 2–5% increase in the optical transmission in the visible spectral range for the parallel component of incident light, while retaining minimum transmission for the orthogonal component. This result is evidently related to the fact that the deposition of CNTs onto the surface of iodinated PVA films modifies the properties of air–film interface and reduces the reflection losses. The losses are decreased due to the Fresnel effect, which is related to a small refractive index of CNTs ($n \sim 1.1$). In a spectral interval of 400–750 nm, the CNT-modified iodinated PVA films ensured the transmission of the parallel component on a level of 55–80%.

An additional feature of a new method to deposit the CNTs onto both surfaces of polarization films is based on the improved mechanical protection of the films. Indeed, the standard approach to mechanical protection of polymeric polarization films against scratching and bending consists of gluing polarizers between plates of K8 silica glass or pressing them into triacetatecellulose. The proposed method of surface nanostructuring increases the surface hardness, while retaining the initial film shape that is especially important in optoelectronic devices for reducing aberrations in optical channels and obtaining undistorted signals in display pixels. The increase in the surface hardness is apparently related to the covalent binding of carbon nano-objects to the substrate surface, which ensures strengthening due to the formation of a large number of strong C–C bonds of the CNTs, which are difficult to destroy. The data supporting the surface mechanical hardness improvement via CNTs used are shown in Table 7.7.

Table 7.7 Comparative data of polarizer microhardness improvement.

Systems/ number of measurements	Microhardness, $\Pi \times 10^9$							Middle value	Coefficient of improve-ment
	1	2	3	4	5	6	7		
Pure polarizing film	0.191	0.154	0.148	0.175	0.182	0.182	0.154	0.1694	2.587
Nanostructured polarizing film	0.492	0.458	0.458	0.402	0.376	0.354	0.558	0.4383	

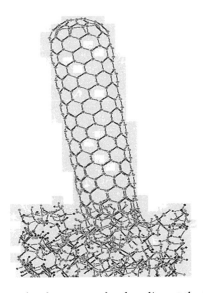

Figure 7.24 Illustration of carbon nanotubes bonding at the PVA surface.

The quantum chemical simulation results illustrate the ability to improve the hardness [50]. The barrier of joining single-wall CNT (10,0) fragment to the surface of PVA film has been calculated, which amounted to ~25 eV. This value is the difference between the energy (10,0)NT+PVA system (PVA 400 atoms fragment of and NTF of 350 C-atoms), when the CNT distance above PVA surface is 1 nm, and the energy of the system in the case of the distance of 0.2 nm, after which CNT could attach to the PVA surface (see Figure 7.24). Using this, it has been estimated that the rate of CNTs bombarding of PVA surface was subsequently introduced in the PVA film. The speed is equaled to ~1000 m/s. It has been estimated the Young's modulus of the composite fragment (12 SWNTs was inculcated one to another in PVA on ~1 nm depth and covalently bonded with PVA molecules) as 10 GPa, that is 10 times more than modulus of pure PVA. All calculations of optimized structures were made by using the molecular dynamics method with GULP package and TBDFT program [51].

The CNT-modified thin film polarizers can be employed in optical instrumentation, laser, telecommunication and display technologies, and medicine. These polarizers can also be used in devices protecting the eyes of welders and pilots against optical damage and in crossed polaroids (polarization films) based on liquid crystals.

7.5   Spectral and Mechanical Properties of the Inorganic Materials via CNTs Application

It is a complex task to modify the inorganic optical materials operated as output window in the UV lamp and laser resonators, as polarizer in the telecommunications, display and medicine systems. Many scientific and technological groups have made some steps to reveal the improved characteristics of optical materials to obtain good mechanical hardness, laser strength, and wide spectral range. Our own steps in this direction have been firstly shown in papers [52, 53]. In order to reveal the efficient nano-objects' influence on the materials surface it is necessary to choose the model system.

It should be noted that magnesium fluoride has been considered as good model system. For this structure the spectral characteristics, atomic force microscopy data, and measurements to estimate the hardness and roughness have been found in good connection. The main aspect has been made on interaction between nanotubes (their C–C bonds) placed at the MgF_2 surface via covalent bonding [53, 54]. Table 7.8 presents the results of surface mechanical hardness of MgF_2 structure after nanotubes placement; Table 7.9 shows the decrease of MgF_2 roughness. One can see from Table 7.8 that the nanostructured samples reveal the better surface hardness. For example, after nanotubes placement at the MgF_2 surface, the surface hardness has been up to 3 times better in comparison with sample without nano-objects. It should be noted that for the organic glasses this parameter can be increased up to one order of magnitude.

Moreover, the roughness of the MgF_2 covered with nanotubes and treated with surface electromagnetic waves has essentially

Table 7.8 Abrasive surface hardness of the MgF_2 structure before and after CNT modifications.

Structures	Abrasive surface hardness (number of cycles before visualization of the powder from surface)	Remarks
MgF_2	1000 cycles	CM-55 instrument has been used. The test has been made using silicon glass K8 as etalon. This etalon permits the obtainment of abrasive hardness close to zero at 3000 cycle with forces on indenter close to 100 g.
MgF_2 + nanotubes	3000 cycles	
MgF_2 + vertically oriented CNTs	more than zero hardness	

Table 7.9 Roughness of the MgF_2 structure before and after CNTs modifications.

Parameters	Materials	Roughness before nanotreatment	Roughness after nanotreatment	Remarks
R_a	MgF_2	6.2	2.7	The area of 5000×5000 nm has been studied via AFM method
S_q	MgF_2	8.4	3.6	

been improved. Indeed, R_a and S_q roughness characteristics have been decreased up to three times. One can see from Table 7.9 that the deposition of the oriented nanotubes on the materials surface decreases the roughness dramatically. Indeed, this process is connected with the nature of the pure materials; it depends on the crystalline axis and the defects in the volume of the materials.

In order to explain the observed increase of mechanical hardness we have compared the forces and energy to bend and to remove the nanotubes, which can be connected with magnesium fluoride via covalent bond MgC. Thus, the full energy responsible for the destruction of the surface with nanotubes should be equal to the sum of W_{rem} (energy to remove the layer of nanotubes) and of W_{destr} (energy to destroy the magnesium fluoride surface). Due to the experimental fact that the nanotubes covering increases drastically the surface hardness of MgF_2 (53,54), the values of W_{rem} and W_{destr} can be close to each other. Under the conditions of the applied forces parallel to the surface, in order to remove the nanotubes from MgF_2 surface, firstly, one should bend these nanotubes, and secondly, remove these nanotubes. In this case, W_{rem} are consisted of W_{elast} (elasticity energy of nanotube) plus W_{MgC} (energy to destroy the covalent MgC binding). For one nanotubes, using the CNTs data [55] and our calculation [54], $W_{elast} = 1.8 \times 10^{-20}$ J and $W_{MgC} = 0.4 \times 10^{-20}$ J. Thus, one should say that in order to break the relief with nanotubes, we should firstly bend the nanotubes with energy that is five times more than the one, which can be applied to simply remove the nanotubes from the surface after destroying the MgC binding. This fact is in good connection with the experimental results.

This calculation can be used to explain the results of the dramatically increased mechanical surface hardness of the MgF_2 covered with nanotubes. The experimental data testified that the surface mechanical hardness of MgF_2 materials covered with nanotubes can be compared with the hardness of etalon based on silicon glass K8. As a result of this process, the refractive index can also be modified, which explains the increase in transparency in the UV. Really, the refractive index of CNTs is close to 1.1, but the refractive index of MgF_2 or other fluoride model materials extends the value to 1.3. Thus, the Fresnel losses can be decreased in comparison with those obtained for MgF_2 without the CNTs bonding at the interface. Moreover, the spectral range saving or increasing in the IR range can be explained based on the fact that the imaginary part of the dielectric constant of carbon nanotubes, which is responsible for the absorption, is minimum (close to zero) in the IR range. The UV-VIS and near IR-spectra of the magnesium fluoride is shown in Figure 7.25(a). The IR spectra for the second model structure, namely, BaF_2 materials, are shown in Figure 7.25(b). Moreover, it should be noted that a drastic increase in the transparency at a wavelength of 126 nm has been observed. Indeed, for the 5 units of MgF_2 sample, the transparency T has been changed after nanotubes deposition as follows: sample No1. $T= 61.8\% \rightarrow T=66.6\%1$ No2. $T=63.6\% \rightarrow T=69\%$; No3. $T=54.5\% \rightarrow T=65.8\%$; No4. $T=58.1\% \rightarrow T=67.5\%$; No5. $T=50.9\% \rightarrow T=65\%$.

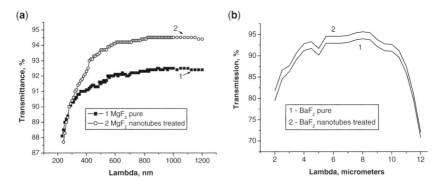

Figure 7.25 UV-VIS-near IR spectra of MgF$_2$ (a) before (curve 1) and after single-wall nanotubes deposition (curve 2). IR spectra of BaF$_2$ (b) before (curve 1) and after single-wall nanotubes deposition (curve 2). The thickness of the samples were close to 2 mm.

Thus, it should be mentioned that the CNTs are good candidates to modify the surface properties of the optical inorganic materials in order to obtain a good advantage in the hardness and spectra.

7.6   Conclusion

Analysis of the obtained results leads to the following summary remarks:

- Doping with nano-objects significantly influences the photorefractive properties of nanobjects-doped organic matrices. An increase in the electron affinity (cf. fullerenes, shungite, QDs) and specific area (cf. QDs, CNTs, nanofibers, graphenes) implies a dominant role of the intermolecular processes leading to an increase in the dipole moment, local polarizability (per unit volume) of medium, and mobility of charge carriers.
- The variations of the length, surface energy, and angle of nano-object orientation relative to the intramolecular donor fragment of matrix organics can significantly change the pathway of charge carrier transfer, which will lead to changes in the electric field gradient and dipole moment.
- Different values of nonlinear optical characteristics in systems with the same sensitizer type and

concentration can be related to a competition between carrier drift and diffusion processes in a nanocomposite under the action of laser radiation.

- The special role of the dipole moment as a macroscopic parameter of a medium accounts for a relationship between the photorefraction and the photoconductivity characteristics. Thus, the photorefractive parameters change can be considered as the indicator of the following dynamic and photoconductive characteristics change.
- The nano-objects-modified interface between solid and LC permits a decrease in the bias voltage, to increase the transparency and to develop both the light addressed spatial light modulator and laser switchers, as well as the new optical elements for displays with vertical orientation of NLC molecules. This method can be considered as an alternative one for MVA-display technology.
- The nano-objects-modified thin film polarizers can be employed in optical instrumentation, laser, telecommunication, and display technologies, and medicine. These polarizers can also be used in devices protecting the eyes of welders and pilots against optical damage and in crossed polaroids (polarization films) based on liquid crystals.
- The nano-objects-modified inorganic materials surface predicts the increase in mechanical hardness and improves the spectral characteristics. It can be useful in the automobile industry, biophysics and biomedicine.

Acknowledgments

The author would like to thank her Russian colleagues: Prof. E.F. Sheka (University of Peoples' Friendship, Moscow, Russia), Prof. L.A. Chernozatonskii (Emanuel Institute of Biochemical Physics, Russian Academy of Sciences, Moscow, Russia), Prof. V.I. Berendyaev (Karpov Research Institute, Moscow, Russia), Prof. A.I. Plekhanov (Institute of Automation and Electrometry SB RAS, Novosibirsk, Russia), Prof. N.M. Shmidt (Ioffe Physical-Technical Institute, St. Petersburg, Russia), Dr. K. Yu. Bogdanov (Lyceum

No.1586, Moscow, Russia), as well as her foreign colleagues, Prof. Francois Kajzar (Université d'Angers, Angers, France), Prof. D.P. Uskokovic (Institute of Technical Sciences of the Serbian Academy of Sciences and Arts, Belgrade, Serbia), Prof. Iwan Kityk (Electrical Engineering Department, Czestochowa University of Technology, Czestochowa, Poland) for their help in scientific discussions. The author would also like to acknowledge her lab colleagues: Dr. V.I. Studeonov, Dr. P. Ya.Vasilyev, PhD students S.V. Serov, N.A. Shurpo, S.V. Likhomanova, P.V. Kuzhakov (Vavilov State Optical Institute, St. Petersburg, Russia) for their participation and help at different steps in this study. Finally, the author would also like to thank Dr. V.E. Vaganov (Vladimir State University) and Dr. I.V. Mishakov (Boreskov Institute of Catalysis, Siberian Branch of Russian Academy of Sciences, Novosibirsk) for presenting the CNTs as types MIG and 65BR.

The basic results have been presented at different international scientific conferences, for example, at the International Symposium on Materials and Devices for Nonlinear Optics, ISOPL'5 (2009, France); at the International Workshop on Nano and Bio-Photonics (IWNBP), St Germain au Mont d'Or (2011, France); at Laser Optics (2010, 2012, Saint-Petersburg, Russia); at YUCOMAT-2011, 2012 (Serbia-Montenegro); a seminar at the Jan Długosz University (2012, Częstochowa, Poland). The presented results are correlated with the work partially supported by the Russian Foundation for Basic Research, grants No.10-03-00916 (2010–2012) and No.13-03-00044 (2013-2015).

References

1. H.W. Kroto, J.R. Heath, S.C. O'Brien, R.F. Curl, R.E. Smalley, *Nature*, Vol. 318, p. 162, 1985.
2. W. Krätschmer, K. Fostiropoulos, D.R. Huffman, *Chem. Phys. Lett.*, Vol. 170, No. 2–3, p. 167, 1990.
3. S. Couris, E. Koudoumas, A.A. Ruth, S. Leach, *J. Phys. B: At. Mol. Opt. Phys.*, Vol. 8, p. 4537, 1995.
4. V.P. Belousov, I.M. Belousova, V.P. Budtov, V.V. Danilov, O.B. Danilov, A.G. Kalintsev, A.A. Mak, *J. Opt. Technol.*, Vol. 64, p. 1081, 1997.
5. J. Robertson, *Mater. Today*, Vol. 7, p.46, 2004.
6. S. Namilae, N. Chandra, C. Shet, *Chem. Phys. Letters*, Vol. 387, p. 247, 2004.
7. O. Buchnev, A. Dyadyusha, M. Kaczmarek, V. Reshetnyak, Yu. Reznikov, *J. Opt. Soc. Am. B*, Vol. 24, No. 7, p. 1512, 2007.

8. N.V. Kamanina, A. Emandi, F. Kajzar, A.-J. Attias, *Mol. Cryst. Liq. Cryst.*, Vol. 486, p. 1, 2008.

9. N.V. Kamanina, S.V. Serov, V.P. Savinov, D.P. Uskokovic, *International Journal of Modern Physics B (IJMPB)*, Vol. 24, Issues 6–7, p. 695, 2010.

10. http://www.ams.org/samplings/feature-column/fcarc-eulers-formula

11. http://moskva.kotoroy.net/histories/34.html

12. 12.http://commons.wikimedia.org/wiki/File:Biosph%C3%A8re_Montr%C3%A9al.jpg?uselang=ru

13. N.V. Kamanina, *Physics-Uspekhi*, Vol. 48, No. 4, p. 419, 2005.

14. N.V. Kamanina and D.P. Uskokovic, *Materials and Manufacturing Processes*, Vol. 23, p. 552, 2008.

15. N.V. Kamanina, S.V. Serov, N.A. Shurpo, S.V. Likhomanova, D.N. Timonin, P.V. Kuzhakov, N.N. Rozhkova, I.V. Kityk, K.J. Plucinski, D.P. Uskokovic, Polyimide-fullerene nanostructured materials for nonlinear optics and solar energy applications, *J. Mater. Sci.: Mater. Electron.*, DOI 10.1007/s10854–012-0625–9, published on-line 26 January 2012.

16. N.V. Kamanina, N.A. Vasilenko, *Opt. Quantum Electron.*, Vol. 29, No. 1, p. 1, 1997.

17. S.A. Akhmanov and S.Yu. Nikitin, *Physical Optics*, Clarendon, Oxford Press, 1997.

18. D.S. Chemla, J. Zyss (Eds), *Nonlinear Optical Properties of Organic Molecules and Crystals*, Orlando, Academic Press, Vol. 2, 1987.

19. V.S. Mylnikov, Photoconducting polymers/metal-containing polymers. In: *Advances in Polymer Science*, Berlin, Springer-Verlag, Vol. 115, 1994.

20. N.V. Kamanina, A.I. Plekhanov, *Opt. Spectrosc.*, Vol. 93, No. 3, p.408, 2002.

21. K. Sutter, J. Hulliger, P. Günter, *Solid State Commun.*, Vol. 74, p.867, 1990.

22. N.V. Kamanina, A.I. Plekhanov, S.V. Serov, V.P. Savinov, P.A. Shalin, F. Kajzar, *Nonlinear Optics and Quantum Optics*, Vol. 40, p. 307, 2010.

23. N.V. Kamanina, N.A. Shurpo, S.V. Likhomanova, D.N. Timonin, S.V. Serov, O.V. Barinov, P.Ya. Vasilyev, V.I. Studeonov, N.N. Rozhkova, V.E. Vaganov, I.V. Mishakov, A.A. Artukh, L.A. Chernozatonskii, The optimization of the composition, structure and properties of metals, oxides, composites, nano- and amorphous materials, *Proceed. 10th Israeli-Russian Bi-National Workshop*, Israel Academy of Science and Humanities and the Russian Academy of Science; 20 June - 23 June, 2011, p.77, 2011.

24. N.V. Kamanina, P.Ya. Vasilyev, S.V. Serov, V.P. Savinov, K.Yu. Bogdanov, D.P. Uskokovic, *Acta Physica Polonica A*, Vol. 117(5), p. 786, 2010.

25. R.J. Collier, C.B. Burckhardt, L.H. Lin, *Optical Holography*, New York and London, Acad. Press, 1971.

26. H. Liu, B. Taheri, W. Jia, *Phys. Rev. B*, Vol. 49, No. 15, p. 10166, 1994.

27. F. Kaizar, C. Taliani, M. Muccini, R. Zamboni, S. Rossini, R. Danieli, *Proceed. SPIE*, Vol. 2284, p. 58, 1994.
28. J. Li, J. Feng, J. Sun, *J. Chem. Phys.*, Vol. 203, p. 560, 1993.
29. W. Krätschmer, L.D. Lamb, K. Fostiropoulos, D.R. Huffman, *Nature*, Vol. 347, p. 354, 1990.
30. P.A. Chollet, F. Kajzar, J. Le Moigne, *Proceed. SPIE*, 1273, p. 87, 1990.
31. J.S. Shirk, J.R. Lindle, F.J. Bartoli, C.A. Hoffman, Z.H. Kafafi, A.W. Snow, *Applied Physics Letters*, Vol. 55, p. 1287, 1989.
32. H.S. Nalwa, T. Saito, A. Kakuta, T. Iwayanagi, *Journal of Physical Chemistry*, Vol. 97, No. 41, p. 10515, 1993.
33. T.C. Wen and I.D. Lian, *Synth. Met.*, Vol. 83, No. 2, p. 111, 1996.
34. N.V. Kamanina, *Optics and Spectroscopy*, Vol. 90, No. 6, p. 867, 2001.
35. N.V. Kamanina, *Synthetic Metals*, Vol. 139, No. 2, p. 547, 2003.
36. R.A. Ganeev, A.I. Ryasnyansky, M.K. Kodirov, T. Usmanov, *Opt. Commun.*, Vol. 185, p. 473, 2000
37. N.V. Kamanina, E.F. Sheka, *Optics and Spectroscopy*, Vol. 96, No. 4, p. 599, 2004.
38. N.V. Kamanina, *Optics and Spectroscopy*, Vol. 90, No. 6, p. 931, 2001.
39. N.V. Kamanina, *Journal of Optics A: Pure and Applied Optics*, Vol. 4, No. 4, p. 571, 2002.
40. A.A. Vasiliev, D. Kasasent, I.N. Kompanets, A.V. Parfenov, *Spatial Light Modulators*, Radio I Svyaz, Moscow, 1987, p. 320.
41. B.S. Lowans, B. Bates, R.G.H. Greer, J. Aiken, *Appl. Opt.*, Vol. 31, pp. 7393, 1992.
42. N.V. Kamanina, L.N. Soms, A.A. Tarasov, *Opt. Spectrosc.*, Vol. 68, p. 403, 1990.
43. V.I. Tsoi, A.V. Tarasishin, V.V. Belyaev, S.M. Trofimov, *Journal of Optical Technology*, Vol. 70, p. 465, 2003.
44. G.M. Zharkova, A.S. Sonin, *Liquid Crystal Composites*, Novosibirsk, Nauka, 1994 [in Russian].
45. N.V. Kamanina, *Tech. Phys. Lett.*, Vol. 22, No. 4, p. 291, 1996.
46. N.V. Kamanina and V.I. Berendyaev, *Proceed. of SPIE*, Vol. 3292, p. 154, 1998.
47. J.S. Gwag, M. Oh-e, K.R. Kim, M. Yoneya, H. Yokoyama, S. Itami and H. Satou, *Nanotechnology*, Vol. 19, p. 395, 2008.
48. N.V. Kamanina, P.Ya.Vasilyev, *Tech. Phys. Lett.*, Vol. 35, No. 6, p. 501, 2009.
49. N.V. Kamanina, P.Ya. Vasilyev, V.I. Studenov, *Tech. Phys. Lett.*, Vol. 36, No. 8, p.727, 2010.
50. N.V. Kamanina, S.V. Likhomanova, P.Ya. Vasilyev, V.I. Studeonov, L.A. Chernozatonskii, V.E. Vaganov, I.V. Mishakov, *Tech. Phys. Lett.*, Vol. 37, No. 12, p.1165, 2011.

51. B. Das, K.E. Prasad, U. Ramamurty, C.N.R. Rao, *Nanotechnology*, Vol. 20. p. 125705, 2009, (http://www.dftb-plus.info/)
52. N.V. Kamanina, P.Ya. Vasilyev, V.I. Studeonov, Yu.E. Usanov, *J. Opt. Technol.*, Vol. 75, No. 1, p. 67, 2008.
53. N.V. Kamanina, P.Ya. Vasilyev, V.I. Studeonov, *J. Opt. Technol.*, Vol. 75, No. 12, p. 806, 2008.
54. N.V. Kamanina, K.Yu. Bogdanov, P. Ya. Vasilyev, V.I. Studeonov, *J. Opt. Technol*, Vol. 77. No. 2, p. 145, 2010.
55. C.P. Poole, Jr., F.J. Owens, *Introduction to Nanotechnology*, New York, Wiley Interscience, 2003.

8

Covalent and Non-Covalent Functionalization of Carbon Nanotubes

Tawfik A. Saleh[1] and Vinod K. Gupta[2,*]

[1]Chemistry Department, King Fahd University of Petroleum & Minerals,
Dhahran, Saudi Arabia
[2]Chemistry Department, Indian Institute of Technology Roorkee, Roorkee, India

Abstract

Currently, there has been a growing interest in investigating the possibility of functionalization of carbon nanotubes to enhance their properties. Carbon nanotube functionalization is one of the most exciting fields of research. The objective of this chapter is to discuss the common covalent and non-covalent functionalization methods used for the chemical modification of carbon nanotubes. The chapter will also discuss the methods used for functionalization of nanotubes with nanoparticles. Also, the activity of the nanotubes-based composites as adsorbents and photocatalysts will be highlighted. The chapter ends with conclusions and recommendations.

Keywords: Carbon nanotube, functionalization, nanocomposites

8.1 Introduction

Functionalization of carbon nanotube (CNT) is one of the most exciting fields of research in the area of nanotechnology. The outstanding properties of CNT, such as high mechanical strength and impressive electrical and thermal conductivity, make it an attractive potential candidate for various applications. Despite the fact that CNT holds great promise for various applications, there are still

Corresponding author: vinodfcy@gmail.com

Ashutosh Tiwari and S.K. Shukla (eds.) Advanced Carbon Materials and Technology, (317–330) 2014 © Scrivener Publishing LLC

some shortcomings prior to its transfer into successful real applications. A frequent problem of carbon nanotubes is the formation of irreversible agglomerates or restacking to form carbon nanotubes bundles via π– π stacking and van der Waals interactions, which are considered to be an imminent concern. Apart from that, the physical handling of the nanotubes is challenging since CNT is not soluble in most solvents.

Chemical functionalization of CNT is one of many solutions to address the above challenges of CNT. Chemical functionalization through synthetic chemistry methods allow for the preparation of CNT-based nanocomposites containing nanoparticles covalently or nano-covalently bonded to nanotube network. The properties can often be controlled by the nanoparticles concentration. The objective of this chapter is to detail the functionalization of CNT with nanoparticles of semiconductor photocatalysts to utilize carbon nanotube in photocatalytic applications.

This chapter starts with the functionalization nature of carbon nanotube to form nanocomposites along with the methods used for synthesis of CNT-based composite. The functionalization methods can be classified into two categories: covalent and non-covalent. The methods investigating the activity of the nanocomposites as adsorbents and photocatalysts will be discussed.

8.2   Functionalization of Carbon Nanotubes

Functionalization of carbon nanotubes (CNTs) can be performed by several methods. These methods can generally be classified into two categories: covalent and non-covalent functionalization. Figure 8.1 presents a scheme of the methods classification.

8.3   Covalent Functionalization

Covalent functionalization is an irreversible process or not reversible attachment of appendage on the nanotube walls or tips. It is based on the formation of a covalent linkage between functional entities and the carbon skeleton of nanotubes. Reactions can be performed at the sidewall, which is called sidewall functionalization, or at the defect sites, which is called defect functionalization, localized usually at the tips. Examples of these types are amidation and

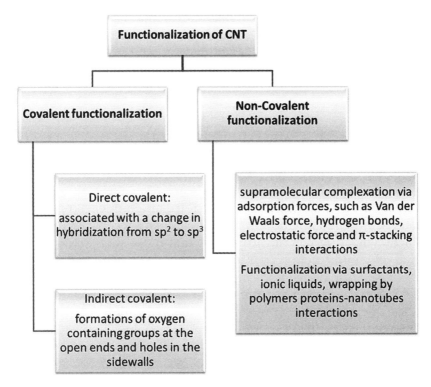

Figure 8.1 Schematic representation of the methods employed for functionalization of carbon nanotubes.

esterification reactions of carboxylic residues obtained on the nanotubes; and the fluorination and cycloadditions [1–12]. Covalent functionalization can be divided into direct covalent sidewall functionalization and indirect covalent functionalization with carboxylic groups on the surface of CNTs.

Direct covalent sidewall functionalization is associated with a change in hybridization from sp^2 to sp^3 and a simultaneous loss of conjugation. Indirect covalent functionalization is achieved by the chemical transformations of carboxylic groups at the open ends and holes in the sidewalls. These carboxylic groups might have existed on the as-grown nanotubes and/or be further generated during oxidative purification and activation by chemical treatment. In order to increase the reactivity of the nanotubes, the carboxylic acid groups usually need to be created. Then, through such functional groups, an esterification or amidation reaction can be performed by converting the oxygen-containing groups into acid chloride. The drawback of covalent functionalization is that the perfect structure of CNTs has to be destroyed, resulting in significant changes in their physical properties. It is feasible that caps or tips are more reactive than sidewalls because of their mixed pentagonal-hexagonal structure.

8.4   Non-Covalent Functionalization

Non-covalent functionalization is based on supramolecular complexation using various adsorption forces, such as van der Waals force, hydrogen bonds, electrostatic force and π-π stacking interactions. Compared to the chemical functionalization, non-covalent functionalization has the advantage that it can be operated under relatively mild reaction conditions. It also has the advantage that it does not perturb the electronic structure of the nanotubesand thus the perfect graphitic structure of CNT can be maintained. However, it has the disadvantage that this functionalization involves weak forces and so some applications are not possible. This system is difficult to control and to characterize as well. The non-covalent functionalization is a reversible process, and reversible de-attachment is possible and can take place easily in the presence of some other interactions or under solvent effects. Examples of non-covalent functionalization are functionalization via surfactants, ionic liquids, and wrapping by polymers proteins-nanotubes interactions [13–25].

8.5   Functionalization of CNT with Nanoparticles

Nanoparticles functionalized CNTs are considered to be one of the hottest areas of research in the field of nanotechnology. There are several methods for the synthesis of CNT/nanoparticles, which are shown in Figure 8.2. These methods include *in situ* reductionof metal precursors onto carbon nanotubes, covalent functionalization and non-covalent functionalization of CNTs with nanoparticles [26].

Synthesis of CNT/nanoparticle composites by covalent method involves various steps. The as-grown CNTs are first oxidized by acid treatment. This can be performed using oxidizing agents such as nitric acid, sulfuric acid, a mixture of sulfuric acid and nitric acid, potassium permanganate, sulfuric acid in presence of potassium permanganate, hydrogen peroxide in presence of nitric acid, hydrogen peroxide, ozone, an oxygen-based atmosphere by an

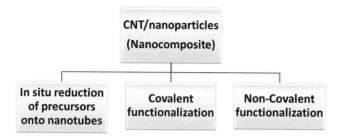

Figure 8.2 Classification of methods for synthesis of carbon nanotube/metal oxides nanocomposites.

inductively-coupled plasma or microwave energy and water [27–32]. The success of the activation with oxygen-containing groups can be confirmed by the means of various tools including transmission electron microscope (TEM), field emission scanning electron microscope (FESEM), energy dispersive X-ray spectroscope (EDX), Raman spectroscope, thermogravimeter (TGA), and Fourier transform infrared spectroscope (FTIR) and X-ray photoelectron spectroscope (XPS). The characterization of the presence of functional groups on the surface of nanotube and the efficiency of functionalization on nanotubes can be confirmed by the interpretation of the characterization results (Figure 8.3).

For the preparation of CNT/nanoparticles, the oxidized CNTs are dispersed in an appropriate solvent like acetone, ethanol, propanol or n-methyl-2-pyrrolidone; tile well homogeneous suspension is formed. The dispersion can be achieved via sonication; some surfactants are used for dispersion. The precursor of the metal oxide is dissolved in a suitable solvent. The latter is drop-wise added into the dispersed o-CNTs and the mixture is sonicated for a while, then, magnetically stirred for some time. After that, the suspensions' mixture is transferred into a round-bottomed flask and refluxed at high temperature; 120–200°C. The system is then allowed to cool. After that, it is filtered and washed with suitable solvent like distilled water and ethanol several times. This is followed by the calcinations process of the composite at temperatures between 250–400°C for 2–4 h.

Characterization by TEM, FESEM, EDX, Raman spectroscope, TGA, FTIR and XPS is to be conducted to confirm the formation of the nanocomposites. As an example, the characterization results of the CNT/titaniana nocomposite are presented and discussed here. Figure 8.4 depicts the FESEM image of MWCNT/TiO$_2$ nanocomposite, which confirms the morphology and microstructure of the composite. The image clearly shows the nodes of titania on both sides

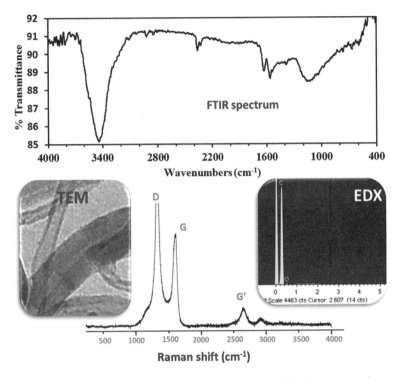

Figure 8.3 Characterization of oxygen-containing groups' carbon nanotubes; CNT oxidized by acid treatment; characterization results of Fourier infrared transform spectrum; high resolution transmission electron microscope image; energy dispersive X-ray spectrum; and Raman spectrum of carbon nanotube.

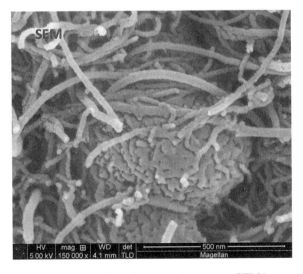

Figure 8.4 Field emission scanning electron microscopy; SEM image of MWCNT/titania nanocomposite.

of the nanotube and on the surfaces and ends of the nanotubes. The presence of titanium oxide on the surface of the nanotube was also confirmed by the EDX spectrum. Figure 8.5 depicts the quantitative analysis of the nanocomposite. It confirms the presence of carbon, oxygen and titanium.

Figure 8.6 depicts the FTIR spectrum of the MWCNT/TiO$_2$ nanocomposite. In the spectrum, the peak at around 700 cm^{-1} is assigned

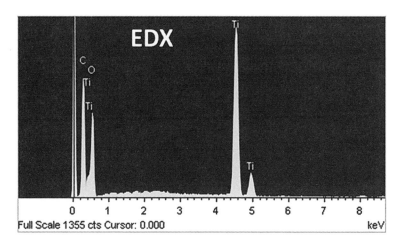

Figure 8.5 Energy dispersive X-ray spectrum of WCNT/titaniana nocomposite.

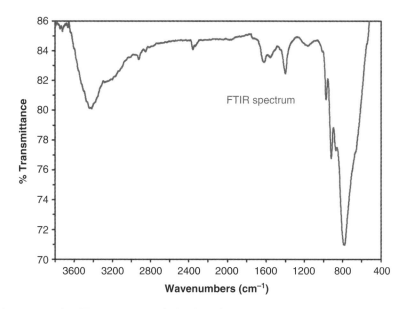

Figure 8.6 The FTIR spectrum of MWCNT/titaniana nocomposite.

to Ti-O and Ti-O-Ti bonding of titania. It also shows the characteristics peaks of carbonyl, hydroxyl and unsaturated carbon bonds in the composite. The band intensity of the $MWCNT/TiO_2$, in the region of the asymmetric carboxylate stretching mode at about 1400 cm^{-1}, is significantly higher than that of MWCNT. This indicates a (C-O-Ti) interaction between the titanium atom and the carboxylate group on the nanotube to form a COO-Ti moiety, in which one of the carboxylate oxygen interacts with the titanium ion.

8.5.1 Applications of the CNT-Based Nanocomposites

The new generation of CNT/nanoparticle composites have interesting properties that combine the properties of CNTs and MOx and hold new properties resulted from the interaction. In addition, the combination of metal oxide nanoparticles into the nanotubes prevents agglomeration. As a result, the composites have attractive wide applications compared with the isolated nanoparticles because CNTs act as carrier to stabilize the nanoparticles, maintaining their integrity. Examples of the fields of application are water treatment, the environment, catalysis and others [33–41].

8.5.2 Nanocomposites as Photocatalysts

Several factors such as surface area, adsorption capacity, light-response range and the recombination time of photogenerated charge carriers play a major role in determining the photocatalytic performance of any photocatalytic material.

A comparison of the photocatalytic activity has been conducted between TiO_2 and TiO_2/CNTs composite for acetone degradation. It has been noticed that the presence of a small amount of CNTs can enhance photocatalytic activity of TiO_2 [42]. This is supported by another study in which photocatalytic activity of TiO_2/SWCNT composite has been investigated for degradation of organic dyes and 2,6-dinitro-p-cresol. High photocatalytic activity of the TiO_2/SWCNT composite has been observed, compared with that of TiO_2 nanoparticles.

Since the change in the surface area is not significant between as-grown CNT and nanocomposites, the variation in adsorption capacity, dominantly determined by the surface area, is not the major reason for the photoactivity enhancement. It can be assumed that the dominant contribution of nanotubes in MWCNT/TiO$_2$ composites is mainly determined by an increase of recombination time for photogenerated electron–hole pairs. This is supported by the delocalized π-structure nature of MWCNT, which facilitates the transfer of photoinduced electrons and can perform as an excellent electron acceptor leading to hole–electron separation. The injected electron is transferred to the oxygen to generate the superoxide anion, which is considered highly active radical that deprecates the dye adsorbed on the nanotube surface [39]. It can be stated that the nanocomposite combines the features of MWCNTs and TiO$_2$, as well as generating new properties, which indicates the presence of interaction between MWCNT and titania, which modify the process of the electron/hole pair formation. The interaction could be by coordination of carbonyl or by ester bonding between carboxyl group of the nanotube and titania [43, 44, 39].

8.5.3   Nanocomposites as Adsorbents

Incorporation of the metal oxides' properties into CNT properties is a good way to have new materials that combine the properties of both components and new properties as a result of the combination. For example, the magnetic property of iron oxide can be incorporated in CNTs. The magnetic adsorbent can be well dispersed in the water and easily separated magnetically.

Magnetic-modified MWCNTs were used for the removal of several pollutants like cationic dyes crystal violet (CV), thionine (Th), janus green B (JG), methylene blue (MB) and heavy metal pollutants [45]. To find the optimum adsorption, the various effects of initial pH, dosages of adsorbent and contact time are to be investigated. Guar gum-grafted Fe$_3$O$_4$/MWCNTs (GG/MWCNT/Fe$_3$O$_4$) ternary composite was prepared and reported for the removal of natural red (NR) and methylene blue (MB) [46, 47]. The adsorption of both dyes was found to follow the pseudo-second-order kinetics and the maximum adsorption of MB and NR reached 61.92 and 89.85 mg/g. The removal efficiency of cationic dyes using GG/MWCNT/Fe$_3$O$_4$ is higher as compared with other adsorbents such as MWCNTs and MWCNT/Fe$_3$O$_4$. The maximum monolayer adsorption capacity of MB and NR by MWCNT–Fe$_3$O$_4$ were reported as 15.87 and 20.51 mg/g [48]. The reported higher adsorption capacity of GG/MWCNT/Fe$_3$O$_4$ could be related to the hydrophilic property of GG, which improved the dispersion of GG–MWCNT–Fe$_3$O$_4$ in the solution, which facilitated the diffusion of dye molecules to the surface of CNTs. Synthesized magnetic MWCNT-starch-iron oxide was used as an adsorbent for removing anionic methyl orange (MO) and cationic methylene blue (MB) from aqueous solutions [49]. The MWCNT-starch-iron oxide exhibits super paramagnetic properties with a saturation magnetization (23.15 emu/g) and better adsorption for MO and MB dyes than MWCNT/iron oxide. While the BET surface area of MWCNT-COOH and MWCNT-starch were 184.3 and 155.36 m^2/g, respectively, the specific surface areas of MWCNT/iron oxide and MWCNT-starch-iron oxide were smaller; 124.86 and 132.59 m^2/g, respectively. However, ternary composites have a higher removal capacity than both the binary composite and the as-grown CNTs. The adsorption capacity of MB and MO onto MWCNTs-starch-iron oxide was 93.7 and 135.6 mg/g, respectively, while for MWCNTs-iron oxide it was 52.1 and 74.9 mg/g [49]. Magnetic composite bioadsorbent composed of chitosan wrapping magnetic nanosized -Fe$_2$O$_3$ and multi-walled carbon nanotubes (m-CS/ -Fe$_2$O$_3$/MWCNTs) was reported to have almost two times higher removal of methyl orange than m-CS/ -Fe$_2$O$_3$ and MWCNTs [50, 51]. The properties of MWCNTs along with that of iron oxide have also been combined in a composite to produce a magnetic adsorbent. The adsorption activities have been tested for the adsorptions of Cr(III). The adsorption capability of the composite is higher than that of MWCNTs and activated carbon [35, 52].

8.6   Conclusion

This chapter has covered the common covalent and non-covalent functionalization methods used for the chemical modification of carbon nanotubes. Compared to the non-covalent functionalization, covalent functionalization has the advantage that it involves strong forces. However, non-covalent functionalization has the advantage of being operated under relatively mild reaction conditions. Thus, it does not perturb the electronic structure of the nanotubes and the perfect graphitic structure of nanotubes can be maintained.

Although there has been a lot of work reported on the covalent and non-covalent functionalization of carbon nanotubes, there is still more to be achieved in this area of research. Apart from developing new methodologies, it is equally important to improve current methods to increase the ease of functionalization. Success in this field would pave the way for future usage of carbon nanotube materials in real applications.

Acknowledgment

Saleh acknowledges the support of the Chemistry Department, Center of Research Excellence in Nanotechnology and King Fahd University of Petroleum and Minerals, (KFUPM) Dhahran, Saudi Arabia.

References

1. A. Hamwi, H. Alvergnat, S. Bonnamy, F. Béguin, Fluorination of carbon nanotubes, *Carbon* 1997; 35(6):723–728.
2. A. Hirsch, Functionalization of single walled carbon nanotubes, *Angewandte Chemie International Edition* 2002; 41:1853–1859.
3. D. Tasis, N. Tagmatarchis, A. Bianco, M. Prato, Chemistry of carbon nanotubes, *Chemical Reviews* 2006; 106:1105–1136.
4. Z.M. Dang, L. Wang, L.P. Zhang, Surface functionalization of multiwalled carbon nanotube with trifluorophenyl, *J. Nanomater.*, 2006; 83583:2006.
5. J.E. Fischer, Chemical doping of single-wall carbon nanotubes, *Acc. Chem. Res.* 2002; 35:1079–86.
6. W. Kim, N. Nair, C.Y. Lee, et al., Covalent functionalization of single-walled carbon nanotubes alters their densities allowing electronic and other types of separation, *J. Phys. Chem. C* 2008; 112(19):7326–31.
7. W.H. Zhu, N. Minami, S. Kazaoui, et al., Fluorescent chromophore functionalized single-wall carbon nanotubes with minimal alteration to their characteristic one-dimensional electronic states, *J. Mater. Chem.*, 2003; 13(9):2196–201.
8. T. Yumura, M. Kertesz, S. Iijima, Confinement effects on site-preferences for cycloadditions into carbon nanotubes, *Chem. Phys. Lett.* 2007; 444(1–3):155–60.
9. Z. Guo, F. Du, D.M. Ren, et al., Covalently porphyrin-functionalized single-walled carbon nanotubes: A novel photoactive and optical limiting donor–acceptor nanohybrid, *J. Mater. Chem.* 2006; 16(29): 3021–30.
10. B. Kim, W.M. Sigmund, Functionalized multiwall carbon nanotube/gold nanoparticle composites, *Langmuir* 2004; 20(19):8239–42.
11. V. Georgakilas, V. Tzitzios, D. Gournis, et al., Attachment of magnetic nanoparticles on carbon nanotubes and their soluble derivatives, *Chem. Mater.* 2005; 17(7):1613–7.
12. D.M. Guldi, G.M.A. Rahman, N. Jux, et al., Integrating single-wall carbon nanotubes into donor–acceptor nanohybrids. *Angew Chem. Int. Ed.* 2004; 43(41):5526–30.

13. S.R. Vogel, K. Müller, U. Plutowski, et al., DNA–carbon nanotube interactions and nanostructuring based on DNA, *Phys. Status Solid B* 2007; 244(11):4026–9.

14. S.R. Shin, C.K. Lee, I. So, et al., DNA-wrapped single-walled carbon nanotube hybrid fibers for supercapacitors and artificial muscles, *Adv. Mater.* 2008 ;20(3):466–70.

15. Q.X. Mu, W. Liu, Y.H. Xing, et al., Protein binding by functionalized multiwalled carbon nanotubes is governed by the surface chemistry of both parties and the nanotube diameter, *J. Phys. Chem. C* 2008; 112(9):3300–7.

16. J.G. Yu, K.L. Huang, S.Q. Liu, et al., Preparation and characterization of polycarbonate modified multiple-walled carbon nanotubes, *Chin. J. Chem.* 2008; 26(3):560–3.

17. N. Nakayama-Ratchford, S. Bangsaruntip, X.M. Sun, et al., Noncovalent functionalization of carbon nanotubes by fluorescein-polyethylene glycol: Supramolecular conjugates with pH-dependent absorbance and fluorescence, *J. Am. Chem. Soc.* 2007; 129(9):2448–9.

18. D.S. Zhang, L.Y. Shi, J.H. Fang, et al., Preparation and modification of carbon nanotubes, *Mater. Lett.* 2005; 59(29–30):4044–7.

19. S. Lefrant, J.P. Buisson, J. Schreiber, et al., Raman studies of carbon nanotubes and polymer nanotube composites, *Mol. Cryst. Liq. Cryst.* 2004; 415:125–32.

20. C.L. Fu, L.J. Meng, Q.H. Lu, et al., Large-scale homogeneous helical amylose/SWNTs complexes for biological applications, *Macromol. Rapid Commun.* 2007; 28:2180–4.

21. A. Satake, Y. Miyajima, Y. Kobuke, Porphyrin–carbon nanotube composites formed by noncovalent polymer wrapping, *Chem. Mater.* 2005; 17(4):716–24.

22. O. Kim, J. Je, J.W. Baldwin, et al., Solubilization of single-wall carbon nanotubes by supramolecular encapsulation of helical amylose. *J. Am. Chem. Soc.* 2003; 125:4426–7.

23. Y.H. Xie, A.K. Soh, Investigation of non-covalent association of single-walled carbon nanotube with amylose by molecular dynamics simulation. *Mater. Lett.* 2005; 59(8–9):971–5.

24. W.Z. Yuan, J.Z. Sun, Y.Q. Dong, et al., Wrapping carbon nanotubes in pyrene-containing poly(phenylacetylene) chains: Solubility, stability, light emission, and surface photovoltaic properties, *Macromolecules* 2006; 39(23):8011–20.

25. G.N. Ostojic, J.R. Ireland, M.C. Hersam, Noncovalent functionalization of DNA-wrapped single-walled carbon nanotubes with platinum-based DNA cross-linkers, *Langmuir* 2008; 24:9784.

26. L. Meng, C. Fu, Q. Lu, Advanced technology for functionalization of carbon nanotubes, *Progress in Natural Science* 2009; 19:801–810.

27. A.K. Cuentas-Gallegos, R. Martínez-Rosales, M.E. Rincón, G.A. Hirata, and G. Orozco, Design of hybrid materials based on carbon nanotubes and polyoxometalates, *Opt. Mater.* 2006; 29:126–133.

28. H. Wang, H.L. Wang, and W.F. Jiang, Solar photocatalytic degradation of 2,6-dinitro-p-cresol (DNPC) using multi-walled carbon nanotubes (MWCNTs)–TiO2 composite photocatalysts, *Chemosphere* 2009; 75(8):1105–1111.

29. B. Smith, K. Wepasnick, K.E. Schrote, H.H. Cho, W.P. Ball, D.H. Fairbrother, Influence of surface oxides on the colloidal stability of multi-walled carbon nanotubes: A structure-property relationship, *Langmuir* 2009; 25:9767–9776.

30. V. Subramanian, E. Wolf, P.V. Kamat, Catalysis with TiO2/gold nanocomposites. Effect of metal particle size on the fermi level equilibration, *J. Am. Chem. Soc.* 2004; 126:4943–4950.

31. S. Wang, X. Shi, G. Shao, X. Duan, X. Yang, and X. Wang, Preparation, characterization and photocatalytic activity of multi-walled carbon nanotube-supported tungsten trioxide composites, *Journal of Physics and Chemistry of Solids* 2008; 69:2396–2400.

32. T.A. Saleh, V.K. Gupta, Functionalization of tungsten oxide into MWCNT and its application for sunlight-induced degradation of rhodamine B, *Journal of Colloid and Interface Science* 2011; 362(2):337–344.

33. V.K. Gupta, I. Ali, T.A. Saleh, A. Nayak, S. Agarwal, Chemical treatment technologies for waste-water recycling – An overview, *RSC Advances* 2012; 2:6380–6388.

34. V.K. Gupta, T.A. Saleh, Sorption of pollutants by porous carbon, carbon nanotubes and fullerene – An overview, *Environ. Sci. Pollut. Res.* 2013; 20:2828–2843.

35. V.K. Gupta, S. Agarwal, T.A. Saleh, Chromium removal by combining the magnetic properties of iron oxide with adsorption properties of carbon nanotubes, *Water Research* 2011; 45(6):2207–2212.

36. V.K. Gupta, S. Agarwal, T.A. Saleh, Synthesis and characterization of alumina-coated carbon nanotubes and their application for lead removal, *Journal of Hazardous Materials* 2011; 185(1):17–23.

37. T.A. Saleh, V.K. Gupta, Characterization of the bonding interaction between alumina and nanotube in MWCNT/alumina composite, *Current Nano* 2012; 8:739–743.

38. T.A. Saleh, V.K. Gupta, Column with CNT/magnesium oxide composite for lead(II) removal from water, *Environ. Sci. Pollut. Res.* 2012; 19(4):1224–1228.

39. T.A. Saleh, V.K. Gupta, Photo-catalyzed degradation of hazardous dye methyl orange by use of a composite catalyst consisting of multi-walled carbon nanotubes and titanium dioxide, *Journal of Colloid and Interface Science* 2012; 371(1):101–106.

40. T.A. Saleh, S. Agarwal, V.K. Gupta, Synthesis of MWCNT/MnO2 and their application for simultaneous oxidation of arsenite and sorption of arsenate, *Applied Catalysis B: Environmental* 2011; 106(1–2):46–53.

41. T.A. Saleh, The influence of treatment temperature on the acidity of MWCNT oxidized by HNO_3 or a mixture of HNO3/H2SO4, *Applied Surface Science* 2011; 257(17):7746–7751.

42. Y. Yu, J.C. Yu, J.G. Yu, Y.C. Kwok, Y.K. Che, J.C. Zhao, L. Ding, W.K. Ge, and P.K. Wong, Enhancement of photocatalytic activity of mesoporous TiO2 by using carbon nanotubes, *Applied Catalysis A: General* 2005; 289:186–196.

43. K. Woan, G. Pyrgiotakis, and W. Sigmund, Photocatalytic-carbon-nanotube–TiO2 composites, *Adv. Mater.* 2009; 21:2233–2239.

44. Y.J. Xu, Y. Zhuang, and X. Fu, New insight for enhanced photocatalytic activity of TiO2 by doping carbon nanotubes: A case study on degradation of benzene and methyl orange, *J. Phys. Chem. C* 2010; 114:2669–2676.

45. V.K. Gupta, R. Kumar, A. Nayak, T.A. Saleh, M.A. Barakat. Adsorptive removal of dyes form aqueous solution onto carbon nanotubes: A review, *Advances in Colloid and Interface Science* 2013; 193–194:24–34.

46. L. Yan, P.R. Chang, P. Zheng, X. Ma, Characterization of magnetic guar gum-grafted carbon nanotubes and the adsorption of the dyes, *Carbohydrate Polymers* 2012; 87(3):1919–1924.

47. V.K. Gupta, R. Jain, M.N. Siddiqui, T.A. Saleh, S. Agarwal, S. Malati, D. Pathak, Equilibrium and thermodynamic studies on the adsorption of the dye rhodamine-B onto mustard cake and activated carbon, *Journal of Chemical and Engineering Data* 2010; 55(11):5225–5229.

48. J.L. Gong, B. Wang, G.M. Zeng, C.P. Yang, C.G. Niu, Q.Y. Niu, et al., Removal of cationic dyes from aqueous solution using magnetic multi-wall carbon nanotube nanocomposite as adsorbent, *J. Hazard Mater.* 2009; 164:1517–1522.

49. P.R. Chang, P. Zheng, B. Liu, D.P. Anderson, J. Yu, X. Ma, Characterization of magnetic soluble starch-functionalized carbon nanotubes and its application for the adsorption of the dyes, *J. Hazard Mater.* 2011; 186:2144–2150.

50. H.Y. Zhu, R. Jiang, L. Xiao, G.M. Zeng, Preparation, characterization, adsorption kinetics and thermodynamics of novel magnetic chitosan enwrapping nanosized γ-Fe2O3 and multi-walled carbon nanotubes with enhanced adsorption properties for methyl orange, *Bioresource Technology* 2010; 101(14):5063–5069.

51. Y. Yao, B. He, F. Xu, X. Chen, Equilibrium and kinetic studies of methyl orange adsorption on multiwalled carbon nanotubes, *Chem. Eng. J.* 2011; 170(1):82.

52. S.G. Wang, W. Gong, W. Liu, Y. Yao, B. Gao, and Q. Yue, Removal of lead(II) from aqueous solution by adsorption onto manganese oxide-coated carbon nanotubes, *Separation and Purification Technology* 2007; 58:17–23.

9

Metal Matrix Nanocomposites Reinforced with Carbon Nanotubes

Praveennath G. Koppad[1,2], Vikas Kumar Singh[3,*], C.S. Ramesh[1], Ravikiran G. Koppad[4] and K.T. Kashyap[1]

[1]*Advanced Composites Research Center, Department of Mechanical Engineering, PES Institute of Technology, Bangalore, India*
[2]*Research and Development, Rapsri Engineering Products Company Ltd., Bangalore, India*
[3]*Department of Industrial Production Engineering, GITAM Institute of Technology, GITAM University, Visakhapatnam, India*
[4]*Department of Electrical and Electronics Engineering, BV Bhommaraddi College of Engineering and Technology, Hubli, India*

Abstract

Carbon nanotubes are the unique nanostructures of carbon having high surface area per unit weight, low density, superior elastic modulus and tensile strength as well as remarkable thermal and electrical properties. On the other hand, considerable research effort has been directed towards the development of metal matrix nanocomposites using a wide range of matrix materials including aluminium, copper, magnesium, nickel and titanium reinforced with nanofiller materials like nanoplatelets, nanoparticles, nanofibers and carbon nanotubes. In this chapter, the recent research and development on the use of carbon nanotubes as reinforcement and their effect on mechanical, thermal and tribological properties of metal matrix nanocomposites are presented. The challenges related to the synthesis of nanocomposites using various processing techniques, dispersion of carbon nanotubes, interfacial bonding and strengthening mechanisms are discussed.

Keywords: Carbon nanotubes, metal matrix nanocomposites, mechanical properties, thermal properties

*Corresponding author: vikas.einstein@gmail.com

Ashutosh Tiwari and S.K. Shukla (eds.) Advanced Carbon Materials and Technology, (331–376) 2014 © Scrivener Publishing LLC

9.1   Introduction

Recent developments in nanoscience and nanotechnology have shown great promise for providing breakthroughs in the areas of materials science and technology, biotechnology, health sector and nanoelectronics. The term nanotechnology is used to cover the design, synthesis/fabrication and application of materials/ structures with at least one characteristic dimension measured in nanometers. The discovery of carbon nanotubes (CNTs) is considered responsible for co-triggering the nanotechnology revolution. Carbon nanotubes are unique nanostructured materials with extraordinary mechanical, thermal and electrical properties. These properties have inspired interest in the development of nanocomposites with extraordinary properties. The prospect of obtaining carbon nanotube reinforced nanocomposites that can satisfy multifunctions at the same time for many applications has attracted many researchers from both academia and industry. The tremendous attention paid to carbon nanotube reinforced nanocomposites is due to their potential applications in fields like aerospace, the automobile industry, sports, microfabrication, tribological and energy storage. The interest in carbon nanotubes and their composites is evident from the fact that the number of publications published by research teams around the world is close to seven papers a day. Several applications of carbon nanotubes reinforced polymer nanocomposites are currently available in the market. For instance, Hyperion offers 15–20 wt% multiwalled carbon nanotube reinforced polymer nanocomposite and Nanocyl has marketed polymer nanocomposites-based products like BIOCYL™ and THERMOCYL™ for barriers against flame and fouling. Other than carbon nanotubes, nanofillers like nanoparticles, nanoplatelets, nanofibers and recently graphene are used to reinforce polymer, ceramic and metal matrices in developing nanocomposites. The advantages of these nanofillers are large interfacial area, high aspect ratio and a small amount of nanofiller can reduce weight compared to that of conventional fillers [1–4].

In the following sections we will briefly discuss carbon nanotubes, their synthesis, and physical, mechanical and thermal properties. This will serve as a fundamental basis for understanding the application of carbon nanotubes as nanofiller in metal matrix nanocomposites. This will be followed by the synthesis of nanocomposites and effect of carbon nanotube addition to the microstructure, mechanical, thermal and tribological properties of nanocomposites. Finally, the challenges faced and need to overcome them for successful realization of commercial applications of carbon nanotube reinforced metal matrix nanocomposites is described. Those intending to get deep into the subject can refer to references [3] and [9].

9.2   Carbon Nanotubes

Carbon is the most unique element and exists in two well-known crystalline forms: diamond and graphite. The arrangement of electrons around the nucleus of the atom makes these allotropic forms either extremely hard or very soft. The new additions to the carbon allotropes family are fullerenes, the closed-cage carbon molecules, for example C_{60} and nanotubules of graphite, popularly known as carbon nanotubes. Discovered by Kroto *et al.* [5] in 1985 at Rice University, the fullerenes are closed-shell configuration of sp^2 hybridized carbon cluster. The famous $C_{60,}$ also known as "Bucky ball," is the most thoroughly studied member among the fullerenes and consists of 12 pentagonal and 20 hexagonal rings fused together forming a cage-like structure. This was followed by the discovery of carbon nanotubes which opened up an era in nanoscience and nanotechnology. Carbon nanotubes were discovered in 1991 by Japanese scientist Sumio Iijima during the examination of carbon soot produced by arc evaporation of graphite in helium atmosphere [6]. The nanotubes observed in high-resolution transmission electron microscopy (HRTEM) were multi-walled and had a diameter in the range of 4 to 30 nm and length up to 1 μm. Carbon nanotubes are the graphene sheets of hexagonal carbon rings rolled into long cylinders with both ends capped by fullerene-like structures. Two years later, Iijima *et al.* [7] and Bethune *et al.* [8] independently reported the synthesis of nanotubes having single walls. Single-walled carbon nanotubes (SWCNTs) are composed of single graphene cylinder, whereas multiwalled carbon nanotubes (MWCNTs) consist of many nested cylinders with an interlayer spacing of 0.34 nm. The carbon nanotubes can have a variety of structures, namely armchair, zig-zag and chiral, by rolling the graphene sheets with varying degrees of twist along its length.

The early carbon nanotubes were synthesized by electric-arc discharge method, in which an arc was triggered between graphite electrodes in helium atmosphere. Since then a number of ways of preparing carbon nanotubes have been explored in which the three main techniques are: electric-arc discharge, laser ablation and decomposition of carbon compounds (hydrothermal synthesis or chemical vapor deposition, CVD process). A simple method to synthesize carbon nanotubes is the electric-arc discharge method. In this process a DC voltage of 20–30 V is applied between two graphite electrodes in helium atmosphere. This results in deposition of carbon atoms from anode to the cathode. Highly crystalline carbon nanotubes can be produced by using liquid nitrogen in the arc-discharge chamber. The transmission electron image of arc-discharge-grown single-walled nanotube is shown in Figure 9.1. In the laser ablation technique, carbon nanotubes are produced by heating a graphite target using high energy laser beams in a controlled atmosphere. In either technique the single-walled carbon nanotubes are produced by impregnating the graphite with metal catalysts like nickel or cobalt. However, both these techniques tend to produce a small quantity of carbon nanotubes. These limitations have led to the development of gas-phase techniques like chemical vapor deposition, in which carbon nanotubes are formed by decomposition of hydrocarbon molecules at temperatures between 350 and 1100°C. High purity single-walled carbon nanotubes of quantity 450 mg/h can be produced using gas-phase catalytic method involving the pyrolysis of $Fe(CO)_5$ and CO, which is known as High pressure CO disproportionation process (HiPco). Using CO as carbon feedstock and $Fe(CO)_5$ as the precursor,

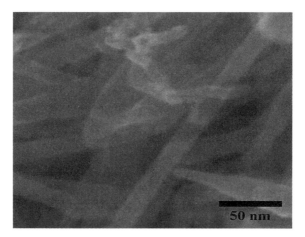

Figure 9.1 Transmission electron microscope (TEM) image of multiwalled carbon nanotubes.

the single-walled carbon nanotubes are produced under continuous flow of CO over the catalytic clusters of iron. The production of nanotubes by the HiPco approach can be 10 g/day, but the yield can still be improved by recirculation of CO using a compressor rather being released into the atmosphere. The unique aspects of the CVD process are good alignment, and the achievement of precise control over the diameter and length of carbon nanotubes. Depending on the heating source, CVD can be classified into plasma-enhanced, thermal- and laser-assisted CVD. Out of these, the laser-assisted CVD technique has higher deposition rates than conventional CVD and offers the possibility to scale-up the nanotube production [9].

Within the plane, three out of four electrons in carbon atom form three σ bonds, a sp^2 hybrid bond. Carbon nanotube with σ bond structuring is considered as the stiffest and strongest fiber to date. A large number of theoretical and experimental studies have been conducted on the estimation of mechanical properties of carbon nanotubes. Studies have confirmed that the axial elastic modulus of carbon nanotube is almost equal to 1.04 Tpa, which is the same as that of in-plane elastic modulus of graphite [10]. A theoretical study on the prediction of Young's modulus of carbon nanotubes using a Keating Hamiltonian with parameters determined from first principles was conducted by Overney *et al.* [11], which gave an estimation of between 1.5 to 5.0 TPa. Elastic and shear modulus prediction of carbon nanotubes by Lu *et al.* [12] using a molecular dynamics approach showed a value of 1.0 TPa and 0.5 Tpa, respectively. The nanotube diameter and the number of walls were found to have a negligible effect on the value of elastic modulus. Kashyap *et al.* [13] have calculated the Young's modulus by assessing the bending of multiwalled carbon nanotubes at the subgrain boundaries in CNT/Al nanocomposites. The value obtained from this analysis was about 0.9 TPa. The tensile strength measurements of both single-walled and multiwalled carbon nanotubes carried out by Yu *et al.* [14, 15] was found to be in the range of 13–52 GPa and 11–63 Gpa, respectively. The tensile strength and fracture studies on 19 multiwalled carbon nanotubes were carried out by using atomic force microscope tip for applying load. The breaking strength obtained was in the range of 11 to 63 GPa and strain at break reached up to 12%. In the case of single-walled carbon nanotubes, about 15 carbon nanotube ropes were tensile loaded in a nanostressing stage operated inside a scanning electron microscope. The average strength values were in the range of 13 to 52 GPa.

Both the allotropes of carbon, namely, graphite and diamond, do exhibit excellent heat capacity and thermal conductivity. The theoretically predicted and measured thermal conductivity values of individual carbon nanotubes are higher than that of graphite. The thermal conductance in nanotubes is due to both phonons and electrons. However, it is still unclear how much of the thermal conductance is due to phonons and due to electrons. But it is presumed that most of the heat is carried by phonons. Using combined results of equilibrium and nonequilibrium molecular dynamics, Berber *et al.* [16] showed that an isolated nanotube (10,10) possesses a thermal conductivity of 6600 W/mK at room temperature. The mesoscopic thermal transport measurements of a single multiwalled carbon nanotube showed a thermal conductivity of 3000 W/mK at room temperature and the phonon mean free path was found to be 500 nm [17]. However, the average thermal conductivity of multiwalled carbon nanotube films with a film thickness of 10 to 50 μm was found to be 15 W/mK at room temperature. The low thermal conductivity value resulted from intertube coupling and defects in the nanotubes causing scattering of phonons [18].

Wetting and capillarity of carbon nanotubes studied by Dujardin *et al.* [19] revealed that the elements with a surface tension below 100 mN/m can wet the nanotubes. The determining factor for wetting was surface tension, and for nanotubes the cutoff point is between 100 and 200 mN/m. The inner cavity of closed nanotubes could easily be filled up with lead compound in an oxidizing environment, whereas the nanotubes with opened tips exposed to molten metal in an inert atmosphere were not drawn in by capillarity. This work showed that the materials with low surface tension could be drawn inside nanotubes by capillary forces. If the pure metals are to be filled in nanotubes, then depending on the diameter of tube and contact angle, an outside pressure must be applied.

Owing to its excellent properties, carbon nanotubes are used in various applications like nanocomposites, chemical sensors, biosensors, nanoelectronics, electrochemical devices, hydrogen storage and field emission devices. For example, the addition of carbon nanotubes to polymers not only reinforces but introduces new electronic properties based on the electronic interaction between the two phases. This enables the polymer nanocomposites to be used as a material for applications in photovoltaic devices and light emitting diodes. Due to its high aspect ratio and physical properties, carbon nanotubes are suitable for field emitter applications, which was successfully demonstrated by Samsung SDI's fabrication of a prototype 4.5-inch color field emission display panel made using single-walled carbon nanotubes. Carbon nanotubes are being considered to replace semiconducting metal oxides for chemical sensor applications. The electrical conductance of semiconducting carbon nanotubes is highly sensitive to the change in chemical composition of the surroundings. This response is attributed to the charge transfer between the p-type semiconducting nanotube and molecules from the gases adsorbed onto the nanotube surface, when electrically charged nanotubes tend to act as actuators and can be used as nano-tweezers for grabbing or releasing the nanoscale structures. Carbon nanotubes are being projected as a tip for scanning probe microscopes because of the promise of ultrahigh lateral resolution due to their smaller diameter (~2 nm). However, technical issues related to growth, morphology and orientation of nanotubes on a wafer must be overcome to realize the commercial applications in industries. The carbon nanotubes are found to be suitable candidates for drug delivery systems given that a wide range of drugs can be adsorbed on their walls. The functionalized carbon nanotubes with more than one therapeutic agent with recognition capacity and specific targeting have created great interest for use in the treatment of cancer [20–24].

Most of the work on nanocomposites has been focused on reinforcement of polymers by carbon nanotubes due to their low processing temperatures and small stresses involved compared to that of metal and ceramic matrices. The incorporation of carbon nanotubes to ceramic matrix has been aimed at enhancing the fracture toughness and improving the thermal shock resistance. The increase in fracture toughness was observed in ceramic matrices like SiO_2, Al_2O_3 and $BaTiO_3$. The addition of carbon nanotubes to Al_2O_3 matrix resulted in high wear resistance and low coefficient of friction by serving as lubricating medium. Research on metal matrix nanocomposites made from carbon nanotubes is receiving much attention and is being projected for applications in structural and electronic packaging. During the last decade a lot of work has been carried out on the synthesis and characterization of metal matrix nanocomposites reinforced with carbon nanotubes. Aluminium is the most sought-after matrix material due to its low density and good mechanical properties, followed by other metals like copper, nickel, magnesium and titanium. The addition of a small amount of carbon nanotubes to these metallic matrices can significantly reduce the weight of nanocomposites when compared to that of microfillers, and increase the wear resistance and mechanical and thermal properties [25, 26]. The several processing routes of metal matrix nanocomposites are explained in detail in the following section.

9.3   Processing and Microstructural Characterization of Metal Matrix Nanocomposites

The past decades have been associated with some remarkable developments in the field of nanofiller-reinforced metal matrix nanocomposites, which have opened up unlimited possibilities for modern science and development. The need for high specific strength, stiffness, good dimensional stability, creep resistance, and high electrical and thermal conductivity have led to the development of metal matrix nanocomposites over conventional metals, alloys and microfiller-reinforced metal composites. Metal matrix nanocomposites are the composites with reinforcement in the form of nanofiber, nanoparticles or nanoplatelets dispersed in a metallic matrix. These custom-made materials are of great interest for applications in various fields like transportation, aerospace, thermal management and sporting goods. As discussed earlier, carbon nanotubes have emerged as potential and promising reinforcements for polymer, ceramic and metal matrices. The work on carbon nanotubes reinforced metal matrix composites has increased in the last few years. Few review papers have been published on this topic accounting for its importance from an application point of view [27, 28].

The widely used processing techniques for nanocomposites are powder metallurgy, electrodeposition, spray co-deposition and melt processing. The selection of processing technique mainly depends on the reinforcement morphology, the matrix material and application. Nanocomposite processing is the most important step because it controls the microstructure, which in turn will decide the various properties. After synthesis, material examination techniques should be used to characterize the quality of nanocomposites both qualitatively and quantitatively. The various tools used for characterization of the nanocomposites include X-ray diffraction (XRD), scanning electron microscope (SEM), transmission electron microscope (TEM), energy dispersed spectroscopy (EDS) and Raman spectroscopy. The SEM and TEM can provide the morphological features of nanocomposites such as carbon nanotubes distribution in the metal matrices, if any chemical reaction products formed at the interface between carbon nanotubes and metal matrix, and can image the defects like stacking fault or dislocations generated due to the addition of carbon nanotubes. X-ray diffraction is used mainly to study the phases present after the processing of nanocomposites, including determination of grain size of the matrix, or in computing the dislocation density. In this section the focus is on the processing and characterization techniques involved in carbon nanotubes reinforced metal matrix nanocomposites.

9.3.1 Powder Metallurgy

Powder metallurgy is a material processing technology in which material is taken in powder form, compacted into the desired shape and sintered to cause bonding between the particles. The main advantage of this technology over others is that it involves very little waste of material and about 98% of the starting material powders are converted into product. For the nanocomposite production the matrix and reinforcement powders are blended ultrasonically or using a ball mill to produce homogeneous distribution. The blending stage is followed by either cold pressing or hot pressing. The cold-pressed nanocomposite compact is subjected to sintering to provide bonding of solid particles by application of temperature. Further, the sintered nanocomposite compacts can be subjected to secondary processing techniques such as extrusion, rolling and forging. Most of the studies on carbon nanotubes reinforced aluminium, copper, magnesium and titanium metal matrices nanocomposites have been prepared using the powder metallurgy technique. Kuzumaki et al. [29] were the first to report the synthesis of carbon nanotubes reinforced aluminum composites by hot press and hot extrusion methods. The carbon nanotubes synthesized by DC arc-discharge were stirred with aluminum powders in ethanol at 300 rpm for 30 minutes. The composite powders were loaded in an aluminum can and hot pressed in a steel die with a pressure of 100 MPa. The hot-pressed samples were hot extruded at a temperature of 500°C into a rod at a speed of 10 mm/min. Microstructural studies revealed the coalescence of carbon nanotubes in the aluminum matrix. The observation at the interface between carbon nanotube and aluminum matrix showed no chemical reaction. This was followed by synthesis of CNT/Cu composite by Chen *et al.* [30]. In this study the authors prepared CNT/Cu copper composite by ball milling, compaction of ball-milled powder at a pressure of 600 MPa at 100°C for 10 minutes followed by sintering at 800°C for 2 hours. The CNT/Cu composite with 16 vol% CNTs showed high porosity which was due to agglomeration of CNTs due to high volume fraction. George *et al.* [31] used single-walled and multiwalled carbon nanotubes synthesized by arc-discharge technique to reinforce commercial purity aluminum. Some of the carbon nanotubes were coated with K_2ZrF_6 to improve the bonding between carbon nanotubes and aluminium matrix. The powders were ball milled at 200 rpm for 5 minutes. The milled composite powder was compacted, sintered at 580°C and hot extruded at 560°C. X-ray diffraction of nanocomposites showed no aluminium carbide formation. The TEM images of nanocomposites revealed the pinning of subgrain boundaries by carbon nanotubes. It was proposed that due to the coefficient of thermal expansion mismatch between carbon nanotubes ($\sim 10^{-6}$ K^{-1}) and aluminium (23.6×10^{-6} K^{-1}), dislocations were generated around the carbon nanotubes. However, it is interesting to note that no free dislocations were observed in the subgrains due to the possibility of dislocations being captured by low angle tilt boundaries at the subgrains.

It is very important for processing techniques to achieve uniform dispersion of carbon nanotubes inside the metal matrices to realize their outstanding mechanical and thermal properties. In order to address the dispersion of carbon nanotubes in the metal matrices, Esawi *et al.* [32] chose mechanical alloying to achieve uniform dispersion of nanotubes in the aluminium powders. About 2 wt% of multiwalled carbon nanotubes and aluminum powder of 75 µm were milled in planetary ball mill up to 48 hours with a ball to powder ratio of 10:1. The morphology of ball-milled powders were analyzed using field emission scanning electron microscope (FESEM). It was observed that in the initial stage aluminum particles were flattened under the impact of balls and then started welding to form large particles. After 24 hours of milling the particles reached a size of 3 mm with a very smooth surface. The carbon nanotubes were found to be intact even after 48 hours of milling and were individually embedded inside the aluminum matrix. No detailed investigation was reported on the mechanical alloying with different weight fraction of carbon nanotubes. However, the authors presumed that carbon nanotubes with higher weight fraction in aluminum could make the composite particles less ductile and result in much lower size of particles.

Koppad *et al.* [33, 34] carried out studies on MWCNT/Cu nanocomposites with varying weight fractions of multiwalled carbon nanotubes. The nanocomposites were fabricated by powder metallurgy technique followed by hot forging. Prior to ball milling the nanotubes were pre-coated with nickel by electroless coating for better interfacial bonding. The ball-milled nanocomposite powder subjected to FESEM studies showed that multiwalled carbon nanotubes were not observed on the surface of copper particles as shown in Figure 9.2. This is due to the incorporation of nanotubes inside the copper particles due to cold welding at the time of ball milling. The microstructural characterization of nanocomposites revealed fair dispersion of nanotubes inside the copper matrix. However, at higher weight fractions the agglomeration of nanotubes was observed as shown in Figure 9.3. This was attributed to the fact that nanotubes tend to form clusters due to van der Waals forces. The grain size analysis of hot-forged nanocomposites carried out using optical microscope revealed that grain size decreased with increasing weight fraction of nanotubes. The structures of nanotubes were found to be intact and did not display any degradation due to high processing temperature. Due to thermal mismatch between copper and nanotubes, dislocations were generated around the nanotubes. In addition to this, stacking faults were also observed at the tips

Figure 9.2 Field emission scanning electron microscope (FESEM) image of ball-milled MWCNT/Cu nanocomposite powder.

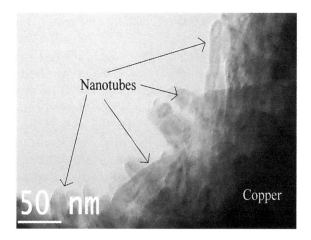

Figure 9.3 Transmission electron microscope (TEM) image of agglomeration of carbon nanotubes in 4 wt% MWCNT/Cu nanocomposites.

of the nanotubes. In another attempt to disperse carbon nanotubes uniformly in copper powders, Chu *et al.* [35] used a novel mixing procedure called particle compositing process. In this process, high speed air caused repeated inter-particle impacting, shearing and friction leading to sticking of fine particles to the large particles. A rotary speed of 5000 rpm and duration of 40 minutes were adopted to mix 5 and 10 vol% of carbon nanotubes and copper particles. Consolidation of nanocomposite powder was carried out in spark plasma sintering with a heating rate of 100°C/min and a pressure of 40–60 MPa was applied from the start to the end of sintering. However, the microstructure of consolidated nanocomposite sample showed the inducement of kinks and twists in the nanotubes. The high carbon nanotube content in nanocomposite showed large pores due to the presence of nanotube clusters.

Apart from pure metals, alloys like aluminum 2024 [36], 6061 [37], 6063 [38] and tin solder alloy [39] were also reinforced with carbon nanotubes. In their work, Choi *et al.* [36] fabricated aluminum alloy (AA) 2024 reinforced carbon nanotubes composites by hot rolling of ball-milled powder. The AA 2024 powder was produced by ball milling the chips at 500 rpm for about 48 hours. To prevent the cold welding among the AA 2024 chips, stearic acid of about 1 wt% was used as a control agent. In order to consolidate the ball-milled powders, the powders were heated up to a temperature of 450°C and then hot rolled with every 12% reduction per pass. After hot rolling, the samples with and without carbon nanotubes were solutionized at a temperature of 530°C for 2 hours and then quenched using cold water.

The grain size of ball-milled powders was analyzed using XRD. The grain size of ball-milled AA 2024 powders after 12, 24 and 48 hours were 92, 50 and 90 nm, respectively. Owing to grain growth, the ball milling duration was limited to 24 hours. The hot-rolled sheets of AA 2024 fabricated using 12 and 24 hours of ball-milled powders had average grain sizes of 250 and 100 nm. It was observed that the dislocation density near carbon nanotubes was very high owing to high stress levels, which stimulate the kinetics of precipitation. The precipitates were nucleated at the tips of carbon nanotubes giving rise to an increase in hardness of composite compared to that of AA 2024. Han *et al.* [39] reported the synthesis of 95.8Sn-3.5Ag-0.7Cu reinforced with nickel-coated multiwalled carbon nanotubes. The starting materials were mixed in a V-type blender for about 10 hours. The nanocomposite powder was subjected to uniaxial compaction and sintering at 175°C for about 2 hours. The compacted sample was extruded at room temperature. The fractographs of nanocomposites showed clustering of nanotubes. The disadvantages of nanotube clusters are that they avoid effective bonding between nanotubes and the tin solder matrix and act as stress concentration sites which promote the formation of cracks.

Most of the methods discussed above have displayed fair dispersion of carbon nanotubes inside the metal matrices. All the research work employed hot extrusion, hot rolling and hot forging as secondary processing after consolidation of ball-milled nanocomposite powders. The secondary processing techniques are employed to improve the density and dispersion of carbon nanotubes and to control the microstructure of nanocomposites. These studies showed that even after hot deformation processing, the integrity of nanotubes was maintained without causing any noticeable damage or degradation of nanotube. However, proper techniques with optimized working parameters are still needed to disperse nanotubes individually in the matrix material. For example, the degree of dispersion of carbon nanotubes in matrix material using ball milling depends on various factors such as type of mill, milling time, milling speed, ball to powder ratio, process control agent and the quantity of nanotubes used.

9.3.2   Electroless and Electrodeposition Techniques

Electroless and electrodeposition are the most important techniques for producing composite materials and thin films to improve the wear/corrosion resistance of machine components and for electromagnetic interference, printed circuit boards and microelectronics applications. The advantages of electroless and electrodeposition techniques include low cost, ease of operation and forming of dense coatings. Electrodeposition of composite coatings can be carried out by either direct current or pulse deposition. Electroless plating is the technique in which a substrate to be coated is dipped in electroless solution containing metallic ions, complexing agent, reducing agent and stabilizer. The potential developed between substrate and electroless solution gives rise to attraction of negative and positive ions towards the substrate surface.

Many attempts were made to synthesize carbon nanotubes reinforced metal nanocomposites by electroless plating and electrodeposition techniques. The electroless plating technique has been successfully employed to develop CNT/Ni and CNT/Cu nanocomposites. Li *et al.* [40] reported the development of CNT/Ni-P coatings on a steel substrate by the electroless plating technique. The carbon nanotubes were initially cleaned by suspending them in hydrochloric acid for 48 hours followed by oxidizing them in potassium bichromate and sulfuric acid for 30 minutes. After drying at 150°C for several days, the nanotubes were introduced into a nickel bath held at 85°C with a pH value maintained at 4.6. The nickel-plated steel substrate was then implanted in the plating bath for 2 hours. The coating thickness of nanocomposite was about 20–30 μm. The SEM images of nanocomposite showed uniform dispersion of nanotubes in nickel matrix with one end protruding from the surface. Synthesis of multiwalled carbon nanotubes reinforced copper nanocomposites with different nanotube volume fractions by electroless technique and spark plasma sintering (SPS) was carried out by Daoush *et al.* [41]. The preparation of electroless CNT/Cu nanocomposite powder included the cleaning, funtionalization and sensitization of nanotubes and introducing them in copper bath. The prepared nanocomposite powder was then subjected to SPS. The compaction pressure of 50 MPa and temperature of 550°C was adopted for consolidation of nanocomposite powder. The SEM analysis showed uniform dispersion of nanotubes without any agglomeration for nanocomposites with low nanotube fraction. With 20 vol% of nanotube in the nanocomposite, the agglomeration was observed and the clusters of nanotubes were present at the grain boundaries of copper matrix. Chen *et al.* [42] reported the synthesis of CNT/Ni by electrodeposition technique in which carbon nanotubes were suspended in an electrolyte held at 48°C with pH of bath maintained at 4.6 and cathode current density in the range of 0.5–3.4 A/dm^2. The nanocomposite coating was deposited on a steel substrate. The nanotubes were also shortened by milling in planetary ball mill for 10 hours and the effect of nanotube length on deposition was studied. The nanotubes with short length were found deposited in greater concentration compared to that of long nanotubes. Long nanotubes tend to form agglomeration easily and with the high concentration the deposition decreases gradually. Agitation of nickel bath was necessary to suspend and transport the carbon nanotubes. The maximum value of nanotube content deposited was obtained when the agitation rate for long nanotubes was 120 rpm and 150 rpm for short nanotubes. The dispersion of nanotubes was found to be uniform in the nickel matrix with nanotubes aligning themselves in the direction perpendicular to the coating surface.

The obtainment of nanocomposite coating with good properties depends on the stable dispersion of carbon nanotubes in the plating bath. The nanotubes tend to form agglomerates due to high surface energy which can lead to non-uniform dispersion in the nanocomposite coatings. The nanotubes can be suspended either by ultrasonication or magnetic stirring. Surfactants are usually used for the stabilization of nanoparticles in the plating bath to facilitate the deposition of coating and to lower the interfacial tension between a liquid and solid surface. Guo *et al.* [43] studied the effect of different surfactants on the electrodeposition of CNT/Ni nanocomposites. The CNT/Ni nanocomposite coating was carried out on steel substrate by maintaining the bath temperature at 54°C, pH at 4 and current density at 4 A/dm^2. The surfactants used were SDS (sodium dodecylsulfate $CH_3(CH_2)_{11}SO_4Na$) and cationic surfactant CTAB (hexadecyltrimethylammonium bromide $CH_3(CH_2)_{15}(CH_3)_3NBr$). It was observed that using CTAB, the carbon nanotubes content in the deposited coating was increased compared to that of the bath using SDS. The CNT/Ni nanocomposite coating prepared using SDS surfactant showed increased hardness and improved adherence of nanotubes with the nickel matrix. Although the nanotubes content in nanocomposite coating was more with the addition of CTAB, there was poor adhesion between nanotubes and matrix. Due to this, hardness and corrosion resistance of coatings were severely affected. The addition of surfactants also influenced the preferred orientations of deposited grains. Without surfactant the preferred orientation was found to be (2 2 0) plane, while with SDS and CTAB the orientations were (2 0 0) and (1 1 1) plane respectively. Ramesh *et al.* [44] have successfully developed CNT/Ni electrocomposites using the sediment co-deposition technique. They have reported enhanced strength of composites coupled with good ductility.

For electroless and electrodeposition techniques the processing parameters that govern the morphology of coatings are different. In the case of electroless coatings, the nature of coating depends upon the type of reducing agent, complexing agent and the surfactant used. The reducing agents, such as hypophosphite baths, are used due to their higher deposition rates and increased stability. Complexing agent, such as ammonium fluoride, improves the deposition rate and buffering capability of Ni-P bath. Surfactant plays an important role for improving the suspension and distribution of nanotubes. In the case of electrodeposition, current density and type of power source plays an important role in uniform deposition of coating. Optimum current density is desired for better and high deposition rate of carbon nanotubes in the metal matrix. The pulse plating is usually considered for uniform deposition of coating over the DC power supply as it provides more nucleation sites for metal ions to deposit on the nanotube surface.

9.3.3 Spray Forming

Gas-turbine engine components are often exposed to hot products of combustion moving at high velocity, while hydraulic turbine components such as guide vanes, runner and needles are often subjected to erosive environment leading to degradation of these components. Much effort has therefore been devoted to the development of surface modification and coating techniques to reduce degradation and prevent failure. Various spray forming techniques such as high-velocity oxy-fuel (HVOF), plasma spray, arc spray, detonation spray and cold spray are used for developing a wide variety of coatings to improve the performance and durability of various components.

Spray forming techniques are being employed in the synthesis of carbon nanotubes reinforced metal matrix composites. Agarwal and coworkers have extensively worked on both thermal and cold spray synthesis of aluminum and its alloy-reinforced carbon nanotubes metal matrix nanocomposites. Laha *et al.* [45] reported the synthesis of freestanding structures of CNT/Al-Si nanocomposites by plasma spray forming technique. In plasma spray forming, the coating powders are heated in argon fed arc formed between the cathode and anode. The coating powders in their molten state were propelled from the gun and projected onto the work piece at velocities of 125–600 m/s. A shielding chamber comprised of plasma gun and component to be coated in inert atmosphere is used to prevent oxidation of coating material. Gas atomized Al-23 wt% Si with carbon nanotubes were initially blended in a ball mill for 48 hours to obtain uniform dispersion. Due to their low density, carbon nanotubes are difficult to spray, and there is a possibility that they can be carried out in gas stream rather than deposited on substrate. To avoid this the Al-Si powders larger micron size were used to carry the nanotubes during spraying. The synthesized nanocomposite had a taper length of 100 mm, 62 mm diameter and thickness of 2 mm. After ball milling nanotubes were found to be uniformly dispersed and resided on the surface of the Al-Si powder particles. The SEM analysis of the nanocomposite coating showed presence of micropores, nanotubes at the edges of splats, splats of both the individual Al-Si powders and Al-Si-carbon nanotube agglomerates. The pores formation at the agglomerated splats was ascribed to the partial melting of agglomerates. The retention of carbon nanotubes at high temperatures is important as there could be a possibility of melting. However in this case, the controlled plasma parameters and partial melting of larger agglomerates of Al-Si-carbon nanotubes lead to their retention in the nanocomposite coating. The XRD analysis also confirmed the presence of graphite phase in the nanocomposite coating. In another attempt, Al-Si alloy with 10 wt% multiwalled carbon nanotubes was fabricated by high velocity oxy-fuel (HVOF) spraying. In HVOF spraying the heat source is a combustion flame produced by combustion of fuel gas and oxygen. The velocity of coating material exiting from the flame is 700–1800 m/s. The cooling rates during solidification of molten splats are normally 10^3–10^5 K/s. The well-blended nanocomposite powder was sprayed on polished aluminium alloy 6061 rotating mandrels of different sizes. The microstructure of HVOF-coated nanocomposite showed dense and layered splat structure with well-defined splat boundaries with nanotubes trapped in between. The strong adherence and low porosity achieved is due to high spray velocity that causes the mechanical bonding between the splats. However, due to high velocity, the carbon nanotubes in the deposited coating were shortened and broken [46].

Other than these thermal spray techniques, a relatively new coating technique known as cold spray is used to produce dense coatings with minimal effects like oxidation or phase changes during the coating process. A high pressure gas flowing through a de Laval type of nozzle is used to accelerate particles to supersonic velocities of 600–1500 m/s. The cold spraying also has some disadvantages such as only plastically deformable metal and alloys can be used as coating materials. Recently this technique was used to developed composite materials with micro- and nanosized filler materials with uniform dispersion. Bakshi *et al.* [47] worked on synthesis of multiwalled carbon nanotubes reinforced Al-11.6%Si-0.14%Fe nanocomposites. Spray drying was used to disperse carbon nanotubes inside the Al-Si eutectic powders. Further, this CNT/Al-Si alloy powder was mixed with pure aluminium powder in a turbula mixer for about one hour. Helium gas was used as a main gas and nitrogen gas kept at 0.1 MPa pressure higher than that of helium gas was used as nanocomposite powder carrier gas. Grit-blasted aluminium alloy 6061 was used as a substrate material for spray deposition. Under the high velocity (600 m/s) impact the spray-dried agglomerates were disintegrated leading to entrapment of Al-Si and nanotubes in between the aluminium particles. Nanocomposite coatings with a thickness of 500 μm were prepared with a carbon nanotubes content of 0.5 and 1.0 wt%. The coatings displayed uniformly distributed carbon nanotubes in between the aluminium splats. A decrease in the length and diameter of the nanotubes was observed due to milling and cold spraying. The TEM analysis of fractured ends showed a fracture of the nanotubes by impact and shear mechanisms. The fracture of the nanotubes was attributed to the impact or shear forces exerted by the striking of incoming and preceding Al-Si particles on the horizontal or inclined surfaces. Peeling of carbon nanotubes layers was also observed due to shearing action between Al-Si particles. The main advantage of this process was the avoidance of oxidation of nanocomposite powders during spraying or any chemical interaction between Al-Si particles and nanotubes after spraying. In a recent study, successful fabrication of multiwalled carbon nanotubes reinforced copper nanocomposites by low pressure cold spray process was reported [48]. Initially the copper powder and carbon nanotubes were blended in a planetary ball mill in argon-protected atmosphere for 20 hours. The MWCNT/Cu agglomerates were cold sprayed onto a pure aluminium substrate using a low air pressure of 0.5 MPa. The SEM analysis of the ball-milled nanocomposite powder particles had a spherical shape due to long milling duration. The surface observation showed the dense microstructure of nanocomposite coating with uniform dispersion of carbon nanotubes. It also was observed that the nanotubes maintained their fibrous form after cold spraying, while some nanotubes suffered shortening in length due to milling operation. The TEM analysis revealed that the grain sizes of copper particles were less than 300 nm. Structural damage to carbon nanotubes due to ball milling and cold spraying was analyzed using Raman spectra obtained from raw nanotubes, ball-milled nanocomposite powders and cold sprayed nanocomposite coating. Ball-milled nanocomposite powders showed relatively broadened peak of D and G bands indicating the damage caused to the nanotubes during processing. The Raman spectra of cold sprayed nanocomposite coating was similar to that of ball-milled powder but showed the appearance of D band, indicating the partial cutting of nanotubes. The TEM analysis also showed that the outer surfaces of carbon nanotubes were firmly attached to the copper matrix with strong compression due to cold spraying. Interfacial studies revealed the absence of any intermediate compounds.

Though the spray forming techniques have been successfully applied to obtain real-life applications of various metals and alloys, some time and effort is needed to realize the carbon nanotube reinforced metal matrix nanocomposites.

9.3.4   Liquid Metallurgy

Liquid metallurgy is used to produce bulk components with near net-shape at a faster rate of processing. This technique is already been employed in large-scale fabrication of metal matrix composites. Of these, the metals such as aluminium, copper, titanium and magnesium are the most popular matrix materials for reinforcements like alumina, SiC, Si_3N_4 and graphite. The reinforcement used to fabricate metal matrix composites by liquid metallurgy are generally in fiber, whisker and particulate form, but in micron size. This technique is now being used to synthesize carbon nanotubes reinforced magnesium and aluminium matrix nanocomposites. However, it is difficult to add the nanotubes directly to the molten metal owing to their low density and possibility of reactivity with the molten metal which can be deleterious to the final nanocomposites.

Li *et al.* [49] have fabricated CNT/Mg alloy AZ91 nanocomposites by a two-step method. In the first step, nanotubes were added to a solution consisting of copolymer dissolved in ethanol and then ultrasonicated for 15 minutes. After this, AZ91 chips were added and the complete suspension was stirred at 250 rpm. The dried nanotube-coated AZ91 chips were melted in a crucible held at 650°C. The melting operation was carried out under inert atmosphere to avoid oxidation. The molten metal was further stirred at a speed of 370 rpm for 30 minutes to obtain homogenous dispersion of nanotubes. The nanotubes dispersed AZ91 chips before casting that were subjected to microstructural studies showed good dispersion and undamaged nanotubes. However, optical micrographs of as-cast nanocomposites have demonstrated that the addition of nanotubes to AZ91 matrix did not contribute to grain refinement. In another attempt, the same authors extended the liquid metallurgy route to synthesize CNT/aluminium alloy Al239D nanocomposites. Multiwalled carbon nanotubes wrapped in an aluminium foil were placed in the entrance of the die and molten aluminium metal was pushed into the die by a high-velocity piston. The turbulent flow of molten metal inside the die led to the dispersion of nanotubes in the melt. Due to low content of carbon nanotubes (0.05 wt%) and limited contact time with the aluminium melt, no carbide formation was observed [50].

9.3.5 Other Techniques

Other than the conventional processing techniques, several novel and modified techniques such as molecular level mixing, friction stir processing and disintegrated melt deposition (DMD) methods have been employed in the synthesis of various metal matrix nanocomposites reinforced with carbon nanotubes.

9.3.5.1 Molecular Level Mixing

Dispersion of carbon nanotubes in the metal matrix using conventional techniques is a major problem during the processing of nanocomposites. Hong *et al.* [51] from Korea Advanced Institute of Science and Technology in Korea have reported a novel technique to synthesize uniformly distributed carbon nanotubes in the copper matrix by a molecular level mixing process. It is known that treating carbon nanotubes with acids, such as nitric and sulfuric acids, can oxidize them and incorporate hydroxyl groups on the end caps. The functionalized carbon nanotubes form a stable suspension in the solvent and strongly interact with metallic ions. A reaction at the molecular level takes place between functionalized carbon nanotubes with metal ions in a solution to form metal complexes. The work of Cha *et al.* [51] stabilized suspension of carbon nanotubes by attaching functional groups to their surfaces. Once the stable suspension was formed, a salt containing copper ions $(Cu(CH_3COO)_2 \cdot H_2O)$ was introduced in the nanotube suspension. A solution of copper ions, suspended nanotubes, solvent and ligands were sonicated for 2 hours to promote a reaction between copper ions and carbon nanotubes. In the next step, the solution was dried by heating at 100–250°C in air during which the copper ions on the surface of the carbon nanotubes were oxidized to form powders. In the last step, the mixture of CNT/CuO was reduced to CNT/Cu by heating in air and reducing under hydrogen atmosphere. Microstructural analysis of both CNT/CuO and CNT/Cu powders displayed homogeneous dispersion of nanotubes located within the powder particles. The morphology of nanocomposite powders was spherical, with carbon nanotubes being reinforced inside the matrix rather than on surface. Further, the nanocomposite powders were consolidated using SPS at 550°C for 1 minute with an applied pressure of 50 MPa. The consolidated nanocomposite showed a homogeneous dispersion with carbon nanotubes forming a network within the copper grains.

9.3.5.2   Friction Stir Processing

Friction stir processing is another technique used to fabricate carbon nanotubes reinforced magnesium and aluminium matrix nanocomposites. Friction stir processing is a solid-state joining process in which coalescence is achieved by frictional heat induced by mechanical rubbing between two surfaces combined with pressure. The friction induced by the rotating tool pin plunged into the surface of metals or alloys to be processed gives rise to temperature at the joint interface to the hot-working range of metals or alloys involved. During the synthesis of composites, the reinforcing particles are placed in a groove on the surface of the matrix material and the rotating tool pin is rastered on the groove at various speeds to disperse the reinforcement in the matrix material. Morisada *et al.* [52] developed the carbon nanotubes reinforced magnesium alloy

31 (AZ 31) nanocomposite. The nanocomposite was prepared by filling the multiwalled carbon nanotubes inside a groove made on the surface of AZ 31 alloy, and this surface was friction stirred by making use of a columnar-shaped friction-stir processing tool rotating at a speed of 1500 rpm with a travel speed of 25 to 100 mm/min. Out of the various travel speeds, the nanocomposite prepared at 25 mm/min showed a good dispersion of carbon nanotubes. On the other hand, nanocomposite prepared at 100 mm/min showed entangled nanotubes inside the AZ 31 matrix. This was ascribed to the fast travel speed of the rotating pin which led to insufficient heat and suitable viscosity in the AZ 31 matrix for distribution of nanotubes. This process has an advantage of producing nanocomposites with fine grain size. The nanocomposites with nanotubes had some grain sizes below 500 nm, while the unreinforced AZ 31 alloy had grain sizes in the range of microns. In another work, Lim *et al.* [53] fabricated multiwalled carbon nanotubes reinforced Al 6111-T4 and Al 7075-T6 alloys. The nanotubes were placed in between these two alloys and were processed using a 10 mm diameter cylindrical tool rotated at 1500 to 2500 rpm. The effect of different rotating speeds and plunge depth on the microstructures of nanocomposites have been reported. The tool rotation speed of 1500 rpm and the shoulder penetration depth of 0.24 mm had a void-free stir zone, while the same rotation speed and a penetration depth of 0.03 mm showed extensive voids in the stir zone. Though the increasing depth showed less voids some cracks were observed in between the lamellae in the stir region. Entangled and fractured nanotubes were observed in these cracks and were not embedded inside the aluminium alloy matrices. However, none of the above parameters employed were successful in uniform dispersion of nanotubes inside the aluminium alloy matrix.

9.3.5.3   Disintegrated Melt Deposition

Gupta and coworkers have extensively worked on CNT-reinforced magnesium and its alloy nanocomposites by disintegrated melt deposition technique (DMD). The magnesium and its alloys are chosen on the basis of their low density (1.74 g/cm³) and good castability. In DMD technique the magnesium turnings and carbon nanotubes placed in a graphite crucible were heated to a temperature of 750°C in inert atmosphere. The molten metal with nanotubes was stirred with a ZIRTEX 25 (86% ZrO_2, 8.8% Y_2O_3, 3.6%

SiO_2, 1.2% K_2O and Na_2O)-coated mild steel impeller at a speed of 450 rpm for 5 minutes. The molten slurry was disintegrated by two jets of argon gas and deposited on a substrate. The ingot obtained from this process was further subjected to hot extrusion at 350°C with an extrusion ratio of 20.25:1. The macrostructural observation of nanocomposites showed defects as such and the absence of macropores indicated good solidification [54]. In another study the magnesium alloy (ZK60A–Mg-Zn-Zr system) with carbon nanotube was processed with the same processing technique. The as-cast and extruded rod macrostructure of nanocomposite showed no shrinkage or macropores. The microstructure of nanocomposites showed equiaxed and smaller grains compared to that of ZK60A alloy. This implies that the carbon nanotubes are not only acting as nucleation sites and obstacles to the grain growth during cooling. It was interesting to note that the intermetallic phase in the nanocomposites was much lower and could not be observed in either XRD or FESEM. This was attributed to the segregation of dissolved zinc at CNT/ZK60A interface leading to lowering of intermetallic phase in nanocomposites [55].

Though these novel techniques have been used in successful fabrication of nanocomposites, there are still problems related to uniform dispersion, wetting and orientation of carbon nanotubes inside the metal matrix that have to be addressed.

9.4   Mechanical Properties of Carbon Nanotube Reinforced Metal Matrix Nanocomposites

Extraordinary mechanical properties like Young's modulus of 1 TPa and tensile strength up to 63 GPa make carbon nanotubes the ultimate high-strength fibers for use as an ideal reinforcement in the development of nanocomposite materials [56]. It is interesting to note that most of the research papers have reported the mechanical properties of the nanocomposites for possible structural applications. In this section the effect of carbon nanotubes on the mechanical properties of different metal matrix nanocomposites will be discussed.

9.4.1   CNT/Al Nanocomposites

Aluminium is the most abundant light-weight metal used in various automotive and aerospace applications. With a low melting point of

660°C, elastic modulus of 69 GPa, yield and tensile strength values of 35 and 90 MPa, it is the first choice for fabricating CNT/Al nano-composites [57]. Most of the work on CNT/Al nanocomposites has used the powder metallurgy route followed by hot extrusion for processing the nanocomposites. Kuzumaki *et al.* [29] studied the mechanical properties of CNT/Al nanocomposites by conducting tensile tests on nanocomposites with 5 and 10 vol% nanotubes. The tensile strength values of nanocomposites were less than that of pure aluminium, however, with an increase in annealing time the strength of aluminium decreases, while that of nanocomposites were stable even after 100 hours of annealing time. The fracture studies showed the agglomeration of carbon nanotubes in the nanocomposite. The TEM studies showed the absence of aluminium carbide at CNT/Al interface, which implies that the nanotubes are chemically stable until they are fractured by external stress. George *et al.* [31] observed an increase of 12% and 23% in the value of Young's modulus of CNT/Al nanocomposites with the addition of 0.5% and 2% volume fraction of multiwalled carbon nanotubes, while the nanocomposites reinforced with single-walled carbon nanotubes did not show any considerable increment. However, the nanocomposite reinforced with K_2ZrF_6-coated single-walled nanotubes showed an increase of 33% in the value of elastic modulus. The elastic moduli of all the nanocomposites except the nanotubes coated with K_2ZrF_6 were in good agreement with the shear lag model. The experimental yield strength values of nanocomposites with multiwalled carbon nanotubes of 0.5% and 2% volume fraction were about 86 and 99 Mpa, while for aluminium it was 80 MPa. The experimentally obtained values of yield strength were in close agreement with those predicted by Orowan looping mechanisms (0.5% - 90.57 MPa and for 2% - 101.14 MPa). The dislocation loops around the nanotubes or any other hard reinforcement particles is indication of Orowan looping [58]. However, no dislocation loops were observed on the carbon nanotubes.

Kwon *et al.* [59] studied the mechanical properties of powder extruded CNT/Al nanocomposites by tensile tests and nanoindentation. For the tensile tests the samples were machined according to EN ISO 6892, while the nanoindentation test was carried out at a load of 50 mN with a loading time of 10 seconds. The tensile strength of 1 vol% carbon nanotubes reinforced aluminium nano-composite (297.7 MPa) obtained by tensile tests showed increment by 2.5 times compared to that of pure aluminium (116.4 MPa). The yield strength of nanocomposites predicted using nanoindentation stress-strain curve were constructed by setting the zero-load and displacement point with the theoretical estimates of the elastic moduli of the samples. The predicted yield strength values from nanoindentation were compared with that of tensile stress-strain curve and were found to be similar. However, the predictable yield strength from nanoindentation tests should be carefully considered because this method is still not standardized. This technique is especially useful in cases where the materials are difficult to machine or in such cases where producing standard tensile specimen is difficult.

Liu *et al.* [60] fabricated the CNT/Al 2009 nanocomposites by powder metallurgy, forging and subjecting them to friction stir processing (FSP). After FSP the samples were solutionized at 495°C for 2 hours, water quenched and then naturally aged for 4 days. The optical and SEM images of nanocomposites revealed that the large nanotube clusters were significantly reduced by FSP and the nanotubes were individually dispersed inside the Al 2009 matrix. It was interesting to note that the forged nanocomposites with 1 and 3 wt% of nanotubes had a tensile strength of 392 and 298 Mpa, which were much less compared to that of unreinforced Al 2009 alloy. However, after subjecting to 4 pass FSP, the tensile strength values of both 1 and 3 wt% nanotubes reinforced nanocomposites were enhanced to 477 and 466 MPa when compared to that of Al 2009 alloy, which was 417 MPa. This was attributed to uniform dispersion and grain refinement by FSP. Lipecka *et al.* [61] reported the effect of temperature on the grain size as well as hardness of carbon nanotube reinforced aluminium alloy S790 (Al-Zn11-Mg2-Cu). The microstructure of extruded nanocomposite was analyzed using a scanning transmission electron microscope and Vickers hardness was measured using diamond indenter with 50 gm of load with force duration of 20 seconds. The nanotubes were successful in retarding grain growth of S790 alloy, as the grain mean diameter at 450°C annealing temperature was 566 nm compared to that of 500 nm at room temperature. With the addition of 3 wt% nanotubes the microhardness of nanocomposite (130 $HV_{0.05}$) after 450°C saw a drop of 15% compared to annealing (150 $HV_{0.05}$). The effect of nanotube morphology and diameter on the mechanical properties of nanocomposites was studied [62]. Multiwalled carbon nanotubes of mean diameter 40 and 140 nm were added to pure aluminium with a content varying from 0.5, 1, 1.5, 2 and 5 wt.%. The tensile test was conducted at a crosshead speed of 1 mm/min to obtain the tensile strength of unreinforced and CNT_{40} and CNT_{140}/Al nanocomposites. The hardness and Young's modulus was measured using nanoindentation at a maximum indentation load of 700 mN. The tensile strength measured for CNT_{40}/Al nanocomposites provided more strengthening than that of CNT_{140}/Al nanocomposites for up to 1.5 wt% nanotubes. The nanoindentation results also showed that the nanotubes with 40 nm mean diameter provided more stiffening and hardness than that of nanotubes with 140 nm mean diameter. This can be due to easy and better dispersion of small diameter nanotubes in the aluminium matrix. The mechanical properties of cold sprayed CNT/Al-Si nanocomposite coatings were measured using nanoindentation by applying a load of 1000 µN and holding for 2 seconds [63]. The elastic modulus and hardness of nanocomposites were found to increase with an increase in nanotube content. With 1 wt% nanotube content in CNT/Al-Si nanocomposite, the elastic modulus and hardness was about 94.8 GPa and 0.64 GPa, while for unreinforced Al-Si alloy it was 45 GPa and 0.47 GPa, respectively. The strengthening of nanocomposites was attributed to the impediment of dislocation motion by nanotubes.

With new techniques like FSP, the breaking of nanotube clusters in the aluminium matrix with uniform dispersion and grain refinement are key factors for developing CNT/Al nanocomposites with better mechanical properties for applications in the automotive and aerospace industries.

9.4.2   CNT/Cu Nanocomposites

Most of the reports on the copper-reinforced carbon nanotubes have focused on the improvement in mechanical and thermal properties. Almost all work on the fabrication of CNT/Cu nanocomposites has been done using the powder metallurgy technique. The compressive tests of CNT/Cu nanocomposites fabricated by novel molecular mixing process were performed on an Instron machine with a crosshead speed of 0.2 mm/min [51]. The uniformly dispersed nanotubes strengthened the copper matrix, which can be observed in the increment of yield strength and Young's modulus values of the nanocomposites. The yield strength of unreinforced copper was 150 MPa while for nanocomposites with 5 and 10 vol% the values were 360 and 455 MPa. The elastic modulus of pure copper was increased from 82 GPa to 112 and 135 Gpa, respectively, for

5 and 10 vol% nanotubes reinforced CNT/Cu nanocomposites. The strengthening in nanocomposites was due to better load transfer efficiency and strong interfacial strength between nanotubes and copper matrix that originated from chemical bonds formed during molecular level processing.

Li *et al.* [64] reported the development of CNT/Cu nanocomposites high strength coupled with good ductility. The mechanical properties of high-pressure torsion processed CNT/Cu nanocomposites were carried in Hysitron Triboindenter nanoindenter. The samples of pillar shape for compression tests were prepared from focused ion beam (FIB) technique. The compression test on a pillar sample of 5 μm diameter and 11 μm length was carried out at a strain rate of 1×10^{-3} s^{-1}. The yield strength of CNT/Cu and copper were 1125 and 738 Mpa, while plastic strains were 28.4% and 30.7%, respectively. In addition to this, the nanotubes located at both grain boundaries and grain interiors lead to grain refinement and narrower grain size distribution. Koppad *et al.* [34] found that the Vickers hardness of multiwalled carbon nanotubes reinforced nanocomposites increased with an increase in nanotube content. The hardness of 4 wt% MWCNT/Cu nanocomposite (141 HV) was 2.04 times higher than that of unreinforced copper (69 HV). Obtaining the good interfacial strength is necessary for better load transfer from matrix to carbon nanotube. In order to obtain a good bonding between copper matrix and carbon nanotube, Koppad *et al.* [33] applied nickel coating to the nanotubes. This was evident in the values of elastic modulus and hardness of the nanocomposites. Both the elastic modulus and hardness of the nanocomposites increased with an increase in nanotube content, and the values also showed better results with nickel-coated nanotubes. The hardness of pure copper was 0.95 Gpa, while that of uncoated and nickel-coated MWCNT/Cu nanocomposites with 3 wt% nanotubes were 1.33 and 1.46 GPa. Similarly, the elastic modulus of pure copper was 98 Gpa, while that of uncoated and nickel-coated MWCNT/Cu nanocomposites with 3 wt% nanotubes were 125 and 133 GPa. Apart from this the dislocation density in nanocomposites increased with the increasing nanotube content, which explains the strengthening by thermal mismatch model. Due to the coefficient of thermal expansion mismatch between copper (17×10^{-6} K^{-1}) and carbon nanotube (1×10^{-6} K^{-1}) the dislocations are generated around the nanotubes as shown in Figure 9.4. This results in punching of dislocations at the CNT/Cu interface leading to work hardening [65]. In this way nanotubes

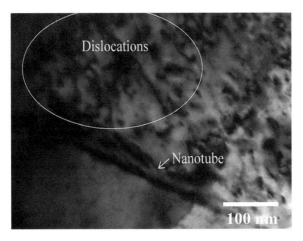

Figure 9.4 Transmission electron microscope (TEM) image of dislocations generated at the vicinity of multiwalled carbon nanotube in 2 wt% MWCNT/Cu nanocomposites.

contribute to the work hardening leading to an increment in the values of hardness of nanocomposites. Singhal *et al.* [66] reported the compressive and flexural strength of CNT/Cu nanocomposites fabricated by molecular level mixing, cold pressing and sintering. The values of compressive and flexural strength for 1.5 wt% MWCNT/Cu nanocomposites were 955 and 306 Mpa, whereas for pure copper the values were 250 and 210 MPa. The increment in mechanical properties of nanocomposite was due to the high density achieved due to high sintering temperature. In one exciting work, CNT/Cu-Sn nanocomposites were compacted into bearings of size 5.4 mm (outside diameter), 2.0 mm (inside diameter) and 2.5 mm (highness) by applying a pressure of 400 MPa. After sintering at 760°C for 2 hours, CNT/Cu-Sn porous oil bearings were obtained. The mechanical properties such as hardness and crushing strength of CNT/Cu-Sn porous bearings with different nanotube content were measured. The microhardness was found to improve with increasing nanotube content. The microhardness and crushing strength of 2 wt% CNT/Cu-Sn porous bearings were 490 HV and 300.3 Mpa, while that of Cu-Sn bearing was 231 HV and 212.2 MPa. The copper coating on the nanotube reacts with Sn forming α-CuSn, which might be useful in good interface adhesion and better stress transfer. The oil storage capacity of CNT/Cu-Sn porous bearings was carried out, which greatly influenced the porosity and the pore size. The oil content of Cu-Sn bearing was increased from

20.6% to 23.42% for 1 wt% nanotube reinforced CNT/Cu-Sn porous bearing [67].

These works clearly demonstrate that carbon nanotubes not only improved the yield strength or hardness of nanocomposites but also the oil storage capacity for lubrication performance of porous oil bearing.

9.4.3   CNT/Mg Nanocomposites

Magnesium and its alloys reinforced with particulate composites are very popular due to their easy fabrication, increased production and low reinforcement costs. However, the limited increment in tensile strength and ductility of particulate magnesium nanocomposites and the call for better reinforcements have led to carbon nanotubes. To study the mechanical behavior of extruded carbon nanotube-added magnesium nanocomposites, macrohardness and tensile tests were conducted. The nanocomposites with 1.3 wt.% nanotube showed a better yield strength of 140 MPa with improved ductility of 13.5% compared to that of pure magnesium, which had values of 126 MPa and 8%. The presence of carbon nanotubes leads to activation of cross slip in non-basal slip planes, which was the main reason why the nanocomposites showed good ductility [54]. Addition of carbon nanotubes to the magnesium alloy ZK60A showed a decrease in the value of microhardness due to the formation of intermetallic phase in the matrix of the nanocomposite. However, an increase in the values of yield and tensile strength from 163 and 268 MPa for unreinforced ZK60A, to 180 and 295 MPa for 1 wt% CNT/ZK60A nanocomposite was observed. The compressive stress of nanocomposite (110 MPa) was found to be lower than that of unreinforced alloy (128 MPa). The drop was attributed to compressive shear buckling of carbon nanotubes in nanocomposite [55]. Li *et al.* [50] conducted the tensile studies on CNT/AZ 91 nanocomposites processed by casting and high pressure die casting. The nanocomposites showed that the compression at failure and compressive strength was enhanced with the addition of nanotubes. No significant increment in strength values were observed with the increasing amount of nanotubes. The mechanical properties of nanocomposites produced by pressure die casting such as tensile strength and elongation at fracture were increased from 8% and 27%, respectively, when compared to that of unreinforced aluminium alloy.

An attempt was made to study the corrosion behavior of AZ31B magnesium alloy reinforced with carbon nanotubes [68]. For this immersion tests were performed to investigate the corrosion resistance of developed nanocomposites. The cylindrical samples of 12 mm diameter and 10 mm length were immersed in 1200 ml of 0.51 M sodium chloride (NaCl) solution at pH of 6.2. A polarization test was also carried out to study whether the nanocomposite samples were prone to Galvanic corrosion. The sample size taken for testing was $15 \times 15 \times 3$, mm^3 and was immersed in 75 ml of 0.017 M NaCl solution at 6.47 pH value and about 1 hour for stabilizing the sample surface. Significant mass loss was observed in CNT/AZ31B nanocomposites after the immersion test in 0.51 M NaCl solution. Corrosion products such as $Mg(OH)_2$ were observed in the vicinity of carbon nanotubes. The white debris of $Mg(OH)_2$ piled around nanotubes was due to strong galvanic corrosion at the interfaces between nanotubes and the matrix. Here the surface potential difference was high enough to form the galvanic cell between the AZ31B alloy and nanotubes. Throughout testing the pH values of the nanocomposites tended to increase, and at one point in time reached a higher value than that of unreinforced alloy. This indeed decreased the corrosion resistance of nanotube reinforced AZ31B alloy.

9.4.4   CNT/Ti Nanocomposites

Titanium and its alloys are widely used in various applications due to their high specific strength and high corrosion resistance. Threrujirapapong et $al.$ [69] reported the synthesis of carbon nanotube reinforced fine (Ti_{Fine}) and sponge titanium (Ti_{Sponge}) powders by spark plasma sintering and hot extrusion at 1000°C. The mechanical properties of hot-extruded CNT/Ti_{Fine} nanocomposites such as yield stress, tensile strength and hardness were 592 MPa, 742 MPa and 285 HV$_{0.05}$ compared to that of 423 MPa, 585 MPa and 261 HV$_{0.05}$ for unreinforced Ti_{Fine}. Similarly the mechanical properties of CNT/Ti_{Sponge} nanocomposites were enhanced due to addition of nanotubes. The increase in tensile strength values was attributed to the dispersion strengthening effect of nanotubes along with the in $situ$ formed fine titanium carbide particles. In another work, Xue et $al.$ [70] reported the elevated compressive properties of multi-walled carbon nanotubes reinforced titanium nanocomposites. The SPS processed nanocomposites showed the formation of TiC at the CNT/Ti interface. The compression test of nanocomposite was carried out at 800, 900 and 1000°C with a strain rate of 1×10^{-3} s^{-1}. With the increasing sintering temperature, the initial compressive yield strength of 143 MPa at 800°C decreased to114 MPa at 900°C, and then a slight rise was observed at 1000°C (129 MPa). The higher sintering temperature of nanocomposites facilitates the reaction between nanotubes and titanium matrix leading to formation of TiC at the interface. The formed TiC, however, had a lower strengthening effect compared to that of nanotubes, so the decrease in yield strength declined at elevated temperatures.

Other than the above mentioned metal matrices, metals like CNT/Ni nanocomposites were developed for better corrosion and wear resistance [71], CNT/Sn nanocomposites for improving the mechanical reliability of solder joints [72] and CNT/Ag nanocomposites for heat dissipating systems [73].

9.5 Strengthening Mechanisms

Several hypotheses have been reported about the possible reinforcing mechanisms that could explain the improved mechanical properties by addition of carbon nanotubes. Strengthening mechanisms of composites have been extensively studied by many researchers [74–78]. To study the direct strengthening of continuous fiber-reinforced composites, Cox et al. [79] developed the continuum shear lag model. This was followed by the prediction of yield strength of particle-reinforced metal matrix composites by considering the dislocations generated by the coefficient of thermal expansion mismatch [80]. The same strengthening mechanisms have been employed to study the strengthening effect of carbon nanotubes on the various properties of nanocomposites. However, it is important to note that the strengthening of nanocomposites can be attributed to the synergistic effect of thermal mismatch between carbon nanotubes and metal matrix, grain refinement due to the addition of carbon nanotubes and load transfer from the metal matrix to the carbon nanotubes. In this section we will discuss the strengthening mechanisms applied to nanocomposites by various authors.

George et al. [31] proposed three mechanisms to study the strengthening mechanism of CNT/Al nanocomposite. The strengthening mechanisms proposed were thermal mismatch, shear lag model and Orowan looping. The difference in coefficient of thermal expansion between the carbon nanotube and aluminium matrix results in generation of dislocations. Increased dislocation density promotes an increase in the yield strength of the nanocomposite. It is interesting to note that this model overestimates the yield strength values compared to that of experimental yield strength. For example, in the case of 2% volume fraction nanotube reinforced aluminium nanocomposite, the experimental yield strength obtained from tensile testing was 99 Mpa, while that by thermal mismatch model was 197.34 MPa. In Orowan looping mechanism the dislocation motion is inhibited by the particles leading to the bending of dislocations around the particles forming dislocation loops. In this case the yield strength predicted (101.14 MPa) was close to the experimental value. However, TEM images showed the absence of the dislocation loops around the nanotubes. This was due to the collapse of dislocation loops at the interface of CNT/Al nanocomposite. The elastic moduli obtained by tensile tests (for 2% CNT/Al – 87.37 GPa) were in good agreement with that predicted by shear lag model (for 2% CNT/Al – 85.84 GPa). In their work, Bhat *et al.* [81] showed that the strengthening in CNT/Cu-Sn nanocomposites was due to the combined effect of thermal mismatch, load transfer and Orowan looping as the experimental values were in close match with the values obtained by these strengthening models. Work on CNT/Al-Cu nanocomposites by Nam *et al.* [82] demonstrated an enhancement of 3.8 times in yield strength and 30% in elastic modulus compared with that of Al-Cu alloy. Two types of nanotubes, one with acid treatment and the other with PVA coating, were used. The difference between them is that the acid-treated nanotubes were shortened, with an aspect ratio of 14:20, and those that were PVA coated were undamaged, with an aspect ratio of 40:60. The yield strength of both acid-treated and PVA-coated CNT/Al-Cu nanocomposites was about 3.8 times higher than that of unreinforced Al-Cu alloy (376 and 384 MPa compared to 110 MPa). These results imply that regardless of nanotube aspect ratio the strengthening effect of nanotubes is the same. The estimated elastic modulus values for nanocomposites by shear lag model were in good agreement with those of measured values. The strengthening mechanisms of nanocomposites varied with the aspect ratio of nanotubes. The fact that the acid-treated short nanotubes behaved as dispersed particles rather than a load transfer constituent suggests the dispersion strengthening mechanism in these nanocomposites. PVA-coated nanotubes with large aspect ratio exhibited the load transfer strengthening mechanism. But it is important to note that the increase in yield strength values is not only due to load transfer but also to dislocation related strengthening mechanisms, which are thermal mismatch and Orowan looping. Hence, it can be concluded that the strengthening of CNT/Al-Cu nanocomposite was mainly by the synergistic effect of all the strengthening mechanisms.

Strengthening of nanocomposites depends on processing techniques used, nanotube dispersion, effect of processing conditions on nanotubes and interface between nanotube and metal matrix. Most of the published data clearly shows that the strengthening of nanocomposites increases for small nanotube content, but at high nanotube concentration the strengthening effect is decreased.

9.6   Thermal Properties of Carbon Nanotube Reinforced Metal Matrix Nanocomposites

It is well known that copper is the second metal after silver that has a very high thermal conductivity of the order of 400 W/mK. Owing to this fact, copper and its alloys are used where high electrical and thermal conductivity are required. The pure copper which is normally soft in nature is strengthened by adding reinforcement like tungsten particles, synthetic diamonds and carbon fibers. Metal matrix composites such as CuW, CuMo, Cu-Gr and Cu-diamond are used as heat spreaders because of their good thermal conductivity and low thermal expansion coefficient. However, these composite combinations have several drawbacks such as wettability, high cost of diamonds in the case of Cu-diamond composite, high density in the case of CuW and low machinability. In this context, carbon nanotubes possess very high thermal conductivity of the order of 3000 W/mK, low thermal expansion coefficient of 1×10^{-6} K^{-1} and a low density of 1.8 g/cm^3. This has led to the development of CNT/Cu and CNT/Al nanocomposites for thermal management applications.

Kim *et al.* [83] have studied the thermal conductivity of nickel-coated carbon nanotubes reinforced copper nanocomposites fabricated by vacuum hot pressing technique. The thermal conductivity was measured using laser flash system with an ND:YAG pulse laser. This method was based on a measurement of the propagation of time of the heat produced by laser pulse and running from bottom to top of the sample. The nanocomposite in the form of a disc was subjected to heating by using the laser beam. The temperature rise at the top of the disc was measured by infrared detector. The thermal conductivity was then calculated using the equation, $k = \alpha \times \rho \times C_p$, where ρ is the density of nanocomposites and C_p is specific heat. The thermal conductivity of pure copper obtained was close to 400 W/mK, but that of sintered Cu-Ni decreased to 150 W/mK, probably due to low thermal conductivity of nickel which is 91 W/mK. The thermal conductivity of nickel-coated CNT/Cu nanocomposite was dropped compared to that of Cu-Ni sample. Though these nanocomposites showed better mechanical properties due to better interfacial bonding provided by nickel coating of the nanotubes, they did not improve the thermal properties. Chu *et al.* [35] reported the thermal conductivity of CNT/Cu nanocomposites fabricated by SPS technique. It is observed that the thermal conductivity of CNT/Cu nanocomposites with 5 and 10 vol% nanotubes decreased slightly, but a large drop was observed in the case of nanocomposites with 15 vol% nanotubes. The drop in the thermal conductivity values was due to the interface thermal resistance between nanotube and copper matrix and poor dispersion and clustering of nanotubes in the copper matrix. The large drop in 15 vol% CNT/Cu nanocomposites was mainly due to large pores due to the clustering of nanotubes. In addition to this, due to processing conditions kinks or twists were produced in the nanotubes. The heat flows along the tube get blocked at these sites and contribute to the low thermal conductivity.

After a few unsuccessful attempts, a few researchers have also reported on the development of CNT/Cu nanocomposites with better and improved thermal conductivity. Chai *et al.* [84] reported an increment of 180% in the thermal conductivity value of nanocomposite. The CNT/Cu nanocomposite was prepared by electrochemical deposition route with a thickness varying from 10 to 50 µm. To measure the thermal conductivity the authors developed a one-dimensional, steady thermal conduction setup. The temperature sensors made up of platinum film were used to measure the temperature gradient formed over the specimen. The measured thermal conductivity of pure copper and CNT/Cu nanocomposite was 339 and 604 W/mK. The obtained conductivity of pure copper was lower than that of bulk copper, which may be due to dimension effects or the crystal structure resulting from the electrochemical deposition process. It was also observed that an increase in the nanotube concentration in the electrolyte increases the thermal conductivity. With the nanotube concentration in the electrolyte increased from 50 mg/L to 250 mg/L, the conductivity of CNT/Cu nanocomposite increased from 441 to 637 W/mK. A possible explanation for this increase in the conductivity values includes good dispersion of nanotubes inside the electrolyte as well as in the deposited coating with good interfacial bonding with the copper matrix. The effect of surface treatment on carbon nanotubes, and in turn on the thermal conductivity of CNT/Cu nanocomposites, was studied by Cho *et al.* [85]. An amorphous carbon layer was observed on the surface pristine nanotubes. This layer was removed by treating the nanotubes with different ratios (2:1 and 3:1) of sulfuric and nitric acid. The pristine and chemically treated nanotubes were added to copper. The thermal diffusivity of SPS fabricated nanocomposites was carried out using laser flash technique at room temperature. The thermal conductivity of pure copper obtained was 348 W/mK. The chemically treated (3:1) MWCNT/Cu nanocomposites (360 W/mK) showed better thermal conductivity then the treated (2:1) MWCNT/Cu nanocomposites (341 W/mK) and pristine MWCNT/Cu nanocomposites (336 W/mK). The low thermal conductivity values in the case of pristine MWCNT/Cu nanocomposites were related to the copper oxide layer on the copper and amorphous carbon layer on nanotubes. The improved conductivity values in the case of treated MWCNT/Cu nanocomposites were due to good interface bonding between nanotubes and copper with low interfacial thermal resistance.

Bakshi *et al.* [86] studied the thermal conductivity of CNT/Al-Si by both computations and experimentation. The experimental thermal conductivity value of plasma sprayed CNT/Al-Si nanocomposite was 25.4 W/mK at 50°C, whereas for Al-Si alloy it was 73 W/mK. The low thermal conductivity values were due to inter-splat porosity and nanotube clusters. The object-oriented finite element method (OOF) was used to compute the thermal conductivity 10 wt% CNT/Al nanocomposite by employing different length scales. The study of microstructure on the small length scale of the order of microns and large length scale (hundreds of microns) was carried out. The OOF computed values were in good agreement with the experimental values. The predicted thermal conductivity values by OOF calculations at micro-length scale showed an increase of 81% in Al-Si matrix due to good dispersion of nanotubes, while at large length scale the thermal conductivity was low due to low thermal conductivity of nanotube clusters.

Still, a substantial amount of work has to be done to utilize these nanocomposites in the thermal management applications, which include uniform dispersion, unidirectional alignment of nanotubes, good bonding between nanotubes and metal matrix.

9.7 Tribological Properties of Carbon Nanotube Reinforced Metal Matrix Nanocomposites

Carbon materials like graphite are excellent solid lubricants. Carbon nanotubes which possess outstanding mechanical properties are expected to provide self-lubrication. The graphene layers of nanotubes may offer lubrication during wear and can lower the coefficient of friction. Carbon nanotubes reinforced polymer and ceramic matrix nanocomposites have already shown reduction in friction and wear rates [87, 88]. An attempt to reduce the coefficient of friction and wear rate in metal matrix nanocomposites by adding carbon nanotubes has been carried out and has resulted in some encouraging results. The carbon nanotubes were added to metal matrices like nickel, copper and aluminium. Tu *et al.* [89] reported the tribological behavior of CNT/Cu nanocomposites. Before adding to copper, carbon nanotubes were coated with nickel by electroless plating method. The friction and wear tests on nanocomposites were carried out using pin-on-disc machine (ASTM G99–90). With a sliding speed of $4.7{\times}10^{-3}$ m/s and loads between 10–50 N, the weight loss was measured after each testing. The values of coefficient of friction for unreinforced copper were in between 0.2 to 0.27, while for nanocomposites they were in the range of 0.096 and 0.19 N. This was attributed to reduced contact between metallic matrix and diamond due to the presence of carbon nanotubes. The nanotubes lubricating property also contributed to a decrease in the values of friction coefficient. However, the nanocomposites with the highest volume percentage (Cu-16 vol% CNTs) showed severe wear due to the presence of high porosity, and wear rate increased rapidly at higher load. The worn surfaces observed by SEM showed that cracking and spalling were the principal wear mechanisms for the nanocomposites with higher nanotube volume percent. In another work, the copper-coated carbon nanotubes reinforced copper matrix nanocomposites fabricated by microwave sintering technique were subjected to a pin-on-disc test to study their tribological characteristics [90]. The loads in the range of 10 to 60 N were applied by maintaining the sliding speed of 2.77 m/s. A gradual reduction in wear rate was observed with the increasing nanotube volume fraction. However, due to agglomeration of nanotubes, a decrease in the value of hardness and increase in wear rate were observed in the case of nanocomposites with nanotubes beyond 15 vol%. Similarly, the coefficient of friction was less in the case of nanocomposites with up to 15 vol% carbon nanotube, beyond which an increase in coefficient friction was observed. With the formation of carbonaceous film at the contact surface, the direct contact between the nanocomposite and steel disc was avoided. Due to the self-lubricating property of the film formed, the adhesion between the contacting surfaces was reduced. At high operating loads, the wear mechanisms observed for low and high nanotube volume fraction nanocomposites were plastic deformation and flake formation-spalling.

A lot of work has been carried out on improving the wear and corrosion resistance of various substrates by electrodepositing CNT/Ni nanocomposite coating over them. For example, aluminium alloys are widely used in a variety of applications, but the matter of their poor wear resistance is of concern. To address this problem, Tu et al. [91] studied the friction and wear properties of CNT/Ni nanocomposite-coated LY12 Al alloy (4.4%Cu-1.5%Mg-0.6%Mn and rest Al). The friction and wear tests were carried out on pin-on-disc tester under dry sliding conditions with a sliding speed of 0.0623 m/s and at a load range of 12 to 150 N. The results revealed that the wear rates of nanocomposite coatings were found to decrease with increasing nanotube volume fraction at lower applied loads. The wear rates of nanocomposites rapidly increased with applied loads of more than 120 N. The worn surface analysis of the nanocomposites with higher nanotube content showed the cracking and spalling at higher applied loads. The coefficient of friction of all the nanocomposites decreased with increasing nanotube content. The authors attributed this to prevention of metal to metal contact by nanotubes which would slide or roll between CNT/Ni-coated disc and the steel pin. Chen et al. [92] studied the friction and wear properties of CNT/Ni-P nanocomposite coating prepared by electroless codeposition using ring-on-block apparatus under lubricated condition. The nanocomposite coating was applied on the test ring and block specimen of medium carbon steel. The wear tests were conducted at a sliding speed of 2.06 m/s. Prior to wear tests some of the nanocomposite coatings were heat treated at 400°C for 2 hours in vacuum. The wear rate of CNT/Ni-P nanocomposite coating was about one-fifth of that of unreinforced Ni-P coating. The wear resistance of nanocomposites further improved after heat treatment due to the formation Ni_3P phase. The worn surfaces of unreinforced Ni-P coating showed some severe wear scars while very few scars were observed in nanotube-reinforced Ni-P coating. The friction coefficient of CNT/Ni-P coatings was greatly reduced due to short nanotubes, which were easy to shear and slide in between the metal-metal friction surfaces. Bakshi *et al.* [93] studied the nanoscratch behavior of 5 and 10 wt% carbon nanotube-reinforced Al-Si nanocomposite coatings by Triboindenter. Scratches of length 20 μm were made at loads 1000, 2000 and 3000 μN at a scratching speed of 0.67 μm/s. The nanoscratch tests carried out on nanocomposite coatings revealed the scratch formation by a ploughing kind of mechanism in which the materials displaced accumulated on the sides of the scratch. It was observed that the width of scratches increased with increasing load. The wear volume, however, was found to decrease with an increase in nanotube content. This is due to the addition of nanotubes which helps to increase the hardness values of nanocomposite coatings. The increased hardness with increase in nanotube content resulted in a reduction in contact depth and wear volume. This was due to uniform dispersion and better strengthening of nanocomposite coatings by the addition of carbon nanotubes.

Although the carbon nanotube reinforced metal matrix nanocomposites have shown good wear resistance and low coefficient of friction, there are some areas which have to be addressed. Optimized parameters of nanotube content and its dispersion are required for development of nanocomposite coating with better properties.

9.8 Challenges

The discovery of carbon nanotubes have opened up a new era in the field of nanotechnology. These nanoscale materials have shown tremendous promise in the development of nanocomposites with multifunctional properties. Either in conducting polymers or high fracture strength ceramics or high strength and high conductivity metal, nanocomposites containing carbon nanotubes are a near-term application. However, there are certain technological issues pertaining to carbon nanotubes which are uniform dispersion, wettability, and alignment in the matrix and the interface.

The first and foremost problem that the carbon nanotubes-based metal matrix nanocomposites are facing is uniform dispersion. Due to large aspect ratio and strong van der Waal force of attraction, carbon nanotubes tend to form clusters. These clusters tend to reduce either the mechanical or thermal properties of the nanocomposites [29, 86]. For example, CNT/Cu nanocomposites have shown a drop in thermal conductivity values due to the agglomeration of carbon nanotubes [35]. Several studies on carbon nanotubes films have reported very low conductivity values. Yang *et al.* [18] investigated the thermal conductivity of carbon nanotube films of thickness varying from 10 to 50 µm. The average thermal conductivity values of these films were about 15 W/mK. Gong *et al.* [94] measured the thermal conductivity of aligned CNT/carbon composite prepared by CVD deposition of carbon on nanotubes. The values measured were in between 10 to 75 W/mK. To address this problem several authors have suggested functionalization by chemical treatment, ball milling, and multi-step secondary processing. Yang *et al.* [95] reported homogeneous dispersion of nanotubes in aluminium powder. The process involved nickel deposition on aluminum powders followed by *in situ* synthesis of nanotubes in these powders by CVD. The microstructural characterization of these powders showed even dispersion of nanotubes on the surface of aluminium powder. Most of the nanotubes had good bonding between the aluminium as most ends of separate nanotubes were implanted into aluminium powder. Singhal *et al.* [96] carried out amino functionalization of carbon nanotubes to improve its dispersion in aluminium matrix. The functionalization was carried out by ball milling carbon nanotubes in the presence of ammonium bicarbonate. The formation of nanosized Al_4C_3 at the interface helped to improve the stress transfer efficiency. In addition to this, the presence of amine and amide groups on the nanotube surface decreased the van der Waals force of attraction between them. The functionalized carbon nanotubes reinforced aluminium nanocomposites showed better dispersion and good mechanical properties compared to that of pure nanotubes reinforced nanocomposites.

The second problem related to nanocomposites is wettability of carbon nanotubes with metal matrix. Work by Dujardin *et al.* [19] has shown that the carbon nanotubes cannot easily wet metals with high surface tension. It is necessary for a reinforcement and matrix material to have good interfacial bonding. The interfacial bonding between carbon nanotubes and metal matrix can be improved by subjecting carbon nanotubes to surface treatments. It has been shown by several authors that the nanotubes coated with either copper or nickel by electroless plating technique could decrease the interface energy between nanotubes and metal matrix, and increase the interfacial bonding strength and the nanocomposites with improved properties [97, 98].

The third factor is alignment of carbon nanotubes inside the metal matrices. It is well known that the thermal conductivity of carbon nanotubes is very high along the tube direction but very low in the radial direction. It is necessary that the nanocomposites for thermal management applications should display the isotropic behavior rather than anisotropic material properties. For this the nanotubes have to be oriented in a particular direction and should avoid random orientation. A research work by Chai *et al.* [99] reported the fabrication of CNT/Cu nanocomposites on silicon substrate by electrochemical plating method. The copper was deposited on the substrates having aligned nanotubes. Both thermal and electrical resistance decreased by increasing copper loading into the aligned nanotubes. The change in the resistance was attributed to improved contact between nanotube and copper metal. The thermal resistance of silicon and aluminium substrate showed very high and higher contact pressure dependence, whereas the copper-filled aligned carbon nanotube nanocomposites showed much smaller interface resistance and were less dependent on the contact pressure.

The last important factor is the interface between carbon nanotubes and metal matrix. The interface plays an important role in load transfer from matrix to the nanotube and crack deflection. The formation of nanosized layers of Al_4C_3 and MgC_2 in CNT/Al and CNT/Mg nanocomposites have shown the better load transfer from the matrix phase to carbon nanotubes by pinning the nanotubes to the matrix material [100, 101]. But in a few cases the observed increment in the mechanical properties could be minimal. Hence, it is necessary to carry out the thermodynamic analysis to predict the interfacial reaction between the two constituents' matrix and nanotubes to ensure better bonding and load transfer [102].

After successfully addressing these problems one can obtain very high quality nanocomposite that will have some real-life applications either in heat sinks or load-bearing members.

9.9   Concluding Remarks

The discovery of the revolutionary material known as carbon nanotubes has stimulated much interest in their use to reinforce metal matrix. The development of nanocomposites with this nanoscale filler material has been a topic of intense research in the past decades owing to its outstanding mechanical, thermal and tribological properties. Already many potential applications of metal matrix nanocomposites have been projected covering the aerospace industry to electronic packaging. However, carrying out a systematic study in this field is broad in scope and includes, for example, uniform dispersion of nanotubes, wettability, alignment in metal matrix and the interface design. New models need to be formulated to predict the mechanical and physical properties by accounting for their agglomeration and defects.

References

1. B. Bhushan, Introduction to nanotechnology, in: B. Bhushan, ed., *Springer Handbook of Nanotechnology*, Springer-Verlag Berlin Heidelberg, pp. 1–6, 2004.
2. P.M. Ajayan, Bulk metal and ceramics nanocomposites, in: P.M. Ajayan, L.S. Schadler, and P.V. Braun, eds., *Nanocomposite Science and Technology*, Wiley-VCH Verlag GmbH & Co. KGaA, Weinheim, pp. 1–75, 2003.
3. A. Agarwal, S.R. Bakshi, D. Lahiri, *Carbon Nanotubes - Reinforced Metal Matrix Composites*, CRC Press, Taylor and Francis Group, LLC, 2011.
4. S.H. Pezzin, S.C. Amico, L.A.F. Coelho, and M.J. de Andrade, Nanoreinforcements for nanocomposite materials, in: C.P. Bergmann and M.J. de Andrade, eds., *Nanostructured Materials for Engineering Applications*, Springer-Verlag Berlin Heidelberg, pp. 119–131, 2011.
5. H.W. Kroto, J.R. Heath, S.C. O'Brien, R.F. Curl, and R.E. Smalley, *Nature*, Vol. 318, p. 162, 1985.
6. S. Iijima, *Nature*, Vol. 354, p. 56, 1991.
7. S. Iijima and T. Ichihashi, *Nature*, Vol. 363, p. 603, 1993.
8. D.S. Bethune, C.H. Kiang, M.S. de Vries, G. Gorman, R. Savoy, J. Vazquez, and R. Bayers, *Nature*, Vol. 363, p. 605, 1993.
9. C.N.R. Rao and A. Govindaraj, *Nanotubes and Nanowires*, The Royal Society of Chemistry Publishing, Cambridge, 2005.
10. R.S. Ruoff and D.C. Lorents, *Carbon*, Vol. 33, p. 925, 1995.
11. G. Overney, W. Zhong, and D. Tomanek, *Zeitschrift fur Physik D: Atoms, Molecules, and Clusters*, Vol. 27, p. 93, 1993.

12. J.P. Lu, *Physical Review Letters*, Vol. 79, p. 1297, 1997.

13. K.T. Kashyap and R.G. Patil, *Bulletin of Materials Science*, Vol. 31, p. 185, 2008.

14. M.-F. Yu, O. Lourie, M.J. Dyer, K. Moloni, T.F. Kelly, and R.S. Ruoff, *Science*, Vol. 287, p. 637, 2000

15. M.-F. Yu, B.S. Files, S. Arepalli, and R.S. Ruoff, *Physical Review Letters*, Vol. 84, p. 5552, 2000.

16. S. Berber, Y.-K. Kwon, and D. Tomanek, *Physical Review Letters*, Vol. 84, p. 4613, 2000.

17. P. Kim, L. Shi, A. Majumdar, and P.L. McEuen, *Physical Review Letters*, Vol. 87, p. 215502, 2001.

18. D.J. Yang, Q. Zhang, G. Chen, S.F. Yoon, J. Ahn, S.G. Wang, Q. Zhou, Q. Wang, and J.Q. Li, *Physical Review B*, Vol. 66, p. 165440, 2002.

19. E. Dujardin, T.W. Ebbesen, H. Hiura, and K. Tanigaki, *Science*, Vol. 265, p. 1850, 1994.

20. S.S. Wong, J.D. Harper, P.T. Lansbury Jr., and C.M. Lieber, *Journal of American Chemical Society*, Vol. 120, p. 603, 1998.

21. W.B. Choi, D.S. Chung, J.H. Kang, H.Y. Kim, Y.W. Jin, I.T. Han, Y.H. Lee, J.E. Jun, N.S. Lee, G.S. Park, and J.M. Kim, *Applied Physics Letters*, Vol. 75, p. 3129, 1999.

22. J. Kong, N.R. Franklin, C. Zhou, M.G. Chapline, S. Peng, K. Cho, and H. Dai, *Science*, Vol. 287, p. 622, 2000.

23. A. Bianco, K. Kostarelos, and M. Prato, *Current Opinion in Chemical Biology*, Vol. 9, p. 674, 2005.

24. R.H. Baughman, A.A. Zakhidov, and W.A. de Heer, *Science*, Vol. 297, p. 787, 2002.

25. S.C. Tjong, *Carbon Nanotube Reinforced Composites - Metal and Ceramic Matrices*, Wiley-VCH Verlag GmbH & Co. KGaA, Weinheim, 2009.

26. W.A. Curtin and B.W. Sheldon, *Materials Today*, p. 44, November 2004.

27. S.R. Bakshi, D. Lahiri, and A. Agarwal, *International Materials Reviews*, Vol. 55, p. 41, 2010.

28. P.G. Koppad, Multiwalled carbon nanotubes in novel metal matrix nanocomposites, *International Conference on Advanced Materials, Manufacturing, Management and Thermal Sciences*, Organized by Siddaganga Institute of Technology, Tumkur, India, p. 57, 2013.

29. T. Kuzumaki, K. Miyazawa, H. Ichinose, and K. Ito, *Journal of Materials Research*, Vol. 13, p. 2445, 1998.

30. W.X. Chen, J.P. Tu, L.Y. Wang, H.Y. Gan, Z.D. Xu, and X.B. Zhang, *Carbon*, Vol. 41, p. 215, 2003.

31. R. George, K.T. Kashyap, R. Rahul, and S. Yamdagni, *Scripta Materialia*, Vol. 53, p. 1159, 2005.

32. A.M.K. Esawi, and K. Morsi, *Composites Part A*, Vol. 38, p. 646, 2007.

33. P.G. Koppad, H.R.A. Ram, and K.T. Kashyap, *Journal of Alloys and Compounds*, Vol. 549, p. 82, 2013.

34. P.G. Koppad, K.T. Kashyap, V. Shrathinth, T.A. Shetty, and R.G. Koppad, *Materials Science and Technology*, Vol. 29, p. 605, 2013.
35. K. Chu, Q. Wua, C. Jia, X. Liang, J. Nie, W. Tian, G. Gai, and H. Guo, *Composites Science and Technology*, Vol. 70, p. 298, 2010.
36. H.J. Choi, B.H. Min, J.H. Shin, and D.H. Bae, *Composites Part A*, Vol. 42, p. 1438, 2011.
37. H.R.A. Ram, P.G. Koppad, and K.T. Kashyap, *Materials Science and Engineering A*, Vol. 559, p. 920, 2013.
38. K.T. Kashyap, K.B. Puneeth, A. Ram, and P.G. Koppad, *Materials Science Forum*, Vol. 710, p. 780, 2012.
39. Y.D. Han, S.M.L. Nai, H.Y. Jing, L.Y. Xu, C.M. Tan, and J. Wei, *Journal of Materials Science: Materials for Electronics*, Vol. 22, p. 315, 2011.
40. Z.H. Li, X.Q. Wang, M. Wang, F.F. Wang, and H.L. Ge, *Tribology International*, Vol. 39, p. 953, 2006.
41. W.M. Daoush, B.K. Lim, C.B. Mo, D.H. Nam, and S.H. Hong, *Materials Science and Engineering A*, Vol. 513–514, p. 247, 2009.
42. X.H. Chen, F.Q. Cheng, S.L. Li, L.P. Zhou, and D.Y. Li, *Surface and Coatings Technology*, Vol. 155, p. 274, 2002.
43. C. Guo, Y. Zuo, X. Zhao, J. Zhao, and J. Xiong, *Surface and Coatings Technology*, Vol. 202, p. 3385, 2008.
44. C.S. Ramesh, M.P. Harsha, K.S. Nagendra, and Z. Khan, Development of nickel-CNT electro-composites, *Society of Tribologists and Lubrication Engineers Annual Meeting and Exhibition 2012*, St. Louis, Missouri, USA, p. 165, 2012.
45. T. Laha, A. Agarwal, T. McKechnie, and S. Seal, *Materials Science and Engineering A*, Vol. 381, p. 249, 2004.
46. T. Laha, Y. Liu, and A. Agarwal, *Journal of Nanoscience and Nanotechnology*, Vol. 7, p. 1, 2007.
47. S.R. Bakshi, V. Singh, K. Balani, D.G. McCartney, S. Seal, and A. Agarwal, *Surface and Coatings Technology*, Vol. 202, p. 5162, 2008.
48. S. Cho, K. Takagi, H. Kwon, D. Seo, K. Ogawa, K. Kikuchi, and A. Kawasaki, *Surface and Coatings Technology*, Vol. 206, p. 3488, 2012.
49. Q. Li, A. Viereckl, C.A. Rottmair, and R.F. Singer, *Composites Science and Technology*, Vol. 69, p. 1193, 2009.
50. Q. Li, C.A. Rottmair, and R.F. Singer, *Composites Science and Technology*, Vol. 70, p. 2242, 2010.
51. S.I. Cha, K.T. Kim, S.N. Arshad, C.B. Mo, and S.H. Hong, *Advanced Materials*, Vol. 17, p. 1377, 2005.
52. Y. Morisada, H. Fujii, T. Nagaoka, and M. Fukusumi, *Materials Science and Engineering A*, Vol. 419, p. 344, 2006.
53. D.K. Lim, T. Shibayanagi, and A.P. Gerlich, *Materials Science and Engineering A*, Vol. 507, p. 194, 2009.
54. C.S. Goh, J. Wei, L.C. Lee, and M. Gupta, *Materials Science and Engineering A*, Vol. 423, p. 153, 2006.

55. M. Paramsothy, J. Chan, R. Kwok, and M. Gupta, *Composites Part A*, Vol. 42, p. 180, 2011.
56. M.-F. Yu, *Journal of Engineering Materials and Technology*, Vol. 126, p. 271, 2004.
57. W.D. Callister Jr., *Fundamentals of Materials Science and Engineering*, John Wiley & Sons, Inc., USA, 2001.
58. D. Lahiri, S.R. Bakshi, A.K. Keshri, Y. Liu, and A. Agarwal, *Materials Science and Engineering A*, Vol. 523, p. 263, 2009.
59. H. Kwon and M. Leparoux, *Nanotechnology*, Vol. 23, p. 415701, 2012.
60. Z.Y. Liu, B.L. Xiao, W.G. Wang, and Z.Y. Ma, *Carbon*, Vol. 50, p. 1843, 2012.
61. J. Lipecka M. Andrzejczuk, M. Lewandowska, J. Janczak-Rusch, and K.J. Kurzydłowski, *Composites Science and Technology*, Vol. 71, p. 1881, 2011.
62. A.M.K. Esawi, K. Morsi, A. Sayed, M. Taher, and S. Lanka, *Composites Part A*, Vol. 42, p. 234, 2011.
63. Y. Chen, S.R. Bakshi, and A. Agarwal, *Surface and Coatings Technology*, Vol. 204, p. 2709, 2010.
64. H. Li, A. Misra, Z. Horita, C.C. Koch, N.A. Mara, P.O. Dickerson, and Y. Zhu, *Applied Physics Letters*, vol. 95, p. 071907, 2009.
65. K.T. Kashyap, C. Ramachandra, C. Dutta, and B. Chatterji, *Bulletin of Materials Science*, Vol. 23, p. 47, 2000.
66. S.K. Singhal, M. Lal, I. Sharma, and R.B. Mathur, *Journal of Composite Materials*, Vol. 47, p. 613, 2013.
67. L. Jun, L. Ying, L. Lixian, and Y. Xuejuan, *Composites Part B*, Vol. 43, p. 1681, 2012.
68. H. Fukuda, J.A. Szpunar, K. Kondoh, and R. Chromik, *Corrosion Science*, Vol. 52, p. 3917, 2010.
69. T. Threrujirapapong, K. Kondoh, H. Imai, J. Umeda, and B. Fugetsu, *Transactions of Joining and Welding Research Institute*, Vol. 37, p. 57, 2008.
70. F. Xue, S. Jiehe, F. Yan, and C. Wei, *Materials Science and Engineering A*, Vol. 527, p. 1586, 2010.
71. X.H. Chen, C.S. Chen, H.N. Xiao, H.B. Liu, L.P. Zhou, S.L. Li, and G. Zhang, *Tribology International*, Vol. 39, p. 22, 2006.
72. E.K. Choi, K.Y. Lee, and T.S. Oh, *Journal of Physics and Chemistry of Solids*, Vol. 69, p. 1403, 2008.
73. C. Edtmaier, T. Steck, R.C. Hula, L. Pambaguian, and F. Hepp, *Composites Science and Technology*, Vol. 70, p. 783, 2010.
74. M. Taya, *Materials Transactions*, Vol. 32, p. 1, 1991.
75. W.S. Miller and F.J. Humphreys, *Scripta Metallurgica et Materialia*, Vol. 25, p. 33, 1991.
76. F. Tang, I.E. Anderson, T.G. Herold, and H. Prask, *Materials Science and Engineering A*, Vol. 383, p. 362, 2004.

77. A.Z. Yazdi, R. Bagheri, S.M. Zebarjad, and Z.R. Hesab, *Advanced Composite Materials*, Vol. 19, p. 299, 2010.
78. S.R. Bakshi and A. Agarwal, *Carbon*, Vol. 49, p. 533, 2011.
79. H.L. Cox, *British Journal of Applied Physics*, Vol. 3, p. 73, 1952.
80. R.J. Arsenault, N. Shi, *Materials Science and Engineering*, Vol. 81, p. 175, 1986.
81. A. Bhat, V.K. Balla, S. Bysakh, D. Basu, S. Bose, and A. Bandyopadhyay, *Materials Science and Engineering A*, Vol. 528, p. 6727, 2011.
82. D.H. Nam, S.I. Cha, B.K. Lim, H.M. Park, D.S. Han, and S.H. Hong, *Carbon*, Vol. 50, p. 2417, 2012.
83. C. Kim, B. Lim, B. Kim, U. Shim, S. Oh, B. Sung, J. Choi, J. Ki, and S. Baik, *Synthetic Metals*, Vol. 159, p. 424, 2009.
84. G. Chai and Q. Chen, *Journal of Composite Materials*, Vol. 44, p. 2863, 2010.
85. S. Cho, K. Kikuchi, and A. Kawasaki, *Acta Materialia*, Vol. 60, p. 726, 2012.
86. S.R. Bakshi, R.R. Patel, and A. Agarwal, *Computational Materials Science*, Vol. 50, p. 419, 2010.
87. W.X. Chen, F. Li, G. Han, J.B. Xia, L.Y. Wang, J.P. Tu, and Z.D. Xu, *Tribology Letters*, Vol. 15, p. 275, 2003.
88. G.D. Zhan, J.D. Kuntz, J.L. Wan, and A.K. Mukherjee, *Nature Materials*, Vol. 2, p. 38, 2003.
89. J.P. Tu, Y.Z. Yang, L.Y. Wang, X.C. Ma and X.B. Zhang, *Tribology Letters*, Vol. 10, p. 225, 2001.
90. K. Rajkumar and S. Aravindan, *Wear*, Vol. 270, p. 613, 2011.
91. J.P. Tu, L.P. Zhu, W.X. Chen, X.B. Zhao, F. Liu, and X.B. Zhang, *Transactions of Nonferrous Metals Society of China*, Vol. 14, p. 880, 2004.
92. W.X. Chen, J.P. Tu, Z.D. Xu, W.L. Chen, X.B. Zhang, and D.H. Cheng, *Materials Letters*, Vol. 57, p. 1256, 2003.
93. S.R. Bakshi, D. Lahiri, R.R. Patel, and A. Agarwal, *Thin Solid Films*, Vol. 518, p. 1703, 2010.
94. Q-M. Gong, Z. Li, X.-D. Bai, D. Li, Y. Zhao, and J. Liang, *Materials Science and Engineering A*, Vol. 384, p. 209, 2004.
95. X. Yang, C. Shi, C. He, E. Liu, J. Li, and N. Zhao, *Composites Part A*, Vol. 42, p. 1833, 2011.
96. S.K. Singhal, R. Pasricha, M. Jangra, R. Chahal, S. Teotia, and R.B. Mathur, *Powder Technology*, Vol. 215–216, p. 254, 2012.
97. B. Lim, C.-J Kim, B. Kim, U. Shim, S. Oh, B.-H. Sung, J.-H. Choi, and S. Baik, *Nanotechnology*, Vol. 17, p. 5764, 2006.
98. M. Schneider, M. Weiser, S. Dorfler, H. Althues, S. Kaskel, and A. Michaelis, *Surface Engineering*, Vol. 28, p. 435, 2012.
99. Y. Chai, K. Zhang, M. Zhang, P.C.H. Chan, and M.M.F. Yuen, *IEEE Electronic Components and Technology Conference*, p. 1224, 2007.

100. H. Kwon, M. Estili, K. Takagi, T. Miyazaki, and A. Kawasaki, *Carbon*, Vol. 47, p. 570, 2009.
101. C.S. Goh, J. Wei, L.C. Lee, and M. Gupta, *Nanotechnology*, Vol. 17, p. 7, 2006.
102. S.R. Bakshi, A.K. Keshri, V. Singh, S. Seal, and A. Agarwal, *Journal of Alloys and Compounds*, Vol. 481, p. 207, 2009.

Part 3

FLY ASH ENGINEERING AND CRYOGELS

10

Aluminum/Fly Ash Syntactic Foams: Synthesis, Microstructure and Properties

Dung D. Luong[1], Nikhil Gupta[1,*] and Pradeep K. Rohatgi[2]

[1]*Composite Materials and Mechanics Laboratory, Mechanical and Aerospace Engineering Department, Polytechnic Institute of New York University, Brooklyn, New York, USA*
[2]*Department of Materials Engineering, University of Wisconsin, Milwaukee, Wisconsin, USA*

Abstract

Increasing demand to improve the fuel efficiency and reduce pollution in the aerospace and automotive sectors, while retaining the integrity of structures, has resulted in the development of new advanced light-weight porous composites. Traditionally, porosity is considered to be an undesired microstructural feature in materials. However, an innovative method of using hollow particles to incorporate porosity in materials has shown great promise. Such composites, known as syntactic foams, have been widely studied in recent years. A number of engineered hollow particles are now available and are used as fillers. Fly ash cenospheres, a waste by-product of coal-fired power plants, are one such class of hollow particles that are commonly used in syntactic foams. Fly ash is generally inexpensive and is considered to be an environmental hazard, thus utilization of fly ash in composites proves to be both economically and environmentally beneficial. In this way, use of fly ash in developing advanced composites is very encouraging for the next generation of advanced light-weight composites. The discussion in this chapter is focused on fly ash filled aluminum matrix composites. Complex composition of fly ash leads to several interfacial reactions and generation of reaction products, which can diffuse to short and long ranges depending on composite processing

*Corresponding author: ngupta@nyu.edu

Ashutosh Tiwari and S.K. Shukla (eds.) Advanced Carbon Materials and Technology, (377–418)
2014 © Scrivener Publishing LLC

conditions (time and temperature). This discussion is focused on synthesis methods, microstructure, mechanical properties, and tribological properties of Al/fly ash syntactic foams.

Keywords: Al/fly ash syntactic foams, synthesis methods, microstructure, mechanical properties, tribological properties

10.1 Introduction

Lightweight composites are in great demand in the transportation sector [1]. Reducing the weight of automobiles can help in increasing their fuel efficiency. Many different approaches are currently tested for automotive weight reduction. For example, smaller turbo-charged engines may be lighter but can provide the same power level as larger engines. Efficient energy absorbers and crumple zones may allow for the reduction of the overall length of the car, which can also save weight. In addition to all of these possibilities, the use of lighter metals and composites can further reduce the structural weight and increase the fuel efficiency of the vehicle. Lightweight metal matrix composites such as Al/fly ash syntactic foams are expected to play an important role in reducing the weight of vehicles in the transportation sector [2].

Modern automobiles make use of a variety of lightweight materials. Seats are made of lightweight polymeric foams, dashboards may be made of plastics, energy absorption zones may contain metal foams, and the seat structure may be made of fiber-reinforced plastics to reduce the overall weight of the automobile. In addition, there is a significant focus on replacing the heavy parts made of steel with either aluminum- or magnesium-based materials. For example, brake discs made of aluminum and aluminum matrix composites are now available; wheel rims made of aluminum and magnesium alloys are also available; and aluminum engine blocks are now widely used.

Metal foams present several unparalleled opportunities in the transportation sector. Metal foams can be classified into three main categories based on their structure:

- Open-cell foams
- Closed-cell foams containing gas porosity
- Syntactic foams containing hollow particles

Open-cell foams contain interconnected porosity. Their primary applications are in the form of core material in sandwich structures and in heat exchange applications. These foams may be comprised of thin ligaments and irregular-shaped pores. The foam density can be modulated by means of a combination of size and number of ligaments in a unit volume. As a general trend, the higher density foams have higher strength and modulus. In these foams, the aspect ratio (length to diameter ratio) of the ligaments is very high (several hundreds), which leads to their buckling and folding under compression. The closed-cell foams containing gas porosity have pores that are not interconnected. Since the pores are not connected with each other, these foams do not absorb water even after developing cracks. The air voids are enclosed inside thin cell walls. The thickness of the cell walls depends on the foam density. Low density foams having thin cell walls may show buckling under compression, while high density foams may show shear failure of cell walls. These foams are also used as core materials in sandwich structures. However, due to the higher mechanical properties of such foams, they may also find structural applications. These foams have lower thermal conductivity than the base metal. The thermal properties of such foams can be tailored by means of size and volume fraction of porosity in the foam structure.

Foams can be reinforced with phases such as micro- or nanoparticles, nanofibers, and nanotubes. A vast body of literature is now available on such reinforced composite foams. For example, SiC-reinforced aluminum foams have been extensively studied [3, 4]. The mechanisms of bubble formation, stabilization, and microstructural effects in the presence of SiC particles have been studied.

In open and closed-cell foams containing gas porosity, the matrix metal is the only solid phase in the microstructure. The foam properties can be tailored by changing the volume fraction and distribution of porosity in the foam microstructure. The range of properties obtained from these modifications is usually not very wide, and mechanical properties are not high enough to lead to their structural applications. In this context, syntactic foams are a class of closed-cell foams that can be very useful.

Syntactic foams are synthesized by embedding hollow particles in a matrix. These foams can also be classified as particulate composites. Porosity is enclosed inside hollow particles in these foams. The presence of thin ceramic particle shells provides a reinforcing effect to each void in these foams, which leads to higher mechanical properties than other closed-cell foams of the same density. Instead of foaming techniques such as gas blowing in a liquid melt, syntactic foams are synthesized by methods used for synthesizing particulate composites. This chapter is focused on discussing the microstructure, synthesis methods, and mechanical properties of aluminum matrix syntactic foams containing fly ash cenosphere hollow particles.

10.2 Hollow Particles

Hollow particles are in demand in many fields, not just in the field of composite materials. They are being used in catalysis, drug delivery, and hydrogen storage, among a wide variety of other applications [5]. The wide variety of applications has resulted in availability of particles of a wide range of materials and also development of several processing methods. In the present case, the discussion will present a brief overview of hollow particles that are of interest in syntactic foams. Two classes of particles are relevant in this regard: engineered ceramic particles specifically developed for applications in syntactic foams and fly ash cenospheres. A brief description of both types of particles is presented below.

10.2.1 Fly Ash Cenospheres

Fly ash is a by-product of coal fired thermal power plants and development of useful applications of such a waste material are of great interest [6, 7]. Combustion of coal results in a variety of materials, which includes ceramic oxides, burnt and unburnt carbon, and several types of gases. The exact nature of the combustion products depends on the composition of the coal and the burning conditions. The chemical composition of fly ash from a number of sources from countries around the world, such as the U.S., the Netherlands, Italy, Turkey, Poland, Greece, and India, has been reported in the literature [8–12].

The ash that is generated as a result of combustion contains ceramic particles. Some of these particles are hollow and have low density. These hollow fly ash particles are called cenospheres. Fly ash can also be recovered from electrostatic precipitators. This type of fly ash usually has higher density than cenospheres and is referred to as precipitator fly ash. Due to their low density, cenospheres are of greater interest in syntactic foam applications because their incorporation in metals may lead to low density composites. Almost half of the fly ash generated in the USA is dumped in landfills, thus utilization of fly ash in any application becomes beneficial for the environment. One of the most common applications of fly ash is its use in cement mixtures. Fly ash utilization also seems very promising for the composite materials sector. Coatings containing fly ash have been developed and are found to have high hardness because of their ceramic content [13, 14].

Since fly ash is formed as a part of a large number of reactions during coal combustion, separation and cleaning of the cenospheres is required. Magnetic separation, flotation in water, cleaning with acids, and sieving are used to obtain useful ceramic microspheres. The Na, K, and Ca can be removed by washing the fly ash with water, whereas washing with hydrochloric acid is able to remove heavy metals such as Pb, Zn, Cd, Fe and Cu [15].

An example of fly ash cenospheres is shown in Figure 10.1. Cenosphere size may range from less than 1 μm to over 1 mm. Most studies have used particles in the range 10–500 μm. Most of the beneficiated cenospheres are spherical in shape. However, some of the particles may be of irregular shape. Figure 10.2 shows examples of defective particles, where cenospheres do not have a uniform smooth surface and impurities may be attached to their surface. Such defective particles may have lower strength compared to the spherical particles and may give rise to stress concentrations in the composite. These particles may become sites for failure initiation. Beneficiation steps are designed to minimize the presence of such particles in the batch that is used for making syntactic foams.

Several methods are available to minimize the effects of the defective particles. In one method, a pressure chamber can be used to pressurize and break the weaker particles and keep only those that have a minimum level of strength. In another method, the particles can be coated with a different material to fill the surface defects and obtain a smooth surface finish. Use of coating materials that are compatible with the matrix alloy can also provide the advantage of having a strong particle-matrix interfacial bonding. For example, nickel and copper coatings on ceramic particles are known to be useful in aluminum matrix composites. Nickel-coated cenospheres have been used for synthesizing aluminum matrix syntactic foams.

(a)

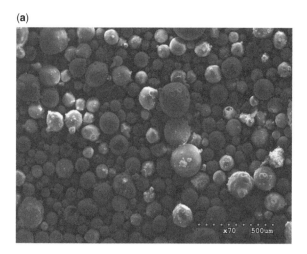

(b)

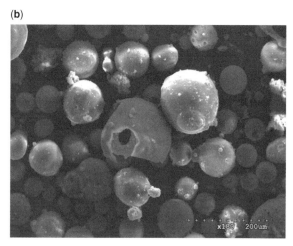

Figure 10.1 A sample of fly ash particles. A wide range of particle sizes can be seen in these figures.

10.2.2   Engineered Hollow Particles

Demand for hollow particles in a variety of fields has resulted in the development of new methods of making hollow particles, and also the development of hollow particles composed of a variety of materials. One of the major challenges in making hollow particles is to have their size and wall thickness uniform. Obtaining consistency in the diameter of the particle is not enough because particles of the same diameter may have different wall thicknesses. Particles of alumina, silica and silicon carbide are now commercially available in

(a)

(b)

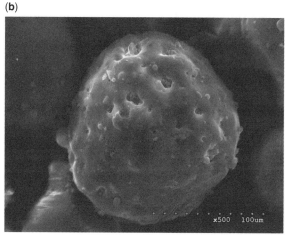

Figure 10.2 All fly ash particles may not be high-quality spheres with smooth surfaces. Some of these particles may have irregular shapes and defects.

sizes ranging from a few microns to several millimeters. An example of SiC hollow particles is shown in Figure 10.3(a). The uniformity in the wall thickness of these particles is shown in Figure 10.3(b). A large part of the particle volume is the void in their center. Such particles have been used in synthesizing aluminum matrix syntactic foams [16]. Particles of 55–60 wt% silica and 36–40 wt% alumina have been used with 1350, 5083 and 6061 aluminum alloys [17]. The high modulus of SiC and alumina compared to that of aluminum helps in obtaining syntactic foams with high mechanical properties, which is what is desired.

(a)

(b)

Figure 10.3 Scanning electron micrographs of engineered SiC hollow particles: (a) uniform size can be observed (b) wall thickness is also uniform in these particles and the walls are free of defects.

Only spherical hollow particles have been used in synthesizing metal matrix syntactic foams. However, particles of other shapes such as cubic and cuboid are also commercially available. With current technologies, it is now possible to fabricate particles of a variety of shapes and sizes. One of the reasons the use of such particles is not common is because any preferential orientation of such particles may lead to directionality in the mechanical properties of syntactic foams, which may be undesired. On the other hand, when directionality in the properties such as strength and

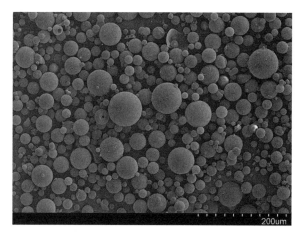

Figure 10.4 Nickel-coated glass hollow particles.

modulus is desired, such particles of different shapes can be very useful.

Engineered hollow particles of glass are available in the size range of less than 1 μm to over 1 mm in diameter, and the density may range from 0.1–0.6 g/cc. It is possible to find higher density glass hollow particles but they do not lead to significant reduction in the density of the composites so their applications are limited in this field. The glass particles can be coated with Ni, Cu, or other metals before incorporating them in aluminum alloys. An example of nickel-coated glass hollow particles is shown in Figure 10.4. Some large-size particles can be seen in the image. The average particle size is about 40 μm. Uniform shape and high surface finish are the advantages of these particles. The modulus of glass is in the range 60 to 80 GPa, which is similar to the modulus of aluminum and aluminum alloys. The effective modulus of particles, due to the void in their center, is usually much lower than that of glass. Therefore, incorporation of these particles in aluminum is not expected to provide much benefit. Rather, it is expected that the modulus of aluminum matrix syntactic foams containing such particles would be low and close to the modulus of closed-cell foams containing a similar level of porosity. These particles can be useful in synthesizing magnesium matrix syntactic foams because magnesium has lower modulus than glass. Considerable interest is currently seen in developing magnesium matrix syntactic foams because of their very low density, which is in the range 1–1.5 g/cc. The existing studies on magnesium matrix syntactic foams have largely used fly ash cenospheres due to their low cost [18–20]. Use of engineered hollow particles can help in obtaining higher mechanical properties at the same level of syntactic foam density because of the high quality of the engineered particles.

10.3   Synthesis Methods

Development of low cost synthesis methods is important for metal matrix composites. Each processing step adds to the production time and cost. Therefore, it is desired to develop methods that can produce the component in the net or near-net shape and reduce the processing time and cost. One of the advantages of metal matrix syntactic foams is that existing foundries can be used for their synthesis with a few small modifications. Two methods of aluminum matrix syntactic foam synthesis are described here. Several modifications of these methods can be found in published literature. The powder metallurgy methods are well developed for synthesizing particle reinforced metal matrix composites [21, 22] but their use for synthesizing syntactic foams is not common because the powder compaction may result in large-scale fracture of fragile hollow particles [23, 24]. Therefore, these methods are not discussed here.

10.3.1   Stir Mixing

Stir mixing, followed by casting, is a well-developed technique for synthesizing particulate composites. In this method, the matrix melt is continuously stirred at a given temperature and particles are slowly added in the vortex formed by the stirrer. Engulfment of particles in the melt and subsequent dispersion is achieved by the stirring. The melt-particle mixture is cast in molds and allowed to solidify. While the synthesis technique is relatively simple, process control is difficult to obtain for several reasons, some of which are outlined below.

Particle preheat temperature: The temperature of pre-heating of cenospheres is an important parameter. If the cenospheres are maintained at room temperature, their addition to the matrix melt would reduce the melt temperature and may lead to partial solidification. Therefore, pre-heating of particles is necessary. The melt superheating temperature is also necessary to control for the same reason. It has been observed that the addition of particles at lower

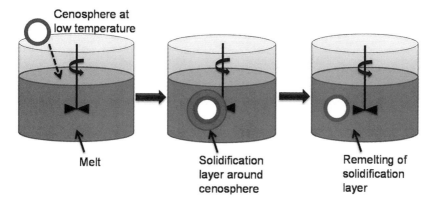

Cenosphere at low temperature

Melt

Solidification layer around cenosphere

Remelting of solidification layer

Figure 10.5 Illustration of formation of solidification layer around particle if the particle pre-heat and melt superheat temperatures are not properly maintained.

temperature may lead to formation of a solidification layer around the particle, which can dissolve as the stirring is continued and the temperature is again homogenized, as illustrated in Figure 10.5.

Rate of cenosphere addition: The rate of cenosphere feeding in the melt needs to be carefully controlled. Generally, the melt and the particles have a temperature difference. Addition of cenospheres can reduce the temperature locally and result in increased viscosity of the melt. Since the mixing is conducted with the intent of developing high shear forces in the melt for particle dispersion, increased viscosity may lead to effects such as poor dispersion and fracture of cenospheres. The rate of cenosphere addition is also important for initial engulfment of particles in the vortex created by the impeller. If the particles are not engulfed, then particle wetting may be incomplete and the overall composite will have poor quality.

Design of impeller: The design of the impeller is an important consideration in the stir mixing process. Impellers that can produce high shear forces in the melt are preferred because they can effectively break particle clusters and disperse them evenly in the melt. Four different impeller designs are shown in Figure 10.6. Selection of an appropriate impeller is even more important for the cenospheres, given that they may fracture during the mixing process.

10.3.2   Infiltration Methods

Infiltration methods rely on filling liquid metal in the interstitial spaces of a packed bed or preform made of particles. Several variants

Figure 10.6 Four different designs of impellers that are used in the stir mixing process.

of such methods are available, of which pressure infiltration and vacuum infiltration are the most widely used. Figure 10.7 schematically illustrates a typical pressure infiltration setup [25]. The metallic melt is pushed into the preform using pressure. The applied pressure depends on a number of factors, including the temperature difference between particles and the melt, particle size and particle packing density. Low pressure would result in incomplete filling of the preform, while high pressure would cause cenosphere fracture. For nickel-coated cenospheres, the threshold pressure to initiate infiltration was experimentally determined to be in the range 0–6.89 kPa [26]. The infiltration length has been calculated using the Washburn equation:

$$L^2 = \left(\frac{pr_e^2}{4\eta} \right) \left(\frac{2\sigma_{LV} \cos \theta}{pr_e} - gL + \frac{P_{app}}{\rho} \right) t \tag{10.1}$$

where L is the infiltration length, ρ is density of liquid aluminum, η is the aluminum viscosity at the processing temperature, g is gravity, P_{app} is the applied pressure, t is the infiltration time, r_e is the interparticle distance, θ is the contact angle and σ_{LV} is the surface tension between liquid and vapor phases.

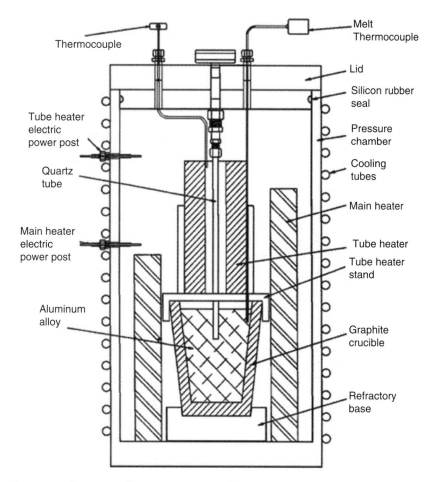

Figure 10.7 Pressure infiltration setup used for Al-alloy/fly ash syntactic foams [25].

Instead of high pressure, a vacuum can be used to promote the infiltration. The wetting characteristics of particles with the melt are extremely important for promoting particle engulfment in the advancing melt front. Poor wetting may lead to undesired effects such as particle pushing and may also result in poor interfacial bonding and higher porosity in the composite.

10.3.3   Comparison of Synthesis Methods

All synthesis methods have some benefits and limitations. Understanding the application domain of each method can help in selecting the best suited method for a given material system and processing conditions. Factors that are considered in selecting the synthesis method are described in this section.

10.3.3.1 Particle Volume Fraction

The stir mixing method is best suited when the cenosphere volume fraction in the syntactic foam is low and particle-to-particle interaction during mixing can be minimized. This method can lead to excessive particle fracture at high volume fractions. On the contrary, the infiltration methods are best suited when syntactic foams are required to have high volume fraction of particles. The infiltration methods require a packed bed of particles, which requires a particle content of at least 40 vol%. At lower volume fractions, particle pushing may be prominent and engulfment of particles in the advancing melt front becomes a problem.

10.3.3.2 Flotation and Segregation

The true particle density of cenospheres is in the range 0.4–0.8 g/cc. In comparison, aluminum alloys have a density of around 2.7 g/cc. The density difference promotes flotation of particles to the top part of the casting. In stir casting methods, the process parameters, especially the rate of solidification, need to be controlled carefully in order to suppress particle flotation. Complete solidification is not required for suppressing flotation. Since the rate of particle rise is a function of the viscosity of the fluid, the temperature can be maintained sufficiently low in addition to continued stirring to obtain a uniform distribution of cenospheres in the composite. The particle flotation is not a major consideration in infiltration methods because the particles do not have freedom to move long distances in a preform.

10.3.3.3 Net Shape Processing

The stir mixing process is comprised of two different steps. The first step involves stir mixing of cenospheres in the aluminum melt. The second step involves casting in molds in required shapes. The mold can be of any shape, and net or near-shaped processing can be conducted. Filling of intricately shaped molds would require the aluminum-cenosphere mixture to have low viscosity. However, if the mixture viscosity is low, then the cenosphere flotation would be promoted and the casting would not have uniform cenosphere distribution along the thickness.

Use of a preform of any desired shape in the infiltration method enables net or near-net shaped processing of Al/fly ash syntactic foam components. The mold and preform temperatures are selected to be high enough to avoid freeze choking and promote complete infiltration.

10.3.3.4   Functionally Graded Structure

The density difference between cenospheres and aluminum alloy promotes particle flotation, which can be used for obtaining a functionally graded structure in syntactic foams at low cost using stir casting methods. Synthesis of functionally graded composites using an infiltration method would require creating a functionally graded preform. The functionally graded structure in hollow particle filled composites can be developed by two different approaches, which include creating a gradient in either the volume fraction or the wall thickness of hollow particles as shown in Figure 10.8 [27, 28]. Creating preforms with any of these approaches will increase the processing cost. Polymer matrix functionally graded syntactic foams have been synthesized and studied [27, 28]. Creating such structures in metal matrix syntactic foams can be of great interest.

10.3.3.5   Ultralight Weight Composites

Synthesis of syntactic foams with high volume fractions of cenospheres is difficult with stir casting methods because of particle collision, resulting in their fracture during mixing. Infiltration methods can provide a much higher volume fraction of particles in composites using innovative approaches such as bimodal or trimodal distributions of particles. Such structures can be designed to provide over 80% filling of the metal matrix syntactic foam with hollow particles. These ultralight composites may be useful in energy absorption or marine structural applications.

10.4   Microstructure of Aluminum/Fly Ash Composites

A number of aluminum alloys have been used as the matrix material in Al/fly ash syntactic foams. Among the alloys that have been used in these studies are A356 [29], A535 [30–32], A4032, and

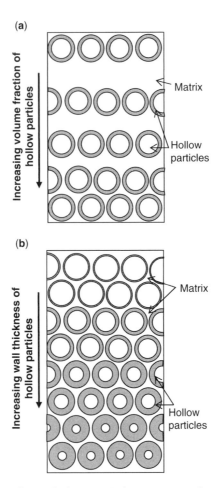

Figure 10.8 Functionally graded syntactic foam structure based on variation in hollow particle (a) volume fraction and (b) wall thickness along the syntactic foam thickness.

AA6061 [33, 34]. Representative microstructures of A4032/fly ash syntactic foams are shown in Figure 10.9. Several features can be observed in these figures. The matrix microstructure, observed in Figure 10.9(a), is usually much more refined in the composite compared to the matrix alloy processed under the same conditions. The matrix grain size depends on the processing time and temperature. The primary and secondary processing time and temperature can be adjusted to achieve the desired grain size. It has been observed that the grain size near and far from the cenosphere particles can be different. Several fly ash particles present in Figure 10.9(a) are undamaged and porosity is present inside them. Some damaged

(a)

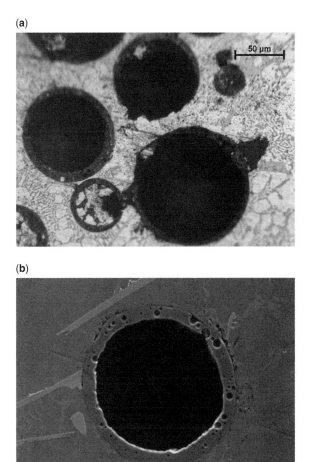

(b)

Figure 10.9 A4032/fly ash composite (a) optical micrograph showing several intact cenospheres and some defective cenospheres infiltrated with the matrix alloy and (b) scanning electron micrograph showing an intact cenosphere that has porosity and defects in its wall.

particles that have been infiltrated with the matrix alloy can also be seen in the micrograph. The proportion of such damaged and infiltrated cenospheres should be low in a syntactic foam specimen. The interface between particle and matrix seems to be continuous in Figure 10.9(b), where the porosity and defects present in the walls of an intact cenosphere can be seen.

Reaction between constituents of the aluminum alloy and fly ash particle is a serious issue [35]. These reactions can not only damage the particle but can also modify the matrix microstructure. The main chemical constituents of fly ash are SiO2, Al2O3, and Fe2O3. Two of the possible reactions are:

$$4\ Al + 3\ SiO_2 = 3\ Si + 2\ Al_2O3 \qquad (10.2)$$

$$2\ Al + Fe_2O3 = 2\ Fe + Al_2O3 \qquad (10.3)$$

These reactions can take place during initial solidification processing or during heat treatment of the syntactic foams [36, 37]. The physical impact of these reactions is schematically illustrated in Figure 10.10 and scanning electron micrographs (SEM) of the syntactic foam showing various stages of cenosphere disintegration

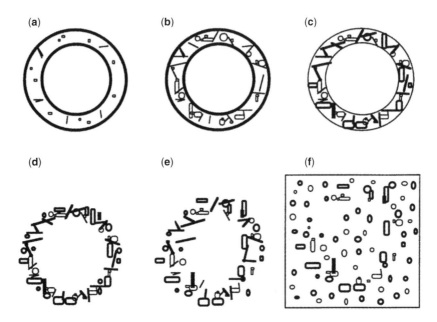

Figure 10.10 Schematic of change in cenosphere structure due to chemical reaction between cenosphere and aluminum: (a) initial intact cenosphere with characteristic cracks and pores in the wall; (b-c) more cracks and pores form progressively due to SiO_2 and Fe_2O_3 reduction by molten aluminum forming Al_2O_3 and releasing Si and Fe in the matrix; (d) remaining unreacted parts of cenosphere as well as newly formed Al_2O_3 particles still maintaining the original shape; (e) small Al_2O_3 particles start to move from the original positions, but the original cenosphere wall shape is still observable; (f) small particles formed by chemical reactions move away from their original position and become uniformly distributed in the matrix given sufficiently high temperature and long times [36].

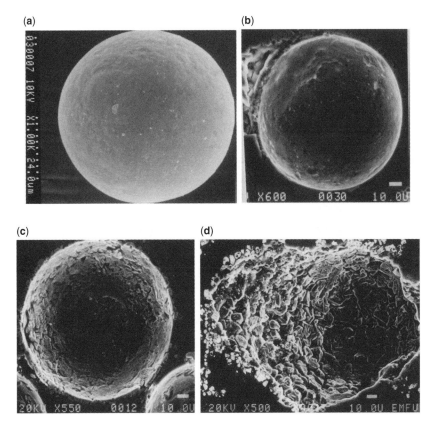

Figure 10.11 SEM micrographs of cenosphere disintegration in molten aluminum due to chemical reactions at 850°C: (a) original cenosphere, (b) soaked for 0.5 h, (c) soaked for 5 h, and (d) soaked for 27 h [36].

are shown in Figure 10.11. These reactions take place at sufficiently high temperatures. In these figures, the temperature of the material system is maintained at 850°C.

The extent of reactions between aluminum and SiO_2 and Al_2O_3 progressively increases with time leading to formation of Fe, Si and Al_2O_3. Elements Fe and Si diffuse in the aluminum melt. The diffusion rate depends on the temperature. The Al_2O_3 particles maintain the shape of the original cenospheres in the initial stages of reactions. However, removal of sufficient Fe and Si results in Al_2O_3 particles, which are not bonded with each other. At this stage, the cenosphere starts to deform and finally the Al_2O_3 particles get distributed throughout the aluminum melt. Processing at lower temperatures can help in keeping the cenospheres intact in the syntactic foam structure, which is required for obtaining a high quality composite of the type observed in Figure 10.9.

There are several other microstructural considerations in Al/ fly ash syntactic foams. Nickel-coated cenospheres have been used for synthesizing aluminum matrix syntactic foams in many studies. Nickel can react with Al to form intermetallic Al-Ni precipitates of several different compositions. Due to the high temperature of composite synthesis, the nickel coating may partially or completely dissolve in the aluminum melt. The intermetallic precipitates are harder than the Al-alloy and can help in improving the mechanical properties of the syntactic foam. The amount and distribution of these precipitates can be controlled by means of heat treatments. However, care must be taken in this process because complete dissolution of the coating will expose the cenospheres to the aluminum melt and cause their degradation.

10.5 Properties of Aluminum/Fly Ash Syntactic Foams

Mechanical properties of Al/fly ash syntactic foams are extracted from the published studies for determining the underlying common trends. Differences in the type of fly ash, matrix material, synthesis methods and synthesis conditions will result in different properties of the composite. However, from the point of view of a given application, the most important considerations are the density and mechanical properties. Therefore, the mechanical properties of syntactic foams are plotted with respect to density in the following discussion. It should be noted that most studies have reported only a few properties, so the comparison between different figures may not be directly applicable. It also must be noted that some of the data are extracted to the best possible accuracy from the figures reported in the respective studies. The standard deviations are not reported here and only the average values are taken for comparison. In some cases, the standard deviations are significantly high and should be a consideration when taking critical decisions based on the mechanical property data.

The porosity present in the microstructure is effective under the compressive loading conditions because the cenospheres can fracture under compression and the composite microstructure can be compacted. Only a few studies on tensile properties are available.

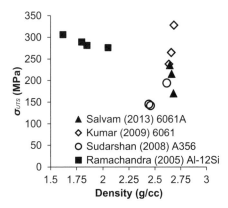

Figure 10.12 The ultimate tensile strength of Al/fly ash syntactic foam with respect to the foam density. The data in this figure are taken from [38,56–58].

Figure 10.12 presents the ultimate tensile strength (σ_{UTS}) of Al/fly ash syntactic foams. Two studies on 6061 alloy matrix syntactic foams are present in Figure 10.12. In one case, the ultimate tensile strength of the composite increases with density, while in two other cases the trend is reversed. The reduction in the strength with increasing density seems to be related to poor wetting and dispersion of cenospheres in the matrix as seen in Figure 10.13 [38]. The compressive modulus data also show a trend that the modulus increases with the composite density, as seen in Figure 10.14. The modulus of syntactic foams is significantly lower compared to that of aluminum. These trends need to be confirmed with additional comprehensive studies.

Compressive properties of syntactic foams have been studied in great detail. A typical compressive stress-strain graph of a typical syntactic foam specimen is illustrated in Figure 10.15. The yield strength (σ_{ys}) is calculated by the strain offset method and is marked as point 1. The subsequent stress peak is marked as point 2 and is called plastic stress (σ_p). The compressive stress-strain graphs of metal matrix syntactic foams show a stress plateau for a large value of strain, as point 3 in Figure 10.15 [16, 39]. This stress plateau provides syntactic foams the ability to absorb energy under compression. The plateau stress is calculated as the average stress over the plateau length. These properties are compared for various types of Al/fly ash syntactic foams with respect to the foam density.

The compressive modulus is found to be reported in only two studies as presented in Figure 10.14. The available results on A356

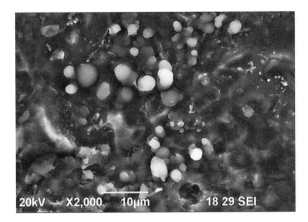

Figure 10.13 SEM micrograph of AA6061/fly ash composite containing 8 wt% fly ash [38]. The wetting and dispersion of fly ash particles do not appear to be uniform.

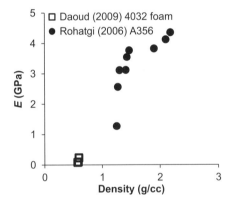

Figure 10.14 The compressive modulus of syntactic foam with respect to the foam density. The data in this figure are taken from [25,59].

alloy matrix syntactic foams show higher density and modulus than the foams of 4032 alloy. It is noticed for A356 alloy matrix syntactic foams that the modulus increases with the foam density. The rapid rise in the modulus in the density range 1.2–1.5 can be very useful in selecting a material with appropriate stiffness for a given application. The general trends in the modulus can be studied by conducting finite element analysis. Figure 10.16 shows a three-dimensional (3D) model containing a random distribution of particles. Such models are able to provide predictions close to the experimental values for syntactic foams [40]. In this case, all of the particles have the

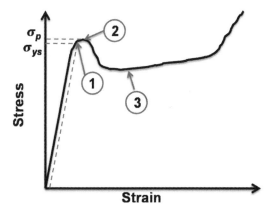

Figure 10.15 Schematic representation of compressive stress-strain diagram of a typical Al/fly ash syntactic foam specimen. The annotations in the diagram refer to 1: yield strength, 2: plastic stress, 3: stress plateau region.

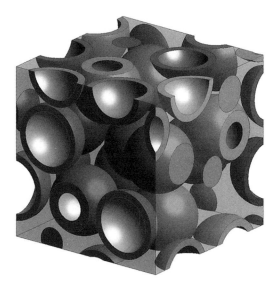

Figure 10.16 A 3D model with random distribution of glass hollow particles in the syntactic foam microstructure. All particles have the same radius and radius ratio of 0.8 in this model.

same size and radius ratio (ratio of inner to outer radius of hollow particle). The model is created by generating a random distribution of particles in space and then defining a cubic unit cell at a randomly selected location. The interparticle space in the unit cell is defined as the matrix material. In this model, the particle volume fraction is not predefined; it depends on the location of the unit cell selection.

Table 10.1 The mechanical properties of aluminum matrix and glass hollow particle materials.

Material	Modulus (GPa)	Poisson's ratio
Aluminum	70	0.3
Glass Microballoon	60	0.21

In the current case, the unit cell has a hollow particle volume of 46%. Compression testing is conducted on this model by applying a strain of 0.1. The material properties assigned to the matrix and particle are given in Table 10.1. The modulus and other properties of cenospheres are not available in the literature, thus glass properties are assigned to the particles. The properties of the cenospheres are expected to be of the same order of magnitude as that of glass because cenospheres mainly have alumina and silica as their constituents. The simulation results are presented in Figure 10.17. The syntactic foam modulus increases with the foam density. The particle radius ratio is selected to be in the range 0.8–1 in this simulation study because this range provides syntactic foams of density 1.5–2.2 g/cc, which is the range observed in many experimental studies. Finite element models of syntactic foams with lower densities can be created by using a higher volume fraction of hollow particles. The modulus of syntactic foams is found to vary in the range of 24–40 MPa within the given parameters. These values are significantly higher than those obtained experimentally. Defects present in cenosphere walls are likely responsible for the low experimental modulus of Al/fly ash syntactic foams. Despite the difference in values, the trend of modulus with respect to density remains the same.

The yield and the plastic strengths of syntactic foams are usually very close to each other and studies have specified only one of these quantities. The yield strength of syntactic foams is presented in Figure 10.18 and the plastic strength is presented in Figure 10.19. The data are available for syntactic foams having commercially pure aluminum (cp-Al), A2014 alloy, A6061 alloy and A4032 alloy. Foams having density in the range 0.5–2.2 g/cc have been tested in these studies. The 4032 foams have the lowest density and plastic strength. The strength of A2014 is as high as 190 MPa, which can enable their structural applications. The other

Table 10.2 Studies on wear testing of Al/fly ash syntactic foams.

Reference	Test condition	Material	Wear rate ×10⁻³ (mm³/m)
[45]	Pin on disc $v = 1\,\mathrm{ms^{-1}}, L = 2.37\,N, t = 300\,s$	A356 alloy	70
		A356-5% fly ash	51.3
[47]	Pin-on-disc $v = 1\,\mathrm{ms^{-1}}, L = 20\,N$	A356	1.2
		A356 Al–6% fly ash	0.9
		A356 Al–12% fly ash	1.0
	Pin-on-disc $v = 1\,\mathrm{ms^{-1}}, L = 20\,N$	A356	1.7
		A356 Al–6% fly ash	3.3
		A356 Al–12% fly ash	1.5
	Pin-on-disc $v = 1\,\mathrm{ms^{-1}}, L = 20\,N$	A356	9.4
		A356 Al–6% fly ash	9.1
		A356 Al–12% fly ash	3.4

(continued)

Table 10.2 (*Cont.*)

Reference	Test condition	Material	Wear rate ×10⁻³ (mm³/m)
[48]	Pin-on-disc $v = 2\ \mathrm{ms^{-1}}, L = 1\ \mathrm{kg}$	Al/fly ash cenosphere syntactic foam	1.4
	Pin-on-disc $v = 3\ \mathrm{ms^{-1}}, L = 1\ \mathrm{kg}$		0.8
	Pin-on-disc $v = 4\ \mathrm{ms^{-1}}, L = 1\ \mathrm{kg}$		0.6
	Pin-on-disc $v = 2\ \mathrm{ms^{-1}}, L = 3\ \mathrm{kg}$		2.2
	Pin-on-disc $v = 3\ \mathrm{ms^{-1}}, L = 3\ \mathrm{kg}$		1.5
	Pin-on-disc $v = 4\ \mathrm{ms^{-1}}, L = 3\ \mathrm{kg}$		0.8
	Pin-on-disc $v = 2\ \mathrm{ms^{-1}}, L = 5\ \mathrm{kg}$		11.5
	Pin-on-disc $v = 3\ \mathrm{ms^{-1}}, L = 5\ \mathrm{kg}$		9.7
	Pin-on-disc $v = 4\ \mathrm{ms^{-1}}, L = 5\ \mathrm{kg}$		2.8

Reference	Test condition	Material	Wear rate ×10⁻³ (mm³/m)
[49]	Slurry erosive wear	Al/fly ash particulate	
[50]	Erosion test	Al/fly ash coating	
[51]	Pin-on-disc $v = 2$ ms⁻¹	Al/cenosphere syntactic foam (dry)	22
	Pin-on-disc $v = 3$ ms⁻¹		16
	Pin-on-disc $v = 4$ ms⁻¹		8
	Pin-on-disc $v = 2$ ms⁻¹	Al/cenosphere syntactic foam (lubricated)	2.5
	Pin-on-disc $v = 3$ ms⁻¹		1.5
	Pin-on-disc $v = 4$ ms⁻¹		1
[52]	Pin-on-disc	AA6351-fly ash	

(continued)

Table 10.2 (*Cont.*)

Reference	Test condition	Material		Wear rate ×10⁻³ (mm³/m)
[53]		Al/fly ash		
[54]	Pin-on-disc $L = 2$ N	A356-60% KFA	0-25 μm	1.5
			25-40 μm	10.3
			40-90 μm	9
			25-40 μm (ground)	5.2
		A356-60% MFA	0-25 μm	13.3
			25-40 μm	15
			40-90 μm	16
			25-40 μm (ground)	3.6
		A356 alloy		4
[55]	Pin-on-disc $L = 40$ N	Al-Mg alloys (8 and 10 wt.% Mg), 0-8 wt.% solid fly ash		–

v = velocity; L = load; t = time; KFA = Kardia fly ash; MFA = Megalopolis fly ash

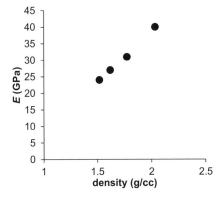

Figure 10.17 The modulus obtained from finite element analysis plotted with respect to density for Al/glass hollow particle syntactic foams.

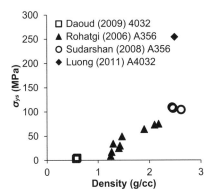

Figure 10.18 The compressive yield strength of syntactic foam with respect to the foam density. The data in this figure are taken from [25,57,59,60].

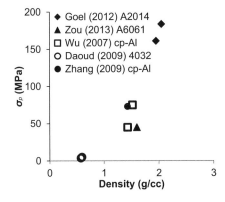

Figure 10.19 The plastic stress of syntactic foam under compression with respect to the foam density. The data in this figure are taken from [59,61–64].

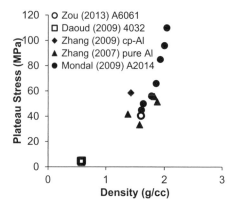

Figure 10.20 The plateau stress of syntactic foam under compression with respect to the foam density. The data in this figure are taken from [59,62,64-66].

foams having much lower plastic strength may be used for energy absorption applications such as foam fillings in automotive body components.

The plateau stress values are plotted in Figure 10.20 for syntactic foams of commercially pure aluminum and various alloys. The plateau stress values vary from about 5 MPa to over 110 MPa. The wide range of values that have been obtained in different syntactic foams is useful in selecting appropriate material as per the design requirements. In this case, the plateau stress also increases with the foam density.

When all tensile and compressive properties show an increase with the foam density, it is desirable to find methods of increasing the mechanical properties without a trade-off on the density of the composite. Detailed studies that clearly demonstrate the effect of particle wall thickness and volume fraction can greatly help in developing the density-mechanical property relationships for syntactic foams. Such studies on polymer matrix syntactic foams are now available and have been very useful in the materials selection and microstructure design to obtain syntactic foams of desired elastic, viscoelastic, and thermal properties [41–44]. Studies are also required to establish the best heat treatment practices for syntactic foams of various aluminum alloys.

The tribological properties of Al/fly ash syntactic foams have been extensively studied. A summary of available studies on the wear properties is provided in Table 10.2. The wear rate of A356 alloy has been reported as 70×10^{-3} mm^3/m [45] compared to 51.3×10^{-3} mm^3/m for A356/fly ash tested under the same conditions. In general, increasing fly ash content in syntactic foams resulted in a reduced wear rate.

Porous materials are known to have high sound and vibration damping capabilities. Syntactic foams have been characterized for their damping properties. The damping capacity of Al/fly ash composites is found to be higher than that of the matrix alloy [34, 46]. Fly ash content as small as 1.5 and 3 wt% has been found to improve the damping capacity of the foam [46]. The particle-matrix interfacial sliding is considered responsible for damping. In cases where the interfacial bonding is poor, the friction during sliding at the interface can dissipate energy and improve the damping capacity of the syntactic foam.

10.6 Applications

Over 80% of all metal matrix composites produced are used in transportation applications. A similar trend is also seen for aluminum matrix syntactic foams. In general, applications for aluminum matrix syntactic foams are not currently widespread at this time, mostly because of lack of design and performance data for components. The development of components may accelerate as research efforts continue and more data become available. Several components of aluminum matrix syntactic foams have been developed in the prototype stages and tested under the conditions of real-world applications. Figure 10.21 shows an

Figure 10.21 Al/fly ash engine intake manifold prototype developed for Ford vehicles.

Figure 10.22 Al/fly ash automotive alternator cover.

Al/fly ash syntactic foam engine intake manifold prototype developed for a car. The weight of this aluminum matrix syntactic foam component is almost 75% lower compared to a similar steel component. This component also shows the possibility of casting a geometrically complex component using Al/fly ash syntactic foam. An alternator cover, another geometrically complex component, is shown in Figure 10.22. The ability to cast geometrically complex components in net or near-net shape can be very useful in developing low cost components that have a realistic possibility of finding applications. In automobiles, several components that do not carry load can be the first target for replacement with such new materials. The thermal expansion and conductivity of aluminum matrix syntactic foams are lower compared to that of matrix alloys, which could be advantage ous in automotive components related to engine or exhaust systems. Some of the initial applications may come from smaller vehicles. For example, a prototype motor mount synthesized by casting methods for Harley Davidson motorcycle is shown in Figure 10.23. Such components can help in building the history of aluminum matrix syntactic foams in automotive components and provide long time performance data that can be used in developing new components for other types of vehicles.

Figure 10.23 Al/fly ash motor mount prototype developed for Harley Davidson motorcycles.

10.7   Conclusion

Aluminum matrix syntactic foams present unique opportunities at a time when weight reduction in a wide variety of industrial applications, especially in automobiles, is desired. A wide variety of hollow particles are now commercially available for use in syntactic foams. The possibility of synthesizing syntactic foams with fly ash filler particles is very attractive because fly ash is an industrial waste material and is available at a very low cost. Replacing the more expensive aluminum with very low cost fly ash in the syntactic foam structure can provide parts that are cheaper than comparable parts made of aluminum. Development of synthesis methods for such composites is a critical issue for minimizing the undesired reaction at the particle-matrix interface and the particle degradation or fracture during processing. Coating the fly ash particles with nickel or other suitable metals has been found useful because it can (a) create a physical barrier and prevent undesired interfacial reactions, (b) improve interfacial bonding between particle and matrix and (c) cover the defects present on the surface of fly ash cenospheres and make them uniform. Aluminum/fly ash syntactic foams have been extensively studied in the published literature. The compressive properties of such composites are especially attractive because fly ash cenospheres can be a load-bearing element in the microstructure under compression. Syntactic foams with commercially pure aluminum as well as with 2014, 4032, 6061, and 356 alloys have been synthesized and characterized. In general, the modulus and yield strength are found to increase with density for syntactic foams. Several compositions showed higher properties compared to the matrix alloy on the basis of per unit weight. It is expected that the availability of results from well-planned detailed studies will increase the confidence of industry in these materials and enable their new applications. Prototype components have been fabricated using syntactic foams and tested under real performance conditions for a number of automotive parts.

Acknowledgments

This work is supported by the Office of Naval Research grant N00014-10-1-0988 and Army Research Laboratory Cooperative Working Agreement W911NF-11-2-0096. The views and conclusions are those of the authors and should not be interpreted as the official policies or position, either expressed or implied, of the ONR, ARL or the U.S. Government unless so designated by other authorized documents. Vasanth Chakravarthy Shunmugasamy is acknowledged for help in preparation of data and figures. Dr. Gary Gladysz and William Ricci of Trelleborg Offshore, Boston provided fly ash samples for imaging. SiC particles were provided for imaging by Mr. Oliver M. Strbik III of Deep Springs Technologies, Toledo, OH.

References

1. P.K. Rohatgi, D. Weiss, and N. Gupta, Applications of fly ash in synthesizing low-cost MMCs for automotive and other applications. *JOM: Journal of the Minerals, Metals and Materials Society*, Vol. 58(11), p. 71–76, 2006.
2. P. Rohatgi, Low-cost, fly-ash-containing aluminum-matrix composites. *JOM: Journal of the Minerals, Metals & Materials Society*, Vol. 46(11), p. 55–59, 1994.
3. O. Prakash, H. Sang, and J.D. Embury, Structure and properties of Al SiC foam. *Materials Science and Engineering: A*, Vol. 199(2), p. 195–203, 1995.

4. W. Deqing, and S. Ziyuan, Effect of ceramic particles on cell size and wall thickness of aluminum foam. *Materials Science and Engineering: A,* Vol. 361(1–2), p. 45–49, 2003.

5. X.W. Lou, L.A. Archer, and Z. Yang, Hollow micro-/nanostructures: Synthesis and applications. *Advanced Materials,* Vol. 20(21), p. 3987–4019, 2008.

6. R.S. Iyer, and J.A. Scott, Power station fly ash – A review of value-added utilization outside of the construction industry. *Resources, Conservation and Recycling,* Vol. 31(3), p. 217–228, 2001.

7. M. Ahmaruzzaman, A review on the utilization of fly ash. *Progress in Energy and Combustion Science,* Vol. 36(3), p. 327–363, 2010.

8. G. Kostakis, Characterization of the fly ashes from the lignite burning power plants of northern Greece based on their quantitative mineralogical composition. *Journal of Hazardous Materials,* Vol. 166(2–3), p. 972–977, 2009.

9. S.V. Vassilev, and C.G. Vassileva, A new approach for the classification of coal fly ashes based on their origin, composition, properties, and behaviour. *Fuel,* Vol. 86(10–11), p. 1490–1512, 2007.

10. S.V. Vassilev, C.G. Vassileva, A.I. Karayigit, Y. Bulut, A. Alastuey, and X. Querol, Phase-mineral and chemical composition of fractions separated from composite fly ashes at the Soma power station, Turkey. *International Journal of Coal Geology,* Vol. 61(1–2), p. 65–85, 2005.

11. I. Narasimha Murthy, D. Venkata Rao, and J. Babu Rao, Microstructure and mechanical properties of aluminum–fly ash nano composites made by ultrasonic method. *Materials & Design,* Vol. 35(0), p. 55–65, 2012.

12. J. Bieniaś, M. Walczak, B. Surowska, and J. Sobczak, Microstructure and corrosion behaviour of aluminum fly ash composites. *Journal of Optoelectronics and Advanced Materials,* Vol. 5(2), p. 493–502, 2003.

13. C.N. Panagopoulos, and E.P. Georgiou, Surface mechanical behaviour of composite Ni–P–fly ash/zincate coated aluminium alloy. *Applied Surface Science,* Vol. 255(13–14), p. 6499–6503, 2009.

14. C.N. Panagopoulos, E.P. Georgiou, A. Tsopani, and L. Piperi, Composite Ni–Co–fly ash coatings on 5083 aluminium alloy. *Applied Surface Science,* Vol. 257(11), p. 4769–4773, 2011.

15. K. Huang, K. Inoue, H. Harada, H. Kawakita, and K. Ohto, Leaching behavior of heavy metals with hydrochloric acid from fly ash generated in municipal waste incineration plants. *Transactions of Nonferrous Metals Society of China,* Vol. 21(6), p. 1422–1427, 2011.

16. D.D. Luong, O.M. Strbik III, V.H. Hammond, N. Gupta, and K. Cho, Development of high performance lightweight aluminum alloy/SiC hollow sphere syntactic foams and compressive characterization at quasi-static and high strain rates. *Journal of Alloys and Compounds,* Vol. 550, p. 412–422, 2013.

17. R.A. Palmer, K. Gao, T.M. Doan, L. Green, and G. Cavallaro, Pressure infiltrated syntactic foams – Process development and mechanical properties. *Materials Science and Engineering: A*, Vol. 464(1–2), p. 85–92, 2007.

18. Z. Huang, S. Yu, J. Liu, and X. Zhu, Microstructure and mechanical properties of in situ Mg2Si/AZ91D composites through incorporating fly ash cenospheres. *Materials and Design*, Vol. 32(10), p. 4714–4719, 2011.

19. Z.Q. Huang, S.R. Yu, and M.Q. Li, Microstructures and compressive properties of AZ91D/fly-ash cenospheres composites. Transactions of Nonferrous *Metals Society of China*, Vol. 20(Supplement 2), p. s458-s462, 2010.

20. P.K. Rohatgi, A. Daoud, B.F. Schultz, and T. Puri, Microstructure and mechanical behavior of die casting AZ91D-Fly ash cenosphere composites. *Composites Part A: Applied Science and Manufacturing*, Vol. 40(6–7), p. 883–896, 2009.

21. B. Ralph, H.C. Yuen, and W.B. Lee, The processing of metal matrix composites – An overview. *Journal of Materials Processing Technology*, Vol. 63(1–3), p. 339–353, 1997.

22. J.M. Torralba, C.E. da Costa, and F. Velasco, P/M aluminum matrix composites: an overview. *Journal of Materials Processing Technology*, Vol. 133(1–2), p. 203–206, 2003.

23. R.Q. Guo, P.K. Rohatgi, and D. Nath, Preparation of aluminium-fly ash particulate composite by powder metallurgy technique. *Journal of Materials Science*, Vol. 32(15), p. 3971–3974, 1997.

24. M. Hrairi, M. Ahmed, and Y. Nimir, Compaction of fly ash–aluminum alloy composites and evaluation of their mechanical and acoustic properties. *Advanced Powder Technology*, Vol. 20(6), p. 548–553, 2009.

25. P.K. Rohatgi, J.K. Kim, N. Gupta, S. Alaraj, and A. Daoud, Compressive characteristics of A356/fly ash cenosphere composites synthesized by pressure infiltration technique. *Composites Part A: Applied Science and Manufacturing*, Vol. 37(3), p. 430–437, 2006.

26. P.K. Rohatgi, R.Q. Guo, H. Iksan, E.J. Borchelt, and R. Asthana, Pressure infiltration technique for synthesis of aluminum–fly ash particulate composite. *Materials Science and Engineering: A*, Vol. 244(1), p. 22–30, 1998.

27. N. Gupta, A functionally graded syntactic foam material for high energy absorption under compression. *Materials Letters*, Vol. 61(4–5), p. 979–982, 2007.

28. N. Gupta and W. Ricci, Comparison of compressive properties of layered syntactic foams having gradient in microballoon volume fraction and wall thickness. *Materials Science and Engineering: A*, Vol. 427(1–2), p. 331–342, 2006.

29. T.P.D. Rajan, R.M. Pillai, B.C. Pai, K.G. Satyanarayana, and P.K. Rohatgi, Fabrication and characterisation of Al–7Si–0.35Mg/fly ash

metal matrix composites processed by different stir casting routes. *Composites Science and Technology*, Vol. 67(15–16): p. 3369–3377, 2007.

30. E. Gikunoo, O. Omotoso, and I.N.A. Oguocha, Effect of fly ash addition on the magnesium content of casting aluminum alloy A535. *Journal of Materials Science*, Vol. 40(2), p. 487–490, 2005.

31. W.A. Uju and I.N.A. Oguocha, A study of thermal expansion of Al–Mg alloy composites containing fly ash. *Materials and Design*, Vol. 33, p. 503–509, 2012.

32. W.A. Uju and I.N.A. Oguocha, Thermal cycling behaviour of stir cast Al–Mg alloy reinforced with fly ash. *Materials Science and Engineering: A*, Vol. 526(1–2), p. 100–105, 2009.

33. P.R.S. Kumar, S. Kumaran, T.S. Rao, and S. Natarajan, High temperature sliding wear behavior of press-extruded AA6061/fly ash composite. *Materials Science and Engineering: A*, Vol. 527(6), p. 1501–1509 2010.

34. G.H. Wu, Z.Y. Dou, L.T. Jiang, and J.H. Cao, Damping properties of aluminum matrix–fly ash composites. *Materials Letters*, Vol. 60(24): p. 2945–2948, 2006.

35. N. Sobczak, J. Sobczak, J. Morgiel, and L. Stobierski, TEM characterization of the reaction products in aluminium–fly ash couples. *Materials Chemistry and Physics*, Vol. 81(2–3), p. 296–300, 2003.

36. R.O. Guo, and P.K. Rohatgi, Chemical reactions between aluminum and fly ash during synthesis and reheating of Al-fly ash composite. *Metallurgical and Materials Transactions B*, Vol. 29(3), p. 519–525, 1998.

37. R.Q. Guo, D. Venugopalan, and P.K. Rohatgi, Differential thermal analysis to establish the stability of aluminum-fly ash composites during synthesis and reheating. *Materials Science and Engineering: A*, Vol. 241(1–2), p. 184–190, 1998.

38. J. David Raja Selvam, D.S. Robinson Smart, and I. Dinaharan, Microstructure and some mechanical properties of fly ash particulate reinforced AA6061 aluminum alloy composites prepared by compocasting. *Materials and Design*, Vol. 49(0), p. 28–34, 2013.

39. N. Gupta, D.D. Luong, and K. Cho, Magnesium matrix composite foams-density, mechanical properties, and applications. *Metals*, Vol. 2(3), p. 238–252 2012.

40. L. Bardella, A. Sfreddo, C. Ventura, M. Porfiri, and N. Gupta, A critical evaluation of micromechanical models for syntactic foams. *Mechanics of Materials*, Vol. 50(0), p. 53–69, 2012.

41. N. Gupta and E. Woldesenbet, Microballoon wall thickness effects on properties of syntactic foams. *Journal of Cellular Plastics*, Vol. 40, p. 461–480, 2004.

42. N. Gupta, R. Ye, and M. Porfiri, Comparison of tensile and compressive characteristics of vinyl ester/glass microballoon syntactic foams. *Composites Part B: Engineering*, Vol. 41, p. 236–245, 2010.

43. N. Gupta and D. Pinisetty, A review of thermal conductivity of polymer matrix syntactic foams—Effect of hollow particle wall thickness and volume fraction. *JOM: Journal of The Minerals, Metals and Materials Society*, Vol. 65(2): p. 234–245, 2013.

44. V.C. Shunmugasamy, D. Pinisetty, and N. Gupta, Viscoelastic properties of hollow glass particle filled vinyl ester matrix syntactic foams: Effect of temperature and loading frequency. *Journal of Materials Science*, Vol. 48, p. 1685–1701, 2013.

45. P.K. Rohatgi, R.Q. Guo, P. Huang, and S. Ray, Friction and abrasion resistance of cast aluminum alloy-fly ash composites. *Metallurgical and Materials Transactions A*, Vol. 28(1), p. 245–250, 1997.

46. Y. Mu, G. Yao, and H. Luo, The dependence of damping property of fly ash reinforced closed-cell aluminum alloy foams on strain amplitude. *Materials and Design*, Vol. 31(2), p. 1007–1009, 2010.

47. Sudarshan and M.K. Surappa, Dry sliding wear of fly ash particle reinforced A356 Al composites. *Wear*, Vol. 265(3–4), p. 349–360, 2008.

48. D.P. Mondal, S. Das, and N. Jha, Dry sliding wear behaviour of aluminum syntactic foam. *Materials and Design*, Vol. 30(7), p. 2563–2568, 2009.

49. M. Ramachandra and K. Radhakrishna, Effect of reinforcement of flyash on sliding wear, slurry erosive wear and corrosive behavior of aluminium matrix composite. *Wear*, Vol. 262(11–12), p. 1450–1462, 2007.

50. S.P. Sahu, A. Satapathy, A. Patnaik, K.P. Sreekumar, and P.V. Ananthapadmanabhan, Development, characterization and erosion wear response of plasma sprayed fly ash–aluminum coatings. *Materials and Design*, Vol. 31(3), p. 1165–1173, 2010.

51. N. Jha, A. Badkul, D.P. Mondal, S. Das, and M. Singh, Sliding wear behaviour of aluminum syntactic foam: A comparison with Al–10 wt% SiC composites. *Tribology International*, Vol. 44(3), p. 220–231, 2011.

52. M. Uthayakumar, S. Thirumalai Kumaran, and S. Aravindan, Dry sliding friction and wear studies of fly ash reinforced AA-6351 metal matrix composites. *Advances in Tribology*, Vol. 2013, Article ID 365602, 6 pages, 2013.

53. G. Withers, Dispersing flyash particles in an aluminum matrix. *Advanced Materials & Process*, Vol. 164(11), p. 39–40, 2006.

54. G. Itskos, P.K. Rohatgi, A. Moutsatsou, J.D. DeFouw, N. Koukouzas, C. Vasilatos, and B.F. Schultz, Synthesis of A356 Al–high-Ca fly ash composites by pressure infiltration technique and their characterization. *Journal of Materials Science*, Vol. 47(9), p. 4042–4052, 2012.

55. S. Zahi, and A.R. Daud, Fly ash characterization and application in Al–based Mg alloys. *Materials & Design*, Vol. 32(3), p. 1337–1346, 2011.

56. P.R.S. Kumar, S. Kumaran, T. Srinivasa Rao, and K. Sivaprasad, Microstructure and mechanical properties of fly ash particle reinforced

AA6061 composites produced by press and extrusion. *Transactions of the Indian Institute of Metals*, Vol. 62(6), p. 559–566, 2009.

57. Sudarshan and M.K. Surappa, Synthesis of fly ash particle reinforced A356 Al composites and their characterization. *Materials Science and Engineering: A*, Vol. 480(1–2): p. 117–124, 2008.

58. M. Ramachandra and K. Radhakrishna, Synthesis-microstructure-mechanical properties-wear and corrosion behavior of an Al-Si (12%)—Flyash metal matrix composite. *Journal of Materials Science*, Vol. 40(22), p. 5989–5997, 2005.

59. A. Daoud, Effect of fly ash addition on the structure and compressive properties of 4032–fly ash particle composite foams. *Journal of Alloys and Compounds*, Vol. 487(1–2), p. 618–625, 2009.

60. D.D. Luong, N. Gupta, A. Daoud, and P.K. Rohatgi, High strain rate compressive characterization of aluminum alloy/fly ash cenosphere composites. *JOM: Journal of The Minerals, Metals & Materials Society*, Vol. 63(2), p. 53–56, 2011.

61. M.D. Goel, M. Peroni, G. Solomos, D.P. Mondal, V.A. Matsagar, A.K. Gupta, M. Larcher, and S. Marburg, Dynamic compression behavior of cenosphere aluminum alloy syntactic foam. *Materials & Design*, .Vol. 42(0), p. 418–423, 2012.

62. L.C. Zou, Q. Zhang, B.J. Pang, G.H. Wu, L.T. Jiang, and H. Su, Dynamic compressive behavior of aluminum matrix syntactic foam and its multilayer structure. *Materials & Design*, Vol. 45(0), p. 555–560, 2013.

63. G.H. Wu, Z.Y. Dou, D.L. Sun, L.T. Jiang, B.S. Ding, and B.F. He, Compression behaviors of cenosphere-pure aluminum syntactic foams. *Scripta Materialia*, Vol. 56(3), p. 221–224, 2007.

64. L.P. Zhang and Y.Y. Zhao, Mechanical response of Al matrix syntactic foams produced by pressure infiltration casting. *Journal of Composite Materials*, Vol. 41(17), p. 2105–2117, 2007.

65. D.P. Mondal, S. Das, N. Ramakrishnan, and K. Uday Bhasker, Cenosphere filled aluminum syntactic foam made through stir-casting technique. *Composites Part A: Applied Science and Manufacturing*, Vol. 40(3), p. 279–288, 2009.

66. Q. Zhang, P.D. Lee, R. Singh, G. Wu, and T.C. Lindley, Micro-CT characterization of structural features and deformation behavior of fly ash/aluminum syntactic foam. *Acta Materialia*, Vol. 57(10), p. 3003–3011, 2009.

11

Engineering Behavior of Ash Fills

Ashutosh Trivedi

Professor, Department of Civil Engineering,
Delhi Technological University, Delhi, India

Abstract

The coal-based thermal power plants often dispose of surplus fly ash or coal ash in a nearby low-lying area normally called ash pond. It is extensively used as a structural fill and as a subgrade for highway embankment. The bearing capacity and settlement of cemented and uncemented ash fills are based upon a quality designation related to RQD and penetration tests, respectively. The cemented ash fill is characterized by a hardening parameter obtained from the joint parameters of a fractured core. Many of these fills in uncemented and uncompacted state show disproportionate settlement due to collapse, piping, erosion and liquefaction. For the use of these fills as foundation soil, the ash should be characterized and compacted. In the present chapter the characterization, hardening, bearing capacity and settlement are analyzed for the ash fills. It is based on relative density and relative compaction for uncemented ash fills. The bearing capacity of compacted ash is a function of relative dilatancy. A plot for settlement and foundation size is utilized to obtain the settlement of compacted ash. The critical values of penetration resistance of standard cone and split spoon sampler in saturated condition are adjudged vulnerable to excessive settlement as shown by the collapse and liquefaction resistance. It is shown that for classified gradation of ashes above a critical degree of compaction and degree of saturation, the ash fill may settle less than the allowable settlements.

Keywords: Ash fills, compaction, RQD, load tests, bearing capacity, settlement

*Corresponding author: prof.trivedi@yahoo.com

Ashutosh Tiwari and S.K. Shukla (eds.) Advanced Carbon Materials and Technology, (419–474)
2014 © Scrivener Publishing LLC

11.1   Background

The burning of coal has been a main source of power generation for ages, but the mass production of power by thermoelectric plants has brought a mounting problem of ash disposal in the past few decades. The ash produced by the coal-fired power plants consists of fly ash composed of particle sizes normally less than 300 µm and bottom ash made up of significantly coarser particles. The mixture of fly ash and bottom ash is eventually disposed of in a slurry containment facility known as ash pond [1]. The combined quantum of ashes, namely fly ash, bottom ash and pond ash, are commonly called coal ash in North American countries. According to one estimate, the world coal consumption of 7.5 thousand million tons in the year 2011 will be closer to 13 thousand million tons in the year 2030 with proportional generation of coal ash. The principal carbon compositions in a variety of coal types, namely anthracite, bituminous, and lignite coal, vary significantly. Consequently, they have ash content ranging from less than twenty, ten and five percent, respectively. On an average, it may require a capacity building to accommodate nearly a hundred to thousand million tons of ash annually. Worldwide, more than 65% of fly ash produced from coal power stations is disposed of in landfills and ash ponds [2, 3]. Therefore, the behavior of ash fill has remained a matter of great interest equally among engineers, scientists, planners and developers, contractors, and owners in the past three decades.

11.1.1   Physico-Chemical Characterization

Normally, the composite ash collected from electrostatic precipitators and the bottom of the hoppers of thermal power plants may be classified as coal ash. The coarse ash collected from the furnace bottom is known as bottom ash. It is around 20 to 25% of the total ash produced. The ash is disposed of in a pond by mixing it with water to form slurry. The slurry usually contains 20% solids by weight. This method of ash disposal is called wet method. There are several environmental issues associated with ash fill [4–7]. The landfill of ash may be used as a construction fill if the suitable ashes are properly characterized. The fine ashes may collapse upon wetting. To avoid excessive settlement upon wetting, suitability of coal ash should be examined as per the criteria of collapse [8, 9] and liquefaction [10, 11]. The chemical and physical characteristics of the ash produced depend upon the quality of coal used, the performance of wash-units, efficiency of the furnace and several other factors. The physical and chemical properties of ash are influenced by the type and source of coal, method and degree of coal preparation, cleaning and pulverization, type and operation of power generation unit, ash collection, handling and storage methods, etc. Ash properties may vary due to changes in boiler load. The choice of furnace type such as stoker fired, cyclone type or pulverized coal furnace is also known to affect the properties of the ash collected.

11.1.2   Engineering Characteristics

Since coal-based power is one of the most reliable means of power generation throughout the world, the continuing practice throughout various countries has been to consider it acceptable to dispose of ash in landfills in areas previously considered wastelands or embankments. The recent studies [12, 5, 13] indicate potential vulnerability of the ash fills to failures [8, 14, 15]. However, land reclamation using coal ash has been viably investigated in various parts of the world. Moreover, it has been a matter of interest among the scientific community to explore the utilization of coal ash [16–18]. During the combustion of coal, minerals are transformed to mullite, magnetite, tridymite, glass, etc., thus forming a composite ash. The main chemical components of coal ash are silica, alumina, iron oxide and other alkalis.

The mineral group present in coal, such as hydrated silicate group, carbonate group, sulphate group and their varying compositions play a major role in determining the chemical composition of ash. As per the source of coal used in different thermal plants, the chemical composition [19] and engineering parameter [20–22] of the ashes are different (Tables 11.1–11.5) compared to natural geomaterials. The design parameters for ash as structural and embankment fill provided in IRC: SP-58: 2001 are shown in Table 11.3.

The ASTM classifications of coal ash are related to the percentage of calcium oxide in ash. The ashes with a high amount of calcium oxide show self-hardening pozzolanic properties in the presence of water. The pozzolanic properties of fly ash have been documented by Mehta and Monterio [23]. Such ashes are designated as class C ash. A typical class C ash is obtained from the burning of lignite coal. The ashes from bituminous coal that do not possess self-hardening properties are called class F ash. The large quantum of ash

Table 11.1 Ash type and coal quality, collection stage, chemical and particle characteristics.

Coal Quality	Stage	Ash Type	Chemical, Mineralogical Composition	Particle Characteristics
Anthracite bituminous and sub bituminous coal	ESP hopper	Low calcium oxide fly ash	Mostly silicate glass containing aluminum, iron, and alkalis. Small amount of crystalline matter present generally consists of quartz, mullite, sillimanite, hematite, and magnetite [23].	15 to 30 percent particles larger than 45μm (surface area 20 to 30 m^2N^{-1}) most of the particles are solid spheres 20μm average diameter. Cenosphere, plerospheres may be present [23].
Lignite coal	ESP hopper	High calcium oxide fly ash/typical ASTM C-618 class C ash	Mostly silicate glass containing calcium magnesium, aluminum, and alkalis. Small amount of crystalline matter present generally consists of quartz, tricalcium aluminates; free lime and periclase may be present. Unburned coal is usually less than 2 percent [23].	10 to 15 percent particles larger than 45μm (surface area 30 to 40m^2N^{-1}) most of the particles are solid spheres less than 20μm in diameter. Particle surface is generally smooth but not as clean as low calcium oxide fly ash [23].
Any coal quality	Furnace bottom	Bottom ash	Residue consisting of silica, mullite, sillimanite, hematite, and magnetite [1].	Sand size particles of rough texture [1].
Any coal quality	Ash pond or ash lagoon	Pond ash	Pond ash is a mixture of ESP ash and furnace bottom ash with water. It contains a wide range of particle sizes, texture and chemical composition depending upon the quality of the charge. It has some aging affects also [1].	

Table 11.2 Chemical composition of coal ashes [1, 8, 24].

Chemical composition %	British ash	American ash	Swedish ash	Polish ash	Indian ash
SiO_2	38–58	30–58	30–53	43–52	48.4–57.5
Al_2O_3	20–40	7–38	14–33	19–34	18.2–27.2
Fe_2O_3	6–16	10–42	10–14	0.7–10.7	11.3–5.4
CaO	2–10	0–13	0.9–6.1	1.7–9.4	11.8–3.1
MgO	1–3.5	0–3	4–6	1–2.9	3.6–0.4
Na_2O, K_2O	2–5.5	0.4–2	1.6–3.5	0.4–0.9	0–0.9
SO_3	0.5–2.5	0.2–1	0.4–1.5	0.3–0.8	0–2.9
Unburned Carbon	–	0–4.8	0.9–3.3	1.9–9.9	1.2–4.1

Table 11.3 Engineering parameters for coal ash [3, 9].

Parameter	IRC	Uncemented Ash	Cemented Ash
Sp. gravity Plasticity Maximum dry unit weight OMC	1.90–2.55 NP 9–16 18–38%	1.70–2.6 P, NP 8–18 20–40%	2.0–2.6 – 10–18 20–40%
Cohesion	negligible	Nil	Max of 1 MPa
Angle of internal friction	30–40	27–44	30–47
Coefficient of consolidation cv (cm^2 sec^{-1})	1.75×10^{-5} to 2.01×10^{-3}	– –	– –
Compression index (cc)	0.05–0.4	0.01–0.006	–
Permeability (cm sec^{-1})	$8 \times 10^{-6} - 7 \times 10^{-4}$	$1-7 \times 10^{-5}$	10^{-7} to 10^{-6}
Particle size distribution Gravel Sand Silt Clay	 1–10% 8–85% 7–90% 0–10%	 5% 10–90% 10–90% 0–5%	Intact to Fragmented
Coefficient of uniformity	3.1–10.7	2–12	

Table 11.4 Engineering parameters for a few uncemented geomaterials [3, 25–27].

Sand Type	c_u	e_{min}	e_{max}	ϕ_c	Reference
Monterey #0 sand	1.6	0.57	0.86	37.0	[27]
Tieino sand	1.5	0.57	0.93	34.8	[26]
Toyoura sand	1.27	0.61	0.99	35.1	[26]
Ottawa sand (round)	1.48	0.48	0.78	29.0	[26]
Sacramento river sand	1.47	0.61	1.03	33.3	[27]
Hokksund sand	1.91	0.55	0.87	36.0	[26]
Yamuna silty sand	2–35	0.31	0.78	25–30	[25]
Coal ash	2–10	0.6–0.8	1.4–2.10	27–30	[3]

Table 11.5 Peak and ultimate angles of friction for a few geomaterials [3, 25, 27, 28].

Material	ϕ'_{peak}	ϕ'_{ult}	Reference
Dense well-graded sand or gravel, angular grains	55°	35°	[27]
Medium dense uniform sand, round grains	40°	32°	
Dense sandy silt with some clay	47°	32°	
Sandy silty clay (glacial)	35°	30°	
Clay-shale, on partings	35°	25°	
Clay (London)	25°	15°	
Yamuna silty sand	46°	25°	[28]
Coal ash	50–53°	27–30°	[3]

produced may be classified as class *F*. Coal ash is normally used in the construction of ash dykes, reclamation of low-lying land, fabricated earth structures such as embankments, road fills, etc. The landfill's intensive utilization of coal ash requires stability analysis of fill.

11.1.2.1 Characterization

X-ray diffraction study is normally carried out to identify the mineral phases present in the ash. A typical X-ray diffraction shows that the ash contains traces of aluminum silicate, quartz and some heavy minerals like hematite and magnetite. Identification of definite crystalline mineral can be based on Bragg's equation [$\lambda = 2$ d sin 2θ] where λ is wavelength of X-ray specific to the Cu target element [= 1.542Å] and d is interplanner spacing. Normally the test is conducted between 0°–70° (2θ), at a rate of 0.8°/sec using the CuKα characteristic radiation of Cu target element. The interplanner spacing of respective peaks on the x-ray pattern are calculated from the corresponding 2θ angle. These peaks are associated with the characteristic minerals. In crystalline form, ash contains traces of aluminum silicate, quartz and some heavy minerals [1].

Figure 11.1 shows an example of a typical X-ray diffraction pattern of an ash sample. The peak near 26.40° is characteristic of alumina-based silicate minerals. Their respective peaks near 33.2° and 35.4° indicate the presence of a heavy mineral like hematite or magnetite. A subordinate amount of $11CaO \cdot 7Al_2O_3$ is evident from peak incidentally close to 18°. The concurrence of a strong peak close to 26.5° indicates that quartz is one of the major constituents along with alumina-based silicate mineral. The crystalline silica sand is

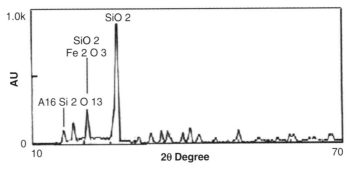

Figure 11.1 X-ray diffraction pattern of a typical ash sample [1].

also characterized by this peak. The crystalline silica sand is the main mineral component of granular fill hydraulically deposited by rivers worldwide. The potential clay minerals may be present or absent in the ash, indicating that ash may or may not have any structural clay cohesion in its natural state. Any peak associated with the hydrated calcium silicate group that is responsible for the development of cohesion due to chemical reaction (marked by the formation of crystals of hydrated calcium aluminium silicate on curing in the presence of water with time) indicates the self-hardening properties of ash. Therefore, the absence or presence of respective peak is classified as cohesionless or cemented ash mass while evaluating its behavior as an engineering fill. The ash samples contain divergent amounts of amorphous phase. The amorphous phase is in the highest amount in the pond among all ash types. This is because of the presence of unburned coal in bottom ash component. Comparing the X-ray diffraction pattern of ash samples with sand (cohesionless) it is implicit that sand has peaks with crystalline quartz, while the ash has peaks of quartz as well as humps of non-crystalline matter. Burning of coal at high temperature and sudden cooling of ash in a short interval produces non-crystalline matter in coal ash. The presence of glassy phase, which is non-crystalline in nature, is around 60 to 88% of ash by weight [23]. The cohesionless soils of similar gradation as that of ash may be characterized as sandy silt to silty sand. The engineering behavior of natural geomaterials is also affected in a similar manner [24–28]. The cemented ash fill may be characterized as weakly cemented rock. These fills have dominant presence of crystalline quartz. The presence of amorphous matter along with crystalline quartz induces significant differences in the engineering characteristics of ash fills compared to the natural geomaterials [1].

11.1.2.2 Chemical Composition

The chemical composition of the ashes is obtained from the non-combustible components produced by burning of the coal. The comparison of a typical range of chemical composition of ashes from different parts of the world along with a typical Indian ash is given in Table 11.2. The main constituent of the ash is silica followed by alumina, oxides of iron and calcium. The presence of sodium and potassium salts is known by the qualitative chemical analysis. The submergence of ashes is critical compared to the other granular soils due to the presence of few soluble matters. The solubility of ash sample is determined separately by boiling in water and then at room temperature. Each sample is thoroughly mixed with the boiling water by a stirrer. The entire experiment is repeated with the cold-water mix at room temperature. This mixture is filtered through the Whatman-42 filter paper. The retained ash is dried in an electric oven at 105° for 24 hours and the percentage soluble in the ash is obtained. The pond ashes have negligible soluble content, while fine ashes obtained directly from electrostatic precipitator have a significant percentage of soluble.

11.1.2.3   Electron Micrographs

The micrographic investigation of a typical ash sample is presented in Figure 11.2. The electron micrograph [2] indicates presence of predominantly coarse grain and finer particles together. It suggests that coarse ash contains rounded spherules, subrounded, and opaque particles. The ash contains superfine that form agglomerates which have a tendency to stick together and appear as larger particle upon pressing. Upon observing ash with a microscope it is seen that ash particles are clear or translucent spherules (siliceous aluminous particles), subrounded and rounded porous grains, irregular agglomerated glass spherule, opaque dark gray and red angular grains of magnetite and hematite, and black porous grains of carbon.

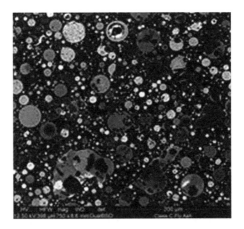

Figure 11.2 Electron micrograph of a typical ash sample [2].

11.1.2.4   Grain Size Distribution

The grain size distribution of varied ash sample indicates particle size in the range of coarse sand, silt to clay. However, the maximum frequency of particle is in the range of fine sand to silt (Figure 11.3). The effective size and mean sizes may control permeability (Figure 11.4), shear strength (Figure 11.5) and compaction (Figures 11.6, 11.7) of ashes. The pond ash which is

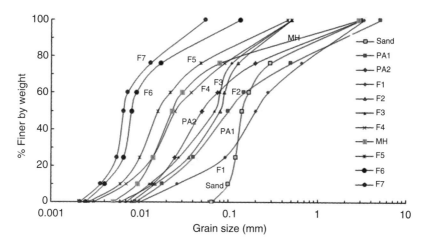

Figure 11.3 Grain size distribution of varied ash samples.

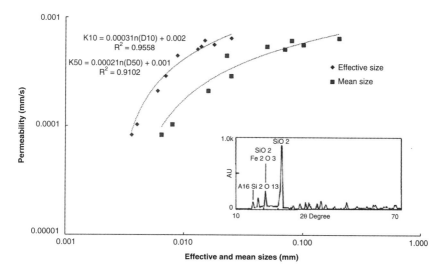

Figure 11.4 Permeability characteristics of ashes.

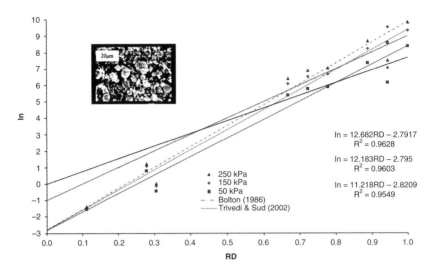

Figure 11.5 Shear characteristics of ashes.

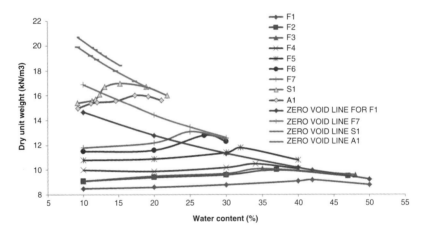

Figure 11.6 Compaction plot for typical ash samples.

examined for mass behavior as a fill may contain 5 to 10% of particles in coarse and medium sand size, 35 to 50 % in fine sand size and 40 to 60% of particles in the range of silt. The presence of superfine (size ~ 0.01 mm) transforms inter-particle friction, agglomeration and formation of pendular bonds in the presence of moisture.

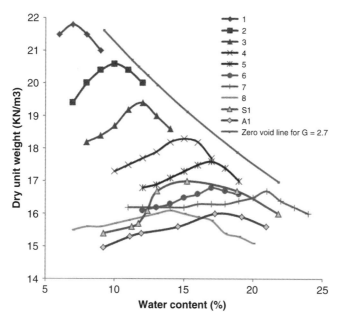

Figure 11.7 Proctor compaction of a few natural geomaterials of similar gradation.

11.1.2.5   Apparent Specific Gravity

The ashes may have lower apparent specific gravity than the natural soils of similar gradation, which is largely composed of α and β quartz, cristobalite and tridymite. The ash contains maximum percentage of silica among all the constituents. A low value of the specific gravity is attributed to the trapped micro-bubble of air in the ash particle and the presence of unburned carbon. The air voids percentage of ash (5 to 15%) is found to be greater than natural soils (1 to 5%) at maximum dry density [29]. It is noticed that as fineness of the ash increases, the specific gravity also increases partly due to the release of entrapped gases. Investigators [30] reported a similar phenomenon in the ash grounded by mortar and pestle, indicating the possibility of the breaking of bigger particles only. The mineralogical composition is one of the other reasons for variation in the specific gravity of the ash relative to soils. The ashes with high iron content tend to have a higher specific gravity. The presence of heavier minerals such as hematite and magnetite results in a higher specific gravity. It is indicated that the bottom ash typically has a higher specific gravity. The pond ash may have a higher specific gravity than the other ashes. This is partly due to the presence of bottom ash in the pond, which contains heavier components of the coal ash. Some of the ash solids contain pores, which are not inter-connected, and hence they possess, on measurement, less specific gravity, although the specific gravity of constituent mineral remains in the usual range. Such cases are referred to as apparent specific gravity, which is based on the weight in air of a given volume of ash solids, which includes the isolated voids [1].

11.1.2.6   Compaction

In the design of ash containment facility, road embankments and ash fills, it is desirable to consider the compaction characteristics of the ashes along with natural geomaterials (Figures 11.6; 11.7). The hydraulically disposed of ash in the ash ponds is normally at a low-density state. In order to improve its engineering properties com-paction is a prerequisite. The coal ash is compacted by vibration if non-plastic in nature. However, owing to the significant percent-age of fines, it is often compacted by impact. A granular material is found in varying states of density, i.e., loosest state to dense states.

The void ratio of ash sample in the loosest state is often obtained by a slow pouring technique. The ash is normally poured in a fixed volume mold from a constant height of fall of 20 mm as the tech-nique applied to loose cohesionless fill. In the vibration test, ash is deposited at varying moisture contents in a standard thick-walled cylindrical mold with a volume of 2,830 cm^3. The ash is vertically vibrated at double amplitude of 0.38 mm for seven minutes in this mold mounted on a vibration table with a frequency of 60 Hz. The difficulties associated with flow of the fines are encountered in using this technique. The capping plate is modified to fit at the top of the mold so that it presses the ash with at least some clearance on the side. The double amplitude of vertical vibration of 0.38 mm is found to be optimum for ash samples [1].

The typical result of Proctor and vibratory compaction of the ashes with varying gradation indicates reduction in water require-ment to achieve maximum density with fineness (Tables 11.6; 11.7; 11.8). The increasing fineness demonstrates a sharp increase in maximum dry density in the Proctor test. Normally, the density in the vibration test is lower than that in the Proctor test on the dry side of optimum due to the rebound action of the spherical ash

Table 11.6 Grain size, specific gravity, Procter density and optimum moisture content.

Ash Type	Cu	D_{50} (mm)	D_{60} (mm)	D_{10} (mm)	Sp Gravity	Max Unit Weight (kN/m³)	OMC (%)
Pond	8.33	0.1	0.15	0.018	1.98	9.50	40
Pond	5.35	0.05	0.075	0.014	2.00	10.3	37.5
Hopper	3.4	0.023	0.03	0.0088	1.90	11.7	33

Table 11.7 Results of vibratory and Proctor compaction.

Sp Gravity	γ_d^{min}	e^{max}	γ_d^{max} (Dry)	γ_d^{max} (moist)	e^{min} (Dry)	e^{min} (moist)	γ_d^{max} Proctor	Void Ratio at γ_d^{max} Proctor
1.98	7.63	1.6	9.56	9.50	1.06	1.08	9.5	1.08
2.00	7.85	1.54	10.56	10.3	0.89	0.94	10.3	0.94

Table 11.8 Gradation and compaction trends for natural geomaterials (Fig.11.7).

S.no	Description	Sand (%)	Silt (%)	Clay (%)
1	Well graded loamy sand	88	10	2
2	Well graded sandy loam	72	15	13
3	Medium graded sandy loam	73	9	18
4	Lean sandy silty clay	32	33	35
5	Lean silty clay	5	64	31
6	Loessial silt	5	85	10
7	Heavy clay	6	22	72
8	Poorly graded sand	92	6	–
9	Sand	99.3	0.63	–
10	Silty sand	96.4	3.55	–

particles at a low degree of saturation. In the vibration test a reduction in the density is observed with moisture content contrary to the Proctor test. This is due to the slacking of ash at a low saturation level. The minimum value of the dry unit weight is observed at critical moisture content. The dry unit weight increases beyond critical moisture due to the contravention of the surface tension force. The maximum dry unit weight is obtained at slightly higher moisture content in the vibration test.

The maximum dry unit weight of coal ash is less than that of the natural soils. This is partly due to a low specific gravity and a high air void content. The maximum dry unit weight by Proctor test is obtained at significantly high moisture content. The maximum dry unit weight by the vibration test is slightly higher (~4%) than the Proctor test. High optimum moisture content (OMC) is normally because of the porous structure of particles. The higher OMC of coal ashes compared to natural soils is due to a large percentage of the water inside the particles in initial stages. It is difficult to work the particles to higher density at lower moisture. The total air in porous structure is hardly expelled to saturate ash up to OMC (Figure 11.6). Hence, the vibratory densification technique resulted in a maximum dry unit weight in dry conditions only. But, the dry state compaction is not very useful in the landfill. By a slight vibration, ash becomes airborne and remains in the air for many hours. Sprinkling a little water helps to get rid of this problem but leads to the bulking of ash. In the standard equipment there is no further improvement in unit weight beyond 5–8 minutes of vibration. Moreover, the finer sample shows less variation in unit weight over time than the coarse sample. This is due to the greater inter-particle friction in fine ashes, as found in the case of powders [31]. All ashes initially show bulking there after they reach a stage of minimum dry unit weight at critical moisture content. The compaction beyond optimum moisture is quite erratic. The presence of porous particle (unburned coal, pelerosphere and cenosphere) increases the optimum moisture content [1].

The ash is often compacted in the field by vibratory rollers. In vibratory compaction, the maximum density in dry state is more than that in wet condition. The air voids remain entrapped among the hollow particles in wet condition. A resulting low density is attributed to surface tension in partly wet condition. The surface tension force hinders the compaction process. The maximum dry density in wet condition also coincides with the maximum dry density as obtained in the Proctor test. The modified Proctor test suggests that there is no perceptible increase in the maximum dry density owing to increased compaction effort. Apparently, it is because ash is a non-plastic material [1].

The usual practice is to compact ash by vibration using a 10 to 20 ton vibratory roller at moisture content close to optimum. Tests indicate that vibratory roller compaction gives test results when soil moisture is slightly higher than optimum moisture content as obtained in the Proctor test [32]. Depending upon the proximity of available moisture with optimum moisture content, degree of compaction may vary from 80% to 100%.

In such a limited area vibratory roller compaction was not possible. Another type of vibratory compaction equipment is the vibrating base plate compactor. It produces results similar to that of vibratory rollers [32].

In field compaction of ash, the use of a hand-operated base plate compactor has been reported near the foundation walls [30]. The use of heavyweight vibrators with low frequency is suggested for gravel. Light- to medium-weight vibrators with high frequency are suggested for sands and finer materials [32]. One of reasons for the selection of high frequency, low weight, and base plate compactor for the ash fill is as sited above. Moreover, any surcharge is found to reduce the amount of densification in the case of ash compacted at constant moisture content [33].

The weight and frequency of the compactor controls the thickness of compacted lift. The lightweight, high-frequency compactors obtain satisfactory densities in thin lifts and heavyweight, low-frequency compactors obtain satisfactory density in thick lifts [32]. A plate compactor of 220 N and a plate size of 152-mm x 390-mm is proposed for vibratory compaction [1]. Vibration is induced on the loose left of 150 to 200 mm. The time of vibration required is settled after several trials. Three passes are required at a frequency of 50 cps. This produces satisfactory results of density at selected moisture contents. The moisture content density data obtained by core cutters at several locations and depth on the test area is plotted along with the data from the laboratory vibration test to provide a guideline for the compaction of ash fill. Ash becomes airborne by slight vibration in dry state and remains suspended in air for a long time. Therefore, compaction below 5% of moisture content is not recommended.

11.1.2.7   Permeability

The permeability of coal ash can be determined at Proctor density by the falling-head method [17]. The permeability of ash depends on particle size distribution and void ratio. Pond ash deposits are loosely stratified, and as a result the values of permeability are higher due to high void ratio. The permeability of ash may be related to effective size or mean sizes at Proctor density (Figure 11.4). The co-relation coefficient for effective sizes was found to be 0.955 compared to 0.912 for mean sizes of ashes. The permeability may be expressed [17] as a function of grain sizes at Proctor density as follows:

$$k = 0.0003 \ln D_{10} + 0.002 \qquad (11.1)$$

$$k = 0.0002 \ln D_{50} + 0.001 \qquad (11.2)$$

where k is permeability in mm/s, D_{10} is effective size and D_{50} is mean size in mm.

Attempts were made to relate permeability of ash samples with void ratio. At higher void ratio, it may be several times that of permeability at Proctor density. However, in loose state, internal erosion plays a greater role than permeability. There is no definite trend observed between permeability and void ratio. The scatter may probably be due to the difference between actual void space available for flow and calculated void ratio. The calculated void ratio includes the blocked space occupied by pores of particle (porous unburnt coal, pelerosphere and cenosphere).

These values are close to the permeability of medium- to fine-grained soil. The permeability of medium fine sand and silt (*SM, SL, SC*) is in the range of 10^{-2} to 10^{-6} mm/sec according to the Unified Soil Classification System.

11.1.2.8   Compressibility

Compressibility is an important parameter to estimate the settlement of a ground in stressed conditions. The compressibility of ash was estimated in a 60-mm diameter and 20-mm thick oedometer ring on reconstituted loose dry samples [17]. Samples were prepared in an oedometer ring by the dry pluviation method. The dry ash was funnelled with zero potential energy in the ring to obtain sample in the loosest density. The sample was then pressed under a surcharge of 1kPa and temping to obtain desired specific volume. All reconstituted ash samples show a common preconsolidation pressure of 100 kPa, probably because of the exposure of ash material to common thermal stresses in the furnace during formation. With increasing coarseness, the compression of coal ash closely resembles sandy soils. There are certain evidences of grain crushing with increasing effort in tests on coarse ashes [17].

At higher consolidation pressures the compression curves of fine ashes may be represented by a unique normal consolidation line and their specific volume may be determined by the current state of stresses. There is a progressive increase in stiffness with increasing stresses and increasing mean sizes [17]. The stiffness is higher for samples having closeness of placement void ratio to their minimum void ratio. Ashes that are predominantly fine show a minimum void ratio significantly lower than coarse ashes. Compression tends to pack them in a denser state and there is negligible rebound on unloading. The specific volume of ashes in an oedometer at varying pressures may be represented by:

$$V = m \, exp \, [-0.4 \, D_{50} / \, D_a] \tag{11.3}$$

Here V is the specific volume and D_{50} is the mean size of ashes and D_a is a reference size parameter. The values of m depend upon the range of pressure. Typically m takes a value of 2.2 to 2.0 in a range of pressure 100 to 1600 kPa [17].

11.1.2.9   Shear Strength

The shear strength of ash in a triaxial test can be estimated at natural densities and confining pressure for different ashes [21]. The Equations 11.4 and 11.5 show the relationship among relative dilatancy and relative density. Bolton [27] reviewed a large number of triaxial and plane strain cases to propose a unique relationship for dependence of frictional properties of cohesion less soil on dilation and effective mean confining pressure (p') in kPa (using p_a as the reference pressure in the same units). In the case of triaxial,

$$In= 0.33(\phi_p - \phi_c) + RD*ln\ (p'/pa) \qquad (11.4)$$

$$In= Q*RD - r \qquad (11.5)$$

The Equations 11.4 and 11.5 contain terms corresponding to peak (ϕ_p) and critical friction angle (ϕ_c). Here $0.33(\phi_p - \phi_c)$ is referred to as the dilatancy index in the triaxial case, with RD being relative density and Q, r being material fitting constants. Bolton [27] suggested that fitting constant Q and critical angle are unique to a granular media that depends upon mineralogical factors. Many subsequent investigators have indicated effects of fines on selected parameters Q and r [26]. The value of Q for ash varied in a wide range for a selected constant value of r (Fig. 11.5). These results are compared with the value of Q (=7.7) and r (= 0) suggested by [3], ignoring low relative density values of Ir corresponding to peak frictional strength (Equation 11.6 and 11.7).

$$0.33(\phi_p - \phi_c) + RD*ln\ (p'/pa)= Q\ RD - r \qquad (11.6)$$

$$0.33(\phi_p - \phi_c) + RD*ln\ (p'/pa) = 7.7\ RD \qquad (11.7)$$

The partly saturated ash exhibits cohesion owing to pendular bonding. The apparent cohesion is lost upon wetting and therefore should not be considered for design purposes. In the highest possible packing, the angle of internal friction may be as high as 60°.

The angle of internal friction of coal ash is compared with that of sand. It may be observed that at low relative density (0–60%) its peak frictional angle is higher than clean quartz sands, while at high relative density, its frictional properties are similar to silty sand [17].

11.2   Engineering Evaluation of Cemented Ash Fill

It has been observed that aged high-lime ash behaves similarly to weakly cemented fill, which consists of fracture under a gradual process of discontinuous depositions and self-setting. The low-lime ash fills under pressure also harden to show behavior similar to cemented fill. The vast assemblage of studies on cemented fills [34–37] indicate that low cemented granular materials show cohesive behavior up to a limiting strain level, and thereafter they take load by friction only. Therefore, the engineering characteristics of cemented ash fill can be well characterized by the techniques applied for weak and fractured rock masses [38, 39].

11.2.1   Measurement of Cemented Ash Characteristics: Application of RQD

The cemented ash fills consist of strong and weak formations similar to the rock masses, which appear to be a continuous deposit. The drilling processes tend to recover a continuous core of strong material. The already existing weaknesses reappear as discontinuity at irregular frequency, which is the average number of joints appearing per unit length.

The *RQD* [40–42] is normally considered as the percentage of a core recovery of spacing length of greater than or equal to 100 mm. The *RQD* also captures the orientation of the discontinuity relative to the scan line, and thus we obtain the orientation parameter [43]. This information should be recorded in the data sheet of *RQD*. The engineering applications usually consider *RQD* as the percentage of the borehole core in a drill run consisting of intact lengths of rock greater than or equal to 100 mm, which is represented numerically as,

$$RQD = 100 \sum_{i=1}^{n} L^i / L^n \%$$ (11.8)

where L^i is the lengths of individual pieces of core in a drill run having lengths more than or equal to100 mm, and L^n is the total length of the drill run.

11.2.2   Concept of Strength Ratio and Modulus Ratio

The strength of weak rock-like-masses, namely cemented ash fill, is recorded in terms of strength ratio (σ_{mr}). The aim of finding the strength relationship with modified joint factor is to readily get the strength and modulus of fractured cemented ash fill by conducting a compression test on both intact cemented ash fill and concrete-forming embedded foundation. The strength ratio (σ_{mr}) is defined as a ratio of strength of a fractured ash mass (σ_m) as compressive strength of least (of the weak material) of its size intact cemented ash sample (σ_r) or concrete sample. If σ_{1m}, σ_{2m}, and σ_{3m} are triaxial principal stresses in the jointed rock and σ_r is uniaxial compressive strength of concrete or intact cemented ash sample, then in the tri-axial state, the strength ratio is defined by:

$$p = (\sigma_{1m} + \sigma_{2m} + \sigma_{3m})/3 \tag{11.9}$$

$$q = [(\sigma_{1m} - \sigma_{3m})] \tag{11.10}$$

$$\sigma_{mr} = [(\sigma_{1m} + \sigma_{2m} + \sigma_{3m})/3]/ [(\sigma_r)/3] \tag{11.11}$$

$$= p/ [(\sigma_r)/3]$$

$$\sigma_{mr} = 3p/ (\sigma_r) \tag{11.12}$$

In unconfined state the strength ratio is:

$$\sigma_{mr} = [(\sigma_m)/3]/ [(\sigma_r)/3] = [\sigma_m]/ [\sigma_r] \tag{11.13}$$

Similarly, confining pressure ratio and shear stress ratio are defined as:

$$p_{mr} = [(\sigma_{1m} + \sigma_{2m} + \sigma_{3m})/3]/ [(\sigma_r)] \tag{11.14}$$

$$q_{mr} = [(\sigma_{1m} - \sigma_{3m})]/ [(\sigma_r)] \tag{11.15}$$

In triaxial conditions, modulus ratio is defined as:

$$E_{mr} = E_m /E_r \tag{11.16}$$

E_r and E_m are tangent modulus of intact and fractured sample, respectively, which may be considered at hardening for s_r and s_m, respectively.

Similarly, in unconfined conditions, modulus ratio is defined as:

$$E_{mr} = E_m (\sigma_{2m} = \sigma_{3m} = 0)/E_r \tag{11.17}$$

The relationship of strength ratio [43] and joint factor is represented by:

$$\sigma_{mr} = exp\,(a_p\,J_{fg}{}^p) \tag{11.18}$$

As per a few of the investigators [38, 39, 42, 44], a_p is a dilatancy-based parameter.

The variation of strength ratio with joint orientation, joint number and friction shows variation of strength ratio according to the experimental observation of various investigators. Further, the modulus ratio and hardening parameter [42] are represented by:

$$E_{mr} = exp\,(C_h\,J_{fg}{}^h) \tag{11.19}$$

$$C_h = \kappa(p_i/\sigma_r)^n \tag{11.20}$$

The pressure dependence of hardening parameter is shown in Figure 11.8. The limiting and appropriate value of hardening

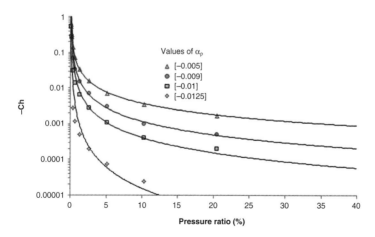

Figure 11.8 Hardening parameter C_h for embedded foundation [42].

parameter is adopted and calculated according to the end conditions as shown in Tables 11.5 and 11.6, respectively.

11.2.3   Evaluation of Joint Parameters

The numbers of joints per unit run of the fill core obtained for estimating RQD is joint number, Jn. Its orientation affects the equivalent number of horizontal joints per unit volume of cemented ash mass. A joint factor captures potential engineering possibilities within a joint, namely number of joints, orientation, and friction and cohesionless infill material. The condition of joints, namely presence of cohesionless infill material, was considered in modified joint factor [44, 38]. The joint number connects RQD, joint factor and modified joint factor. According to several investigators [41, 43–46], the joint factor is a significant joint mapping parameter in relation to the strength ratio of rock-like masses. A concept for model behavior of cemented ash mass with joint inclination, joint number and friction considers the number of joints, joint orientation and friction in the expression as:

$$J_f = J_n / n_\beta r \tag{11.21}$$

where J_n is the number of joints per unit length in the direction of loading (joints per meter length of the sample); n_β is orientation parameter corresponding to the angle of inclination of joint (β) to the load direction and; r is reference roughness parameter.
It is presented in nondimentional form as:

$$J_f = J_n L_{na} / n_b r \tag{11.22}$$

where L_{na} is a reference length = 1 meter.
   On the basis of experimental data of several investigators [43], the joint factor was modified [44] to incorporate varied engineering possiblities amid cemented ash mass as:

$$J_{fg} = c_g J_f \tag{11.23}$$

where c_g is a modification factor for the condition of the joint, water pressure and the cohesionless infill material,

$$c_g = J_{dj} J_t / (g_d J_w) \tag{11.24}$$

where J_{dj} is the correction for the depth of joint (joint stress parameter); J_t is the correction for the thickness of cohesionless infill material in joint (thickness parameter); J_w is the correction for ground water condition and; g_d is the correction factor depending upon the compactness or relative density of cohesionless infill material in joint, equal to unity for fully compacted joint fill.

For clean compact joints, c_g is equal to unity. These discontinuities may consist of fragments of the ash material to a varied extent of thickness, density and orientation. An observation of cemented ash core recovery captures g_d in terms of designated discontinuity condition or precisely by assigning a packing density of fragments in the discontinuity. The packing in the discontinuity tends to compact, dilate or crush during the process of loading.

The progressive compression of the discontinuities significantly influences the strength ratio and deformation of the cemented ash mass. The granular material in the joint's rupture zone undergoes shear deformation depending upon relative density of the cohesionless infill material. The relative density (R_D) is conventionally considered as a ratio of difference of natural state void ratio (e_n) and minimum void ratio (e_{min}) from maximum void ratio (e_{max}) of the infill material as:

$$R_D = (e_{max} - e_n)/ (e_{max} - e_{min}) \qquad (11.25)$$

The effect of pore pressure (u) is adjusted so that mean effective confining pressure (p') is equal to mean confining pressure (p) $[p = (\sigma_1 + \sigma_2 + \sigma_3)/3$, where $\sigma_1, \sigma_2, \sigma_3$ are principal stresses].

11.2.4 Relationship of RQD and Joint Parameters

Several investigators [40] showed that *RQD* value changes with increasing difference between size, i.e., joint spacing. In fact, changing block size modulates stress intensity on discontinuity, hence upon the volume of joint material. Such an observation calls for an adjustment in *RQD* vs joint properties, namely spacing, volume, friction, gouge material, ground water and internal pressure. The volumetric joint count tends to have an exponential relation with *RQD* as block size increases.

$$J_{fg} = a \; exp \; (b \; RQD) \qquad (11.26)$$

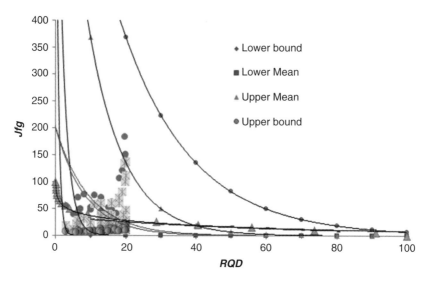

Figure 11.9 Relationship of modified joint factor and RQD [42].

where a, b are fitting constants, which take a value according to the rock block characteristics, discontinuity and friction in relation to RQD.

The volumetric effects on strength and deformation of jointed rocks namely spacing, orientation, volume, friction, gouge material, ground water and internal pressure are considered in modified joint factor (J_{fg}).

The factors, namely a and b, are readily estimated from core drilling data of the rock mass and geotechnical site investigation report. The variation of J_{fg} with RQD is shown in Figure 11.9. Zhang [41] and Trivedi [42] related rock mass modulus with RQD. We use the following relation [42] to find deformation as described by:

$$E_{mr} = exp\ [a\ C_h\ exp\ (b\ RQD)]$$ \hfill (11.27)

11.2.5   Steps to Obtain Deformations from the Present Technique

The following steps should be followed to find out deformation of cemented ash mass.

- Prepare bore log sheet containing important information, namely, depth of sampling at requisite interval, unit weight, overburden pressure, joint number appear in coring, orientation of joints, orientation parameter (n_β), maximum, minimum and mean thickness of infill, density of infill, friction factor, water table, RQD, and columns for calculated inputs of J_{fg}, a_p, C_h, E_{mr}, and σ_{mr}.

- Based upon inputs of J_{fg} and n_β decide values of a_p.
- Evaluate the overburden and operate upon C_h (Tables 11.9, 11.10; Fig.11.8).
- Estimate values of C_h (as per Fig. 11.8 and Eqs. 11.19 and 11.20) and a_p find out E_{mr}.
- Draw a relationship between J_{fg} and RQD (Fig. 11.9) for each data point to find fitting parameters a and b.
- Using the Equation 11.25, E_m can be found out to estimate settlement of the footing embedded in rock mass as per the selected cases.

Table 11.9 Values of empirical parameters 'a_p' and 'C_h.'

Failure Mode	End Conditions	Limited Hardening	Applications
	a_p	C_h	
Rotation	−0.025	−0.050	Slopes in cemented ash mass
Sliding	−0.018	−0.036	Slopes in cemented ash mass
Splitting	−0.0123	−0.025	Vertical cuts in cemented ash mass
Shearing	−0.011	−0.022	Vertical cuts in cemented ash mass
Shearing	−0.009	−0.018	Shallow depth fractured cemented ash mass
Shearing	−0.008	−0.016	Shallow depth moderately fractured cemented ash mass
Shearing	−0.005	−0.010	Shallow depth lowly fractured cemented ash mass
Shearing	−0.003	−0.006	Fully cemented fractured cemented ash mass

Table 11.10 Value of hardening parameter for end conditions [42].

End conditions	Coefficient (κ) and power (η) for the pressure ratio	
a_p	κ	η
–0.005	–0.040	–1.00
–0.009	–0.020	–1.25
–0.0100	–0.011	–1.42
–0.0125	–0.001	–2.00

11.3 Problems of Uncemented Ash Fill

11.3.1 Collapse, Piping and Erosion, Liquefaction

The uncemented ash fills are potentially vulnerable to excessive settlement [6, 47], collapse [48], flow failures [8], piping [49], erosion and liquefaction [10, 50]. Their potential vulnerability to failure has a relation to their grain size parameter [48].

Terzaghi [49] described a piping failure as essentially induced by an excess head of water so that the fill adjoining the ash embankment remains in equilibrium, provided the hydraulic head h_1 is smaller than a certain critical value h_c. However, as soon as this critical value is approached the discharge increases more rapidly than the head, indicating an increase of the average permeability of the fill. Simultaneously the surface of the fill rises within a belt with a width of approximately n.D (n=0.5, as considered by Terzaghi [49] for sheet pile embedded to a depth of D) as depth of embedment D of the screen line of embankment. Finally a mixture of fill material and water breaks through the space located below the screen embankment. This phenomenon is called piping, and the hydraulic head at which piping takes place is the critical head. The piping beneath an ash fill is likely to cause a failure of the ash fill retention system or ash pond embankment. The ash fill needs a factor of safety with respect to piping of the ash fill retention system after the water level has been lowered within the dam to a depth below the outside water level.

The process of piping is initiated by an expansion of the fill material between the buried portion of the impermeable ash embankment and a distance of about n.D downstream from the screen line. This expansion is followed by an expulsion of the infill out of this zone. No such phenomenon occurs unless the water pressure overcomes the weight of the fill located within the zone of expulsion. It can be assumed that the body of fill, which is lifted by the water, has the shape of a block with a width n.D and a horizontal base at some depth m.D below the surface [m~1, for sheet pile embedded to a depth of D]. The rise of the block is resisted by the weight of the block and by the friction along the vertical sides of the block. At the instant of failure the effective horizontal pressure on the sides of the block and the corresponding frictional resistance are negligible. Therefore, the block rises as soon as the total water pressure on its base becomes equal to the sum of the weight of the block, fill material and water combined. The head h_c at which the body is lifted is the critical head. The elevation of the base of the body is determined by the condition that h_c should be a minimum because piping occurs as soon as the water is able to lift a block of fill regardless of where its base is located. The factors, namely "n and m," depend upon friction, liquid limit, specific gravity and particle size similar to collapse. The critical hydraulic gradient (~0.3) for coefficient of uniformity is nearly equal to 10 for pond ash, making it potentially vulnerable to erosion.

Singh [50] analyzed the liquefaction characteristics of similar gradation of silty sand as that of ash fills [10]. The magnitude of lateral acceleration required for liquefaction of ash fill compared with sand (Table 11.11) in the same-sized horizontal vibration table (100 cm long, 60 cm wide, 60 cm deep) is significantly lower than sand at a low (70%) relative density (0.31 to 0.14%), while at the higher (85%) relative density (0.51 to 0.31%). The increase in the liquefaction resistance of ash is reported [10] to be (1.84 to 1.70 times) higher compared to that in sand (1.10 times) for a similar increase in relative density (70 to 85%). In fact, the settlement of the ash fill was significantly higher than sand.

The cyclic mobility-liquefaction characteristics of ash fill (obtained from the same source) in vibration table studies was evaluated by Trivedi *et al.* [10] and compared to clean sand. The ash fill settled nearly twice as much as sand for the equal volume contained in the vibration chamber (1x0.6x0.6 m^3) for nearly half of the dynamic disturbance. Further tests are recommended to verify predicted settlements of large-size footings on dynamic loads.

Table 11.11 Range of parameters for liquefaction analysis [10].

Materials (Sand and Coal Ash)	Range	Remarks
Relative density	65 to 85%	By weight volume relation
Amplitude	1to15 mm	Vibration table
Pore pressure at 5, 15, 25cm depth	At varied density and amplitude	At sinking of metallic coin
Settlement	5 to 20 mm	At sinking of metallic coin
Number of cycles	At varied density and amplitude	At sinking of metallic coin

11.3.2 Collapse Behavior of Ash Fills

The general characteristics of collapsing fills are a sudden and a large volume decrease at a constant stress when inundated with water. According to Lutenegger and Saber [51] the collapse is associated with the meta-stable structure of a large open and porous fabric of the material. The earthen structures such as embankments, road fills and structural fills may collapse when the placement moisture content is dry of optimum. The infiltration of the rainfall may be sufficient to reduce the matric suction within the ash to a value low enough to trigger a shallow failure [52]. It has been observed that in a wide range of placement parameters the ash remains vulnerable to the collapse on submergence in working stress range. A slip failure was reported at the ash dump of Vijayawada thermal power plant resulting in the destruction of several houses and the swamping of land with fly ash. The sudden failure of a large fly ash disposal dump after rainfall and the associated mudflow at Panki, Kanpur was reported [53]. Such failures are not quite representative of conventional failures. Several studies have indicated that compaction control of coal ash in the field by usual methods is often poor. It adds to the vulnerability of ash fill to a wetting-induced collapse. The soils that exhibit collapse have an open type of structure with a high void ratio as expected in the case of ashes. According to Barden *et al.* [54] the collapse mechanism is controlled by three factors: a potentially unstable structure, such as the flocculent type associated with soils compacted dry of optimum or with less soils, secondly, a high applied pressure which further increases the instability, and a high suction which provides the structure with only temporary strength which dissipates upon wetting. As per an empirical study [1], the dry unit weight and water content are generally considered as important parameters that control the collapse of metastable structure of soils, if the dry unit weight is less than 15 kNm³. The tentative dry unit weight of the coal ashes was often found to be less than 10 kNm³, suggesting a possibility of collapse.

Once a granular arrangement takes up a loose packing under a favorable condition of pressure and moisture the collapse of ash material occurs. This tendency is quantified in terms of the contact separation parameter (D_{50}/D_a) defined in the Figures 11.10–11.12 and distance of placement void ratio to the minimum void ratio. Therefore, the collapse potential (C_p) and collapsibility factor (F) are:

$$C_p = \Delta h/h \qquad (11.28)$$

$$F = (e_i - e_{min})/e_{min} \qquad (11.29)$$

where Δh is change in the sample thickness (h) upon inundation. Alternatively, e_i is placement void ratio of granular materials and

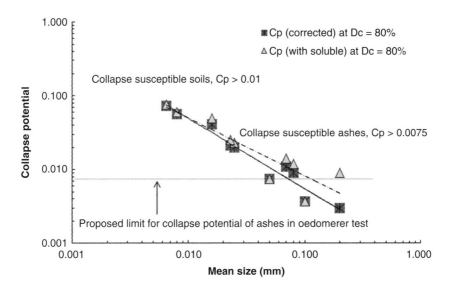

Figure 11.10 Effect of grain sizes on collapse potential [8].

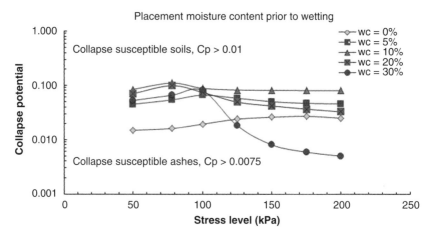

Figure 11.11 Effect of moisture content on collapse potential [8].

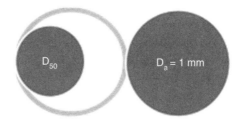

Figure 11.12 Definition of contact separation parameter D_{50}/D_a [48].

e_{min} is void ratio corresponding to maximum dry density in Proctor compaction. The variation of maximum and minimum void ratio of sand and ashes is empirically related with the grain sizes [15]. The void ratio extent defined by a difference of maximum and minimum void ratio drops with the increasing grain sizes [3]. This implies that the collapsibility increased by decreasing sizes. Therefore, the larger the value of F, the more the granular materials are predisposed to collapse [8]. Figure 11.13 shows a reduction in collapsibility factor F, with mean size in the loosest and the compacted states.

In the loosest state when grains are in progressive contact, a = 0.2 and b = 0.5.

$$F = a(D_{50}/D_a)^{-b} \qquad (11.30)$$

On compaction the negative exponent "b" of grain separation parameter goes on reducing from 0.5 to one.

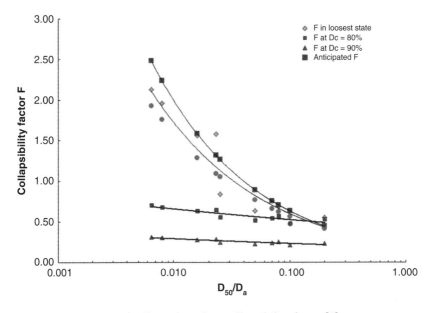

Figure 11.13 Criteria of collapse based on collapsibility factor [8].

However, compared to the loosest state, all the granular materials reach nearly a common collapsibility level in a compacted state. At 90% degree of compaction, a collapsibility level is arrived at, which is associated with small volume change upon collapse that does not reflect collapse. Moreover, it has the practical problem of precise measurement of the volume change. Thus, the variations in the measured collapse at 90% degree of compaction may forbid interpretation of any trend.

The collapsibility factor allows for assessment of the probable collapse. The collapse occurs if the sample attains a minimum void ratio on inundation. The maximum probable collapse potential is computed by:

$$C_{pr} = (e_i - e_{min})/ (1+e_i)$$
(11.31)

where C_{pr} is the maximum probable collapse potential, e_i is void ratio in a lcose state and e_{min} is void ratio corresponding to a maximum dry density in Proctor compaction. The collapse potential shows that the decreasing mean size tends to reduce the difference between the maximum probable and the observed collapse at 80% degree of compaction. While at 90% degree of compaction a significant scatter of the data is observed.

Because of the above observations the classification of granular materials at 80% degree of compaction was found to be appropriate for the evaluation of collapse. The mean particle size was seen to control the collapse of granular materials. If the mean size was greater than 1mm the granular materials were non-collapsible and others were collapsible under specific conditions. The value of collapse potential in the critical range of stress and moisture was 3 to 6 times that of the corresponding dry condition (Figure 11.11). This suggested susceptibility of a non-collapsible dry granular material to the collapse in partly saturated condition. In order to obtain the value of collapse potential of partly wet granular materials, a multiplier may be applied. The vulnerability of fine ash fill is validated by conducting triaxial tests on partly saturated ash compacted at maximum dry density and saturated ash. The deformation modulus of saturated ash is significantly lower than the partly saturated ash (Figure 11.14).

The collapsible granular materials were further divided into the granular materials of low, medium and high collapsibility based on their collapse potential. The collapsible and the non-collapsible granular materials were identified using a model collapse test on selected samples. Normally, the weight of a particle of a natural soil

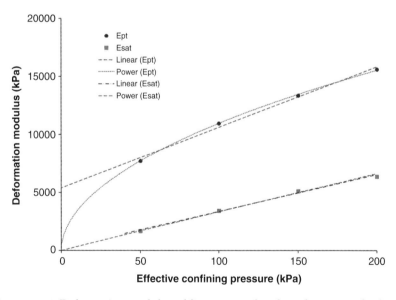

Figure 11.14 Deformation modulus of fine saturated and partly saturated ash based on triaxial test.

of similar grain size is 1.5 to 1.3 times that of ash materials. These soils remain stable at or less than 1% volume change ($C_p = 0.01$). Being light in weight, the ash material has a propensity to be unstable in the presence of buoyancy, which plays a role in the model and the field conditions. Therefore, among the lightweight granular materials particles, 0.75% volume change ($C_p = 0.0075$) triggered collapse failure in the field. Coincidentally, 1% volume change of soils is 1.3 times that of the limit recognized for the collapsible ash materials.

It was observed that the sand and the coarse ash had very close value of the median size. It being a granular material collected dry, having around 25% particles in silt range, it had a higher collapse potential than the sand. It was recognized that all the collapsible granular materials had relatively more fines. Among coarse-grained granular materials, a scatter in collapse potential was observed. A relationship between corrected collapse potential at 200 kPa (Dc = 80%) and mean particle size is obtained with a satisfactory coefficient of determination. The collapse potential is expressed by:

$$C_p = n \ (D_{50}/D_a)^m \tag{11.32}$$

where C_p is collapse potential of a granular material, D_{50} is mean particle size in mm, and D_a is reference size = 1 mm, and m and n are fitting constants for the granular materials.

11.4   Ash as a Structural Fill

There is only scant data available on the interpretation of the load-bearing behavior of ash fills. The penetration test results analyzed by Cousens and Stewart [12, 57] and Trivedi and Singh [17] showed the scope of development of new correlations for evaluation of foundation settlements on coal ash. Leonards and Bailey [30] favored the use of plate load test results for coal ash. Trivedi and Sud [20, 22] examined the evaluation of bearing capacity and settlement of ash fills. The present work reviews the plate settlement on coal ash for working out a strategy for evaluation of foundation settlement. Some of the case studies are reported on the investigation and assessment of load-bearing behavior of coal ash, which is also relevant to its uses in embankment fills. These tests are namely penetration and plate load tests as described below.

11.4.1 Penetration Test

The standard and cone penetration devices are used to evaluate the stability of a landfill. The standard penetration test is used in different parts of the world with a slight variation in the version of its use. It involves an estimation penetration value of 300 mm run of a split barrel of 50(±2) mm external and 35(±1) mm internal diameter under impact of 63.5 kg hammer The results of standard penetration tests (SPTs) conducted on hydraulically deposited Illinois ash were reported. The penetration was observed to be of several stretches of 30 mm under the weight of drill rods to 9 blows per 300 mm of penetration. The SPT value for compacted Kanawha ash was observed among 10 and 31 and had an average of 19.5 for seventeen independent tests conducted in four borings, excluding the values obtained for lenses of bottom ash. Dry density values determined for five of the nine Shelby tube samples taken from these bore holes suggest hardly any correlation with the N-values (field density 95% to 100% of maximum dry density ~14 to 16 kN/m³). The average cone penetration resistance is related to SPT value (= 200 x N kNm²), where N is the SPT value. The ashes normally in silt size, especially in loose condition, may liquefy under the tip of penetrometer below the water table resulting in a lower penetration record. For the compacted ash fills, the density projection and SPT (N) number are presented in Table 11.12. These low

Table 11.12 Variation of SPT value for ash deposits.

Ash Type	D_c (%)	N-value	Reported by
Compacted Kanawaha ash	95–100	10–31	[9]
Hydraulically deposited Illinois ash	Loose state	Zero	[9]
Well compacted Ontario flyash	85–100	10–55	[56]
Bottom ash	–	75	[56]
Compacted ash dyke	95	4–27	[24]
Hydraulically deposited ash	Loose state	Zero	[24]
Well-compacted fly ash in a valley-Pittsburg	–	75	[57]
Hydraulically deposited ash	Loose state	Zero-1	[1]
Hydraulically compacted ash	Dense state	1–10	[1]

values of N are partly associated with a low unit weight and partly with a high percentage of fines and a high moisture condition. A high value of N is associated with the presence of lenses of cemented ash and partly with a higher compaction and aging.

An average friction ratio for sleeve and ash is from 3 to 5 times higher than the value cited by Schmertmann [55] for clay silt sand mixes, silty sands, silts and sands. Toth *et al.* [56] quoted a wide variation in SPT value ranging from 10–55 in fly ash with angle of internal friction ranging from 35 to 36°. The empirical co-relation between SPT values and ϕ (angle of internal friction) for natural soils in this range of SPT value (N = 10 to 55) indicated angle of internal friction ranging from 30 to 45°. The investigations carried out by Cousens and Stewart [12] for the range of cone resistance and the friction ratio (200 kPa and 8%, respectively) indicated grain sizes in the range of silt (60–80%) and clay (5–10%). For a target relative density (50 to 85%), variation in standard cone resistance ranges from 2000 to 6000 kPa. However, the average friction (between sleeve/cone and ash material) ratio was observed to be 3 to 5% [15, 9]. The settlement of these ash fills on the basis of the Schmertmann method was found to be a non-conservative estimate.

11.4.2   Load Test

On the basis of a case study on Indianapolis ash, Leonards and Bailey [30] suggest that the load-settlement relation for the foundation on compacted ash cannot be inferred from standard penetration or static cone penetration tests. This is largely attributed to the inadequacy of penetration tests to sense the effect of compaction related to pre-stressing of the coal ash. The predicted settlements for a selected footing 2.1-m wide at design pressure of 239 kPa on well-compacted ash from the data of standard penetration (SPT), cone penetration (CPT) and plate load (PLT) tests are presented in Table 11.13.

Table 11.13 Predicted settlement for well-compacted area [30].

Testing Technique	CPT	SPT	PLT
Settlement (mm)	15	25.4	5.08

blows/305 mm penetration) indicated that ash materials are significantly less compressible in the pressure range of interest. At 100 kPa, compacted ash may settle less (0.6 mm) compared to the settlement of the same plate on sand (approximately 1.2 mm).

Toth *et al.* [56] reported a case study on the performance of Ontario ash (a typical class *F* ash). During the compaction of ash for landfill, it was observed that the densities being achieved in the field were normally below 95% of maximum Proctor density. On the basis of the good bearing capacity observed in the plate load test, 95% of maximum Proctor density is recommended as the target density for the fly ash landfills. Toth *et al.* [56] obtained short- and long-term test results for circular plates of 0.3-m and 0.6-m in diameter. The settlements for long-term tests occurred within the first hour of load application.

The results of investigations by various investigators are given in Table 11.14. The tests conducted on a common degree of compaction (93.4%) and plate size (600-mm diameter), had different settlement records at a stress level of 100 kPa (2.61 and 3.59 mm). Trivedi and Sud [3] have shown that the variation in grain sizes

Table 11.14 Settlement of test plate on compacted ash fill.

Plate size (mm) and shape	Degree of Compaction (%)	Moisture Content	(S/B)% at 100 kPa	Interpolation Data
900, square	85.24	Wet of Critical	0.56	[22]
600, square	85.24	Wet of Critical	0.63	[22]
300, square	85.24	Wet of Critical	0.45	[22]
300, square	90.29	Wet of Critical	0.35	[22]
300, square	85.24	Dry of Critical	1.03	[22]
300, square	81.55	Dry of Critical	1.56	[22]

Table 11.14 (Cont.)

Plate size (mm) and shape	Degree of Compaction (%)	Moisture Content	(S/B)% at 100 kPa	Interpolation Data
600, square	90.29	Wet of Critical	0.40	[22]
900, square	90.29	Wet of Critical	0.34	[22]
600, circular	93.40	Wet of Critical	0.43 0.50	[56]
600, circular	98.20	Wet of Critical	0.23	[56]
300, circular	98.20	Wet of Critical	0.15	[56]
300, square, Long term	–	Wet of Critical	0.15	[56]
600, square	< 95%	Wet of Critical	0.22	[30]
300, square	< 95%	Wet of Critical	0.23	[30]
300, square	Sand, N = 50	–	0.36	[30]

of ashes may result in different settlement characteristics even at a common degree of compaction. Trivedi and Singh [9] reported a higher load-bearing capacity of ash fills than actually estimated by cone resistance.

11.4.3 Test Setup for Ash Fills and Testing Technique

The ash is normally deposited in loose lift of 150 mm in a trench of plan dimension of 1.5 m × 1.5 m. It is compacted by a pre-calibrated plate vibrator mounted on a flat rectangular plate (152 mm x 390 mm). The rating of the plate vibrator is kept at 3000 rpm. A constant magnitude of vibration is required to achieve the desired relative

Table 11.15 Summary of load tests used for prediction.

Ash type	Test conditions	Size (m)	Shape	Dc(%)	No. of tests	Max. Pressure
Sand	Dry	0.1	Strip	–	2	Failure
A1	Dry of Critical	0.1	Strip	–	2	Failure
	Wet of Critical	0.1	Strip		2	Failure
A2	Dry of Critical	0.1	Strip	–	2	Failure
	Wet of Critical	0.1	Strip		2	Failure
A2	Dry of Critical	0.3 0.6	Square	81.55 85.24	4 4	Failure Failure
	Wet of Critical	0.3 0.6 0.9	Square	85.24 90.29	4 4 4	200 kPa 160 kPa 100 kPa

density. The trench is filled up in layers maintaining constant density throughout. The density checks are applied at regular intervals using thin core cutter sampling and penetration of an 11-mm diameter needle penetrometer under constant pressure.

The plate load test is conducted on the compacted ash fill. A few model tests (Table 11.15) are carried out on surface footings (0.1 to 1 m wide plates) in dry as well as submerged conditions for different ashes and a sand to check the reproducibility of the results. Additionally, on site density checks and laboratory shear tests are also carried out. The displacement of the plate is monitored using pre-calibrated settlement gauges of least count 0.01 mm. The total assembly including hydraulic jack, proving ring and the plate are aligned with the help of a plumb bob to attain verticality.

The load capacity of ash fill is estimated by conducting load tests using different plates on ashes at varying degree of compaction. A summary experimental program is considered. An average of at least two tests is considered to reach a common load settlement plot

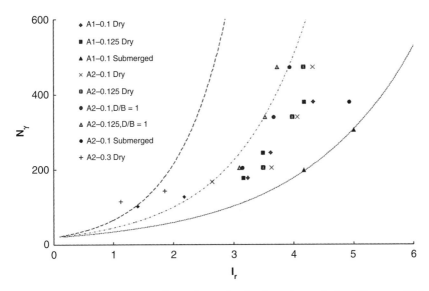

Figure 11.15 Bearing capacity factor for coal ashes based on relative dilatancy [20].

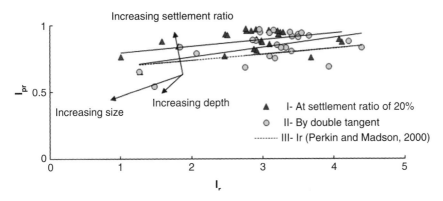

Figure 11.16 Progressive failure index and relative dilatancy for ashes [20].

if the values are within the range of 10%. The results evaluated from typical pressure settlement plots are shown in Figures 11.15–11.17.

A plate of the desired size is placed on the ash fill. A leveled 10-mm thick layer of dry ash is spread on compacted ash to ensure relatively complete and uniform contact between the bearing plate and compacted ash. The plate is loaded with the hydraulic jack against a reaction truss. After application of seating load, the load is increased in regular increments. Bearing plate settlement is measured with an accuracy of 0.01 mm.

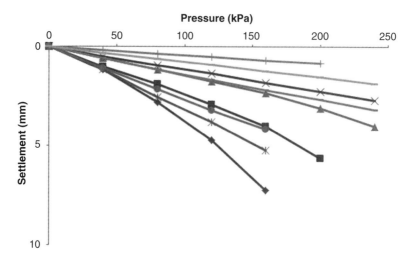

Figure 11.17 Pressure settlement plot interpreted for ashes from varied sources.

Each load increment is maintained on the bearing plate as long as no change in the settlement is observed for two hours in succession. Maximums of load required for failure of the deposit were estimated by the bearing capacity factors obtained from small-scale tests. A sharp increase of bearing plate settlement is considered as an indication of the beginning of the ash failure phase. The settlement observation under the final load is taken to the maximum of 24 hours.

11.4.4   Bearing Capacity of Ash Fill

The generally agreed upon bearing capacity equation for shallow depths uses bearing capacity factors N_c and N_q. However, substantial differences have been reported in the semi-empirical bearing capacity factor for shallow foundations N_γ in numerous studies [9, 20, 59].

The bearing capacity equation for smooth strip foundations on the surface of a fill with surcharge, is given by:

$$q_{ult} = C\,N_c + \sigma_{ov}'\,N_q + 0.5\,N_\gamma\,\gamma'\,B \qquad (11.33)$$

where σ_{ov}' is the effective surcharge pressure acting at footing base expressed in terms of effective stress, γ' is buoyant unit weight and B is footing width; N_c, N_q and N_γ are the bearing capacity factors.

The value of N_q is obtained using a simple plasticity theory for a weightless soil;

$$N_q = \tan^2(\pi/4 + \phi'/2)\, e^{\pi \tan \phi'} \qquad (11.34)$$

The bearing capacity obtained from the Equation 11.33 does not increase linearly with the width of footing or overburden. This phenomenon was frequently termed as scale effect by De Beer [59], who attributed it to the nonlinear shape of soil failure envelope resulting in secant measure of friction angle, which decreases with mean normal effective stresses. With increasing confinement, dense and loose cohesionless soils have much less of a marked difference in peak angle of internal friction. This effect is pronounced in the geomaterials, namely ash fills, which suffer from progressive crushing.

Experimentally, N_γ for a surface footing without surcharge is obtained as:

$$N_\gamma = q_{ult}/(0.5\, B\, \gamma\, S\gamma) \qquad (11.35)$$

where $S\gamma$ = shape factor, which is not required in the relative dilatancy approach, and B = width of the strip footing, and γ is the unit weight of the ash. Since ϕ varies as the state of stress, density and material characteristics of the soil, the concept of stress dilatancy developed by Bolton [27] is utilized. He proposed the empirical equation:

$$\phi_{peak} = \phi_{cr} + A\, I_r \qquad (11.36)$$

$$I_r = RD\,(Q - \ln p') - r \qquad (11.37)$$

where A is an empirical constant and has a value of 3 for triaxial case; I_r is relative dilatancy index; p' is effective mean confining pressure in kPa; RD is relative density; Q and r are empirical material fitting constants with a value of 10 and 1, respectively, for clean silica sand. Incorporating the triaxial test data, Bolton [27] suggested that progressive crushing suppresses dilatancy in the soils of weaker grains, i.e., limestone, anthracite, and chalk, which show "Q" values of 8, 7, and 5.5, respectively. The ash containing substantial amount of crystalline fine silica grains followed by alumina and oxides of iron, calcium and magnesium may have Q as low as 7.7 [20]. This occurs mainly because of the reduction of critical mean confining pressure beyond which increase in mean confining pressure for a relative density does not increase peak angle above critical angle. Perkins and Madson [59] proposed to integrate this approach of progressive failure with the bearing capacity of shallow foundations on sand. This approach is extended to meet the requirements of the plate load test on an ash fill.

The ultimate bearing capacity of the surface footing on ash fills is proposed to be estimated by:

$$q_{ult} = 0.5 \, N_\gamma \, \gamma' \, B \tag{11.38}$$

where B is the width of the footing and γ' is effective unit weight of the footing.

The value of bearing capacity factor N_γ (Fig. 11.15) may be estimated [20] by:

$$N_\gamma = 20e^{\zeta \, Ir} \tag{11.39}$$

The typical value of $\zeta = 0.5$ to 1, which depends upon progressive failure I_{pr} (Fig. 11.16) obtained from relative dilatancy, and ϕ_{peak} and ϕ_{cr} are the peak and constant volume angle of internal friction. A=5 and 3 for plain strain and triaxial conditions respectively.

$$Q \, RD \text{-}r = (\varphi'_p - \varphi_c)/A + RD\ln(p'/pa) \tag{11.40}$$

where $Q = 10$ for sand and 7.7 for coal ash. Rd is relative density and p' is effective mean confining pressure below the footing.

The ultimate bearing capacity of ash fills was observed among the estimates of critical and peak friction angles. The bearing capacity of ash fill evaluated is always lower than that obtained by the use of peak friction angle. Figure 11.16 shows the variation in the index of progressive failure (I_{pr}), with relative dilatancy index for a surface and an embedded footing. The index of progressive failure is hereby defined as:

$$I_{pr} = [q_{ult \, (at \, \phi' \, peak)} - q_{ult \, (from \, cone \, penetration \, test)}] / [q_{ult \, (at \, \phi' \, peak)} - q_{ult \, (at \, \phi \, critical)}] \tag{11.41}$$

If I_{pr} takes a value of 1, it implies that ultimate bearing capacity of ash fill is governed by critical friction angle, while a value of zero indicates the peak angle of friction is fully mobilized. The concurrence of a relatively high value of factor $(Q - \ln p)$ at peak cone resistance in ash fills leads to higher values of relative dilatancy index among ashes.

11.4.5   Settlement of Ash Fills by PLT

Trivedi and Sud [22] presented an evaluation of the settlement of ash fills. In order to investigate settlement characteristics of compacted coal ash, field plate load tests were analyzed for ash compacted to varying degrees of compaction and plate sizes (Table 11.16).

It is observed that ash may be compacted to the same degree of compaction at two moisture contents, one dry of critical and the other wet of critical moisture content. The critical moisture content is defined as the moisture content or range of moisture content in which vibratory effort becomes ineffective and ash bounces back to a loosest packing corresponding to which dry unit weight of ash is minimum in the presence of moisture.

On the dry side of critical, ash packing is very sensitive to moisture. Within the limitation of workability in field, different degrees of compaction were selected (i.e., 85 and 80%). The observations of the moisture density relationship in a field are similar to that in a laboratory vibration test. The increasing moisture content from 5 to 10% decreases the degree of compaction from 85 to 80%. Further, the settlement of 300-mm x 300-mm test plate increases from 3 to 5 mm at 100 kPa.

The coal ash is compacted to a higher degree of compaction (90%). By increasing degree of compaction from 85 to 90% on the wet side of critical, settlement was reduced from 1.5 to 1 mm. That

Table 11.16 Effect of plate size on settlement at wet side of critical [47].

Plate Size (mm)	Settlement at 100 kPa in mm	
	$D_c = 90\%$	$D_c = 85\%$
300 × 300	1.05	1.45
600 × 600	2.4	3.8
900 × 900	3.1	5.1

is an improvement in the degree of compaction by 5% (from 85 to 90%), and the settlement is reduced by one-third.

11.4.6   Settlement on Ash Fills by PLT, CPT and SPT

The case studies [30, 56, 20, 22] have shown that standard penetration test results might overestimate settlements of ash fill as high as five times that of predicted value by plate load test, while the cone penetration test overestimated settlements as high as three times [30]. The plate load tests tend to give a more precise indication of actual settlements of larger sizes. The observed data of several investigators [30, 56] is given along with the results of the investigations in Table 11.17.

From the analysis of data of the present investigation and that published by Leonards and Bailey [30], it is understood that well-compacted ash for the footing size and stress level of interest, has settlement directly proportional to pressure up to 200 kPa. Therefore the settlements at 100 kPa are interpolated from the data published [30, 56, 20, 22].

The critical moisture content is defined as moisture content at which ash attains minimum density when compacted by vibration. The moisture-density curve is almost symmetrical about this moisture content. Therefore, ash may be compacted at two different moisture contents: one dry of critical and the other wet of critical.

Table 11.17 Predicted settlements (mm) using Terzaghi and Peck formula [22].

B_f (m)	$D_c \sim 85\%$ [22]	$D_c \sim 90\%$ [22]	$D_c < 95\%$, [30]	$D_c \sim 98.2\%$, [56]
0.60	2.56	1.85	1.24	0.79
0.90	3.26	2.36	1.575	1.07
1.2	3.71	2.68	1.79	1.15
1.5	4.01	2.90	1.94	1.24
1.8	4.25	3.07	2.05	1.31
2.1	4.43	3.21	2.14	1.37

Adding moisture beyond this critical moisture content, apparent cohesion develops which impedes the deformability of ash. This apparent cohesion is destroyed gradually by addition of water beyond optimum moisture. It seems that owing to the development of apparent cohesion, ash becomes far less deformable above a constant dry density (D_c = 85%).

The settlement record at 100 kPa suggest that, on the dry side of critical, settlement is more than two times that of the wet side of critical. The percentage increase in settlement by compacting ash at dry side of critical, instead of wet side of critical, at 100 kPa on 0.3m x 0.3m square plate at D_c of 85% is 115%.

There is a significant impact of degree of compaction on the dry side of the critical. For a drop in the degree of compaction from 85 to 80% (decrease in degree of compaction is 4.32%) there is an increase of settlement at 100 kPa from 3.1 to 4.7 mm (increase in settlement is 51.61%). Similarly on the wet side of critical, for a decrease in the degree of compaction from 90 to 85% (percentage decrease in degree of compaction is 5.6%) there is an increase in settlement from 1.05 to 1.45 mm (percentage increase in settlement is 38%).

The scale effects are clearly visible in the settlement of ash deposits. Plots are drawn from experimentally observed and predicted settlement for 0.3m x 0.3m, 0.6m x 0.6m and 0.9m x 0.9m square size plates, respectively. Using settlement of 0.3m x 0.3m plate as plate settlement, settlement of footing is estimated by the formula [58]:

$$s_f = S_p \left[\frac{B_f \left(B_p + 0.3 \right)}{B_p \left(B_f + 0.3 \right)} \right]^2 \tag{11.42}$$

where B_p = width of plate in meter; B_f = width of footing in meter; S_p = settlement of plate in mm; and S_f = settlement of footing in mm.

The predicted settlements underestimated actual settlement at all degrees of compaction, even at a high degree of compaction (98% and at less than 95%). The settlement of 0.6 m (least dimension) plate has been underestimated by 45% and 7.5%, respectively. In the investigation [47] the predicted settlement from the actual settlement of 300-mm plate underestimated the actual settlement of

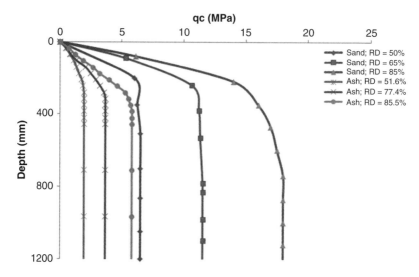

Figure 11.18 Cone resistance plot interpreted for ash and sand.

0.6 m plate (least dimension) at 90 and 85% degree of compaction by 56 and 62%, respectively.

The settlement of ash fill at varying relative densities is obtained by CPT (Figure 11.18) using the Meyerhof [1] method modified by Trivedi and Singh [9]. The following relation gives the settlements of ash fill which is less than fifty to less than ten percent of the settlement obtained by the Meyerhof [1] method.

$$Sc= \Delta pB/[n^*RD + m)]q_c \qquad (11.43)$$

where Δp = net foundation pressure. The cone resistance (q_c) is taken as the average over a depth equal to the width of the footing (B) and RD is relative density, while n and m are constants which take a value ~ 10 to 3.5, respectively.

11.4.7   Settlement of Footings on Ash Deposit

The predicted settlements according to the Terzaghi and Peck [58] extrapolation is not in agreement of settlement of footings larger than 0.6 m (least dimension) on compacted ash fill (Figure 11.19). The predicted settlements based on actual settlement of 300-mm square plate seriously underestimate the observed settlements using method by D'Appolonia *et al.* [60].

The mean value of ratio of predicted settlement according to the Terzaghi and Peck extrapolation and experimentally observed settlements was found to be 0.3. Table 1.18 presents percentage of underestimation of (0.6m x 0.6m) footing settlement by the Terzaghi and Peck formula at varying degrees of compaction.

The predicted settlements according to the criterion suggested by D'Appolonia *et al.* [60] is estimated at 100 and 200 kPa. The experimental data for varying sizes of footing at probable degree

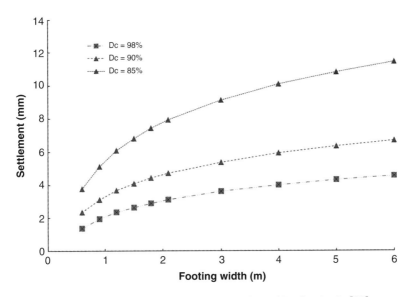

Figure 11.19 Predicted settlements as per Terzaghi and Peck criteria [22].

Table 11.18 Underestimation of settlement by the Terzaghi and Peck formula [47].

D$_c$ (%)	% Under estimation of settlement	Interpreted from the reference
98.2	44.36	[56]
< 95	7.46	[30]
90.29	56.25	[47]
85.24	61.84	[47]

of compaction is plotted in Figures 11.19 and 11.20. The expected settlement, at a pressure of 100 kPa, indicates least possibility of exceeding the allowable limit of settlements in the probable degree of compaction.

The settlement estimate corresponding to the PLT values obtained for the ash fills is estimated for 1m wide footing as per Terzaghi and Peck method as shown in Figures 11.19 and 11.20, which shows excessive rigidity of ash fills to settlement in partly saturated condition. The settlement estimate corresponding to the SPT values obtained for the ash fills is estimated for 1m wide footing at 100 kPa as per Meyerhof [61] and Burland and Burbage [62] methods (shown as Mh and B&B, respectively, in Figure 11.21), which may not show excessive vulnerability of ash fills to settlement in saturated condition. The settlement estimate corresponding to the CPT values obtained for the ash fills is estimated for 1m wide footing at 100 kPa (Figure 11.22) which shows comparison of ash fills to settlement with PLT (Table 11.19).

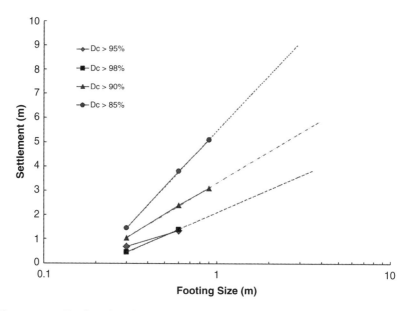

Figure 11.20 Predicted and observed settlement at varying degrees of compaction and footing width [22].

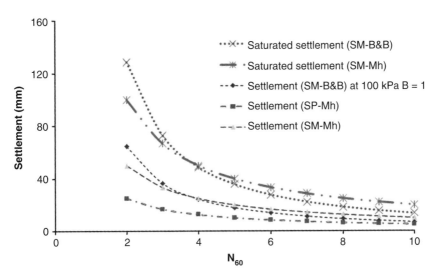

Figure 11.21 Predicted settlements as per *SPT*.

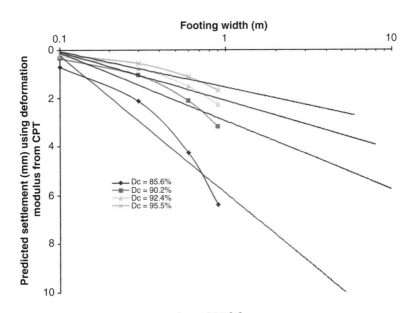

Figure 11.22 Predicted settlement from *CPT* [9].

Table 11.19 Predicted settlements according to modified criteria [47].

Degree of Compaction D_c (%)	Settlement (in mm) at 100 kPa 3m wide footing	Interpreted from the reference
85.24	9.3	[47]
90.29	5.6	[47]
98.20	3.7	[56]

11.5   Conclusions

The analysis in this chapter comprehensively provides a design focused upon the engineering behavior of ash fill so that it can fulfill its intended function from a structural and utilization point of view. The applicability of RQD and its correlation with strength parameters of geomaterials is available in the literature. The use of RQD techniques for the evaluation of cemented ash having refusal to penetrate standard cone and SPT may provide a useful insight into the behavior of cemented fill. The bearing capacity and settlement of footing on weakly compacted ash is relatively critical compared to that degree of compaction at the wet side. The extent of progressive failure below the footings at varied degrees of compaction can be estimated from the relative dilatancy considerations.

The settlement of fill is worked out as per load tests, SPT and CPT at the favorable degree of compaction, for intended footing size and at desired stress level for the ash compacted on the wet side of the critical. The pressure corresponding to safe settlement may also be ascertained from the available data.

From the estimates of safe loads and safe settlements, allowable bearing pressure may be estimated.

The submergence of fine ash deposit is critical to its stability against piping, collapse and liquefaction, hence, effects of water on ash fill (finer than 75μ) need to be critically examined for loose uncemented ash fills.

Salutations, Acknowledgement and Disclaimer

The present chapter is compiled from the work of numerous investigators on varied geomaterials. The conclusions drawn from different resources are interpretive in nature and collectively provide a useful insight into the engineering behavior of ash fill. This framework is drawn from laboratory and field data that are predominantly available in published form; however, further endorsement through a comprehensive testing program is recommended.

References

1. A. Trivedi, Engineering behavior of coal ash, Ph.D. Thesis, Dept. Civil Eng., TIET, Patiala, 1999.
2. http://en.wikipedia.org/wiki/Fly_ash, 2013.
3. A. Trivedi and V.K. Sud, Grain characteristics and engineering properties of coal ash, *Granular Matter*, Vol. 4, No. 3, pp. 93–101, 2002.
4. *Federal Register*, Vol. 65, No. 99, pp. 32214–32237, 1989.
5. K. Prakash and A. Sridharan, Beneficial properties of coal ashes and effective solid waste management, *Practice Periodical of Hazardous, Toxic, and Radioactive Waste Management*, Vol. 13, No. 4, pp. 239–248, 2009.
6. A. Trivedi and V.K. Sud, Effect of grain size on engineering properties of coal ash, *Indian Journal of Environmental Protection*, Vol. 23, No. 5, pp. 525–553.
7. M.G. Ayoola and V.O. Ogunro, Leachability of compacted aged and fresh coal combustion fly ash under hydraulic flow conditions, in: *GeoCongress 2008*, American Society of Civil Engineers, 2008, pp. 692–699.
8. A.Trivedi and V.K.Sud, Collapse behavior of coal ash, *J. Geotech. Geoenviron. Eng.*, Vol. 130, No. 4, pp. 403–415, 2004.
9. A. Trivedi and S. Singh, Cone resistance of compacted ash fill, *Journal of Testing and Evaluation*, Vol. 32, No. 6, pp. 429–437, 2004.
10. A. Trivedi V.K. Sud and R. Pathak, Liquefaction characteristics of coal ash, in: *Fly Ash Characterization and Its Geotechnical Applications*, 1999, pp. 189–194.
11. A. Dey and S.R. Gandhi, Evaluation of liquefaction potential of pond ash, in: *Geotechnical Engineering for Disaster Mitigation and Rehabilitation*, H. Liu, A. Deng, and J. Chu, Eds. Springer Berlin Heidelberg, 2008, pp. 315–320.
12. T. Cousens and D. Stewart, Behaviour of a trial embankment on hydraulically placed pfa, *Engineering Geology*, Vol. 70, No. 3–4, pp. 293–303, 2003.
13. N.A. Lacour, Engineering characteristics of coal combustion residuals and a reconstitution technique for triaxial samples, Virginia Polytechnic Institute and State University, 2012.
14. S. Sharma, Ash dam damaged, NTPC's production from plant hit, *Economic Times*, Sept 27, 2011.

15. A. Trivedi, S. Singh, and C. Singh, Characterization and penetration resistance of ash fill, *Indian Journal of Environmental Proctection*, Vol. 23, pp. 768–773, 2003.
16. A. Sridharan, N.S. Pandian, and S. Srinivas, Compaction behaviour of Indian coal ashes, *Ground Improvement*, Vol. 5, No. 1, pp. 13–22, 2001.
17. A. Trivedi and S. Singh, Geotechnical and geoenvironmental properties of power plant ash, *Journal of the Institution of Engineers. India. Civil*, Vol. 85, pp. 93–99, Aug. 2004.
18. A.K. Choudhary, J.N. Jha, and K.S. Gill, Laboratory investigation of bearing capacity behaviour of strip footing on reinforced flyash slope, *Geotextiles and Geomembranes*, Vol. 28, No. 4, pp. 393–402, 2010.
19. D.P. Mishra and S.K. Das, A study of physico-chemical and mineralogical properties of Talcher coal fly ash for stowing in underground coal mines, *Materials Characterization*, Vol. 61, No. 11, pp. 1252–1259, 2010.
20. A. Trivedi and V.K. Sud, Ultimate bearing capacity of footings on coal ash, *Granular Matter*, Vol. 7, No. 4, pp. 203–212, 2005.
21. R. Gupta and A. Trivedi, Effects of non-plastic fines on the behavior of loose sand-An experimental study, *Electronic Journal of Geotechnical Engineering*, Vol. 14, pp. 1–15, 2009.
22. A. Trivedi and V.K. Sud, Settlement of compacted ash fills, *Geotechnical and Geological Engineering*, Vol. 25, No. 2, pp. 163–176, 2007.
23. P.K. Mehta and P.J.M. Monteiro, *Concrete: Microstructure, Properties, and Materials*. McGraw-Hill, 2006.
24. U. Dayal and R. Sinha, *Geo Environmental Design Practice in Fly Ash Disposal & Utilization*. Allied Publishers, 2005.
25. S. Ojha and A. Trivedi, Shear strength parameters for silty-sand using relative compaction, *Electronic Journal of Geotechnical Engineering*, Vol. 18, No. Bund A, pp. 81–99, 2013.
26. R. Salgado, P. Bandini, and A. Karim, Shear strength and stiffness of silty sand, *J. Geotech. Geoenviron. Eng.*, Vol. 126, No. 5, pp. 451–462, 2000.
27. M. Bolton, The strength and dilatancy of sands, *Geotechnique*, Vol. 36, No. 1, pp. 65–78, 1986.
28. S. Ojha, P. Goyal, and A. Trivedi, Non-linear behaviour of silty sand from catchment area of Yamuna River, in: *Indian Geotechnical Conference, 2012*, pp. 277–280.
29. S.K. Rao, L.K. Moulton, and R.K. Seals, Settlement of refuse landfills, in: *Proceedings of the Conference on Geotechnical Practice for Disposal of Solid Waste Materials*, University of Michigan, Ann Arbor, Michigan: June 13–15, 1977.
30. G.A. Leonards and B. Bailey, Pulverized coal ash as structural fill, *J. Geotech. Eng. Div., ASCE*, Vol. 108, 1982.
31. A.R. Cooper Jr and L.E. Eaton, Compaction behavior of several ceramic powders, *Journal of the American Ceramic Society*, Vol. 45, No. 3, pp. 97–101, 1962.

32. J.W. Hilf, Compacted fill, in: *Foundation Engineering Handbook*, Springer, 1991, pp. 249–316.
33. Y.S. Chae and J.L. Snyder, Vibratory compaction of fly ash, in: *Geotechnical Practice for Disposal of Solid Waste Materials*, 1977, pp. 41–62.
34. S.K. Saxena and R.M. Lastrico, Static properties of lightly cemented sand, *Journal of the Geotechnical Engineering Division*, Vol. 104, No. 12, pp. 1449–1464, 1978.
35. A. Puppala, Y. Acar, and M. Tumay, Cone penetration in very weakly cemented sand, *Journal of Geotechnical Engineering*, Vol. 121, No. 8, pp. 589–600, Aug. 1995.
36. F. Schnaid, P. Prietto, and N. Consoli, Characterization of cemented sand in triaxial compression, *J. Geotech. Geoenviron. Eng.*, Vol. 127, No. 10, pp. 857–868, Oct. 2001.
37. N. Consoli, A. Viana da Fonseca, R. Cruz, and K. Heineck, Fundamental parameters for the stiffness and strength control of artificially cemented sand, *J. Geotech. Geoenviron. Eng.*, Vol. 135, No. 9, pp. 1347–1353, Feb. 2009.
38. A. Trivedi and N. Kumar, Foundation settlement for footings embedded in rock masses, in: *UKIERI Concrete Congress - Innovations in Concrete Construction*, 2013, pp. 2031–2039.
39. A. Trivedi and N. Kumar, Strength of jointed rocks with granular fill, in: *ISRM International Symposium - 6th Asian Rock Mechanics Symposium*, October 23 - 27, 2010, New Delhi, India, 2010.
40. Z. Sen, RQD models and fracture spacing, *Journal of Geotechnical Engineering*, 1984.
41. L. Zhang, Method for estimating the deformability of heavily jointed rock masses, *J. Geotech. Geoenviron. Eng.*, Vol. 136, p. 1242, 2010.
42. A. Trivedi, Estimating In Situ Deformation of Rock Masses Using a Hardening Parameter and RQD, *Int. J. Geomech., ASCE*, 13(4), 348–364, 2013.
43. T. Ramamurthy and V. Arora, Strength predictions for jointed rocks in confined and unconfined states, *International Journal of Rock Mechanics and Mining Sciences*, Vol. 31, No. 1, pp. 9–22, 1994.
44. A. Trivedi, Strength and dilatancy of jointed rocks with granular fill, *Acta Geotechnica*, Vol. 5, No. 1, pp. 15–31, Aug. 2010.
45. V.K. Arora and A. Trivedi, Effect of Kaolin gouge on strength of jointed rocks, in *Asian Regional Symposium on Rock Slopes*, 7–11 December 1992, New Delhi, India: proceedings, 1992, pp. 21–26.
46. A. Trivedi and V.K. Arora, Discussion of 'Bearing Capacity of Shallow Foundations in Anisotropic Non-Hoek–Brown Rock Masses' by M. Singh and K.S. Rao, *J. Geotech. Geoenviron. Eng.*, Vol. 133, Issue 2, pp. 238–240, 2007.
47. A. Trivedi and V.K. Sud, Settlement of compacted ash fills, *Geotechnical and Geological Engineering*, Vol. 25, No. 2, pp. 163–176, Oct. 2007.

48. A. Trivedi, R. Pathak, and R. Gupta, A common collapse test for granular materials, in: *Indian Geotechnical Conference*, 2009, pp. 47–51.

49. K. Terzaghi, *Theoretical Soil Mechanics*. John Wiley & Sons, Inc.USA, 1943.

50. S. Singh, Liquefaction characteristics of silts, *Geotechnical and Geological Engineering*, Vol. 14, No. 1, pp. 1–19, 1996.

51. A.J. Lutenegger and R.T. Saber, Determination of collapse potential of soils, *ASTM Geotechnical Testing Journal*, Vol. 11, No. 3, 1988.

52. A.B. Fourie, D. Rowe, and G.E. Blight, The effect of inÆltration on the stability of the slopes of a dry ash dump, *Geotechnique*, Vol. 49, No. 1, pp. 1–13, 1999.

53. B. Indraratna, P. Nutalaya, K.S. Koo, and N. Kuganenthira, Engineering behaviour of a low carbon, pozzolanic fly ash and its potential as a construction fill, *Canadian Geotechnical Journal*, Vol. 28, No. 4, pp. 542–555, 1991.

54. L. Barden, A. McGown, and K. Collins, The collapse mechanism in partly saturated soil, *Engineering Geology*, Vol. 7, No. 1, pp. 49–60, 1973.

55. J.H. Schmertmann, Static cone to compute static settlement over sand, *J. Soil Mech. Found. Div., ASCE*, Vol. 96, No. SM3, pp. 1011–1043, 1970.

56. P.S. Toth, H.T. Chan, and C.B. Cragg, Coal ash as structural fill, with special reference to Ontario experience, *Canadian Geotechnical Journal*, Vol. 25, No. 4, pp. 694–704, 1988.

57. R. Turgeon, Fly ash fills a valley, *Civil Engineering—ASCE*, Vol. 58, No. 12, pp. 67–68, 1988.

58. K. Terzaghi and R.B. Peck, *Soil Mechanics in Engineering Practice*, Wiley, New York, pp. 64–65, 1948.

59. S.W. Perkins and C.R. Madson, Bearing capacity of shallow foundations on sand: A relative density approach, *J. Geotech. Geoenviron. Eng.*, Vol. 126, No. 6, p. 521, 2000.

60. D.J. Dappolonia, E.E. Dappolonia, and R.F. Brissette, Settlement of spread footings on sand, *Journal of Soil Mechanics & Foundations Div*, Vol. 94, No. SM3, pp. 735–758, 1968.

61. G.G. Meyerhof, Penetration tests and bearing capacity of cohesionless soils, *Journal of the Soil Mechanics and Foundation Division*, Vol. 82, No. 1, pp. 1–19, 1956.

62. J.B. Burland, M.C. Burbidge, E.J. Wilson, and Terzaghi, Settlement of foundations on sand and gravel, *ICE Proceedings*, Vol. 78, No. 6, pp. 1325–1381, 1985.

12

Carbon-Doped Cryogel Thin Films Derived from Resorcinol Formaldehyde

Z. Marković[1], D. Kleut[1], B. Babić[1], I. Holclajtner-Antunović[2], V. Pavlović[3] and B. Todorović-Marković[1,*]

[1]Vinča Institute of Nuclear Sciences, University of Belgrade, Belgrade, Serbia
[2]Faculty of Physical Chemistry, University of Belgrade, Belgrade, Serbia
[3]Joint Laboratory for Advanced Materials, Serbian Academy of Sciences and Arts, Belgrade, Serbia

Abstract

This chapter presents the results of structural properties of carbon-doped cryogel thin films derived from resorcinol formaldehyde (RF). The RF cryogels were doped by single-wall carbon nanotubes (SWCNTs), graphene and graphene quantum dots (GQDs). Different characterization techniques were used to investigate structural properties of these films: atomic force microscopy (AFM), scanning electron microscopy (SEM), Fourier transform infrared spectroscopy (FTIR) and Raman spectroscopy. Raman spectroscopy was performed using two excitation laser energies in the visible range in order to investigate the effect of ceramic glass substrate on the Raman features of investigated carbon-doped RF cryogel thin films. The particle size of these films was determined by atomic force microscopy and scanning electron microscopy.

Keywords: Single-wall carbon nanotubes, graphene, graphene quantum dots, resorcinol formaldehyde, thin films, Raman spectroscopy, Fourier transform infrared spectroscopy, atomic force microscopy, scanning electron microscopy

Corresponding author: biljatod@vin.bg.ac.rs

Ashutosh Tiwari and S.K. Shukla (eds.) Advanced Carbon Materials and Technology, (475–486)
2014 © Scrivener Publishing LLC

12.1 Introduction

Carbon aerogels represent novel porous carbon materials with inter-connected structure and higher electrical conductivity than other carbons [1–3]. This kind of carbon material exhibits relatively large specific area (500–1000 m^2/gm), low mass densities and an intricate structural morphology of macropores (> 50 nm), mesopores (2–50 nm) and nanopores (< 2 nm). In this way they could be used in various new technological advances. The possible applications of carbon cryogels could be achieved as adsorbents, molecular sieves, filters, thermal insulates and electrodes [2–5 icom]. Due to their con-trollable porous structure, carbon cryogels are considered ideal elec-trode materials for supercapacitors and rechargeable batteries [4, 5].

This chapter represents for the first time resorcinol formaldehyde (RF) thin films doped by single-wall carbon nanotubes, graphene and graphene quantum dots [6–8]. In this work we tried not only to understand the structure of these disordered carbon materials, but also to see how the structure of these materials relates to the pres-ence of dopants in different concentrations.

12.2 Experimental Procedure

In the present work, RF gels were synthesized by polycondensation of resorcinol, ($C_6H_4(OH)_2$) (R), with formaldehyde, (HCHO) (F), according to the method proposed by Pekala *et al.* [9]. Sodium car-bonate, (Na_2CO_3) (C), was used as a basic catalyst. The concentra-tion of starting solutions was 5 wt% while the molar ratio between R and F was 0.5 and catalyst ratio between R and C was 200. RF solution was mixed with carbon-based nanomaterials with concen-tration that varied in the range of 0.5 to 2%.

A detailed procedure for the preparation of stable carbon-based col-loids is given in our earlier investigations [6–8]. The homogeneously mixed solution was poured between two glass plates and dried. The thin films were heat treated in a (1,1,1)-cycle, where the numbers stand for the days at room temperature, 50°C and 90°C, respectively. The RF cryogels were prepared by freeze drying according to the procedure of Tamon *et al.* [10–13]. The RF gels were immersed in a 10-times volume of t-butanol, p.a. quality (Centrohem-Beograd), for more than one day and rinsed to displace the liquid contained in the gels with t-butanol. The rinsing with t-butanol was repeated twice.

The samples were prepared by freeze-drying using laboratory set-up. First, the samples were pre-frozen at -30°C for 24 hours. After that, they were freeze dried in the acrylic chambers with shelf arrangements mounted directly on top of the condenser of the freeze dryer. The vacuum during twenty hours of freeze-drying was around 4 mbar.

Carbon cryogels were prepared by carbonization of the cryogels in a conventional furnace, in a nitrogen flow, at 850°C for 2 h. After pyrolysis, the furnace was cooled at room temperature.

Fourier transform infrared spectra of carbon-doped RF cryogel thin films were measured at room temperature in the spectral range from 400 to 4000 cm^{-1} on a Nicollet 380 FTIR, Thermo Electron Corporation spectrometer.

Raman spectra of carbon-doped RF cryogel thin films deposited on a glass substrate were obtained by DXR Raman microscope (Thermo Scientific) using 532 and 633 nm excitation lines with constant laser power of 5 mW [14, 15]. The spectral resolutions were 0.5 and 1 cm^{-1}, respectively. We used a 50x objective to focus the used excitation laser lines with spot sizes of 0.7 and 2.5 μm, respectively. Raman spectra of deposited carbon-doped cryogel thin films were recorded at room temperature. Each sample was analyzed in five different points and mean spectrum was presented in the manuscript. At each point Raman spectra was acquired three times. Acquisition time was 10 sec with 20 scans. We failed to detect any differences among subsequent Raman spectra.

The microstructure and morphological changes of carbon-doped RF cryogel thin films were recorded by atomic force microscope (Quesant) operating in tapping mode in air at room temperature [16]. Standard silicon tips (purchased from NanoAndMore GmbH) with force constant 40 N/m were used. Surface morphology was also investigated by scanning electron microscopy (JSM-6390 LV, JEOL, Tokyo, Japan).

12.3   Results and Discussion

12.3.1   FTIR Analysis

Figure 12.1 shows the FTIR spectrum of the carbon-doped RF cryogel thin films. Two sharp peaks at 2852, 2923 cm^{-1} and peak at 1471 cm^{-1} represent CH_2 stretching vibrations. Broad band at 1587 cm^{-1}

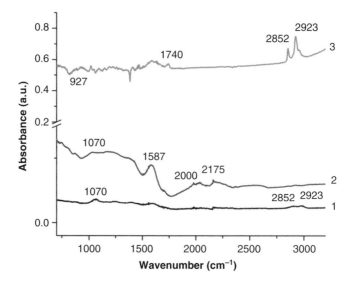

Figure 12.1 FTIR spectrum of carbon-doped RF cryogel thin films by: 1-SWCNTs, 2-graphene and 3-GQDs. Concentration of carbon-based material was 2%.

represents the aromatic group, while the C–O–C stretching vibrations of methylene ether bridges between resorcinol molecules could be found between 1000 and 1300 cm^{-1} (several peaks appear in this interval) [17]. Peak at 1740 cm^{-1} stem from C=O stretching vibrations of aldehyde, while peak at 927 cm^{-1} could be assigned to CH$_2$ bending vibrations [18]. Peaks at 2000 and 2175 cm^{-1} could be assigned to C-C asymmetric stretching vibrations [19].

12.3.2 Raman Analysis

Raman spectroscopy is a powerful tool for determination of carbon nanomaterial structure [20]. A key requirement for carbon research and applications is the ability to identify and characterize all the members of the carbon family, both at the lab- and mass-production scale. Raman spectroscopy as a nondestructive characterization tool with high resolution gives the maximum structural and electronic information about carbon nanomaterials. Analysis of the observed Raman spectra and the resonance condition provides precise information on the electronic states, phonon energy dispersion, and electron-phonon interaction in sp^2 carbon nanostructures.

The Raman spectra of all carbon systems show only a few prominent features. In Figure 12.2, Raman spectra of doped resorcinol

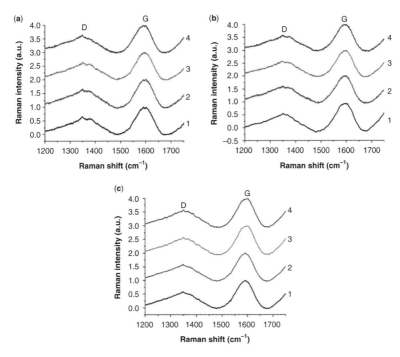

Figure 12.2 Raman spectra of RF thin films doped by (a) SWCNTs, (b) graphene and (c) GQDs. Concentrations of carbon-based materials were changed in the range of 0.5 to 2%.

formaldehyde thin films by three different carbon materials are presented. The concentration of carbon materials were changed in the range from 0.5 to 2%. In Figure 12.2(a) Raman spectra of RF cryogel thin films doped by SWCNTs are presented, while in Figure 12.2(b,c) Raman spectra of RF cryogel thin films doped by graphene and GQDs are presented, respectively. For measurements of these spectra we used lasers with 532 nm wavelength. Curve 1 in all graphs presents Raman spectrum of RF cryogel thin films with carbon concentration of 0.5%, while curves 2, 3, and 4 present carbon concentration of 1, 1.5 and 2%, respectively. As can be seen from these figures, there are two prominent bands assigned by G and D bands. The existence of G band corresponds to the E_{2g} mode while D band has been attributed to in-plane A_{1g} zone edge mode.

Raman intensity mainly depends on the exciting line and the polarizability tensor of the chemical bond [21]. We used two different exciting laser lines in order to specifically analyze the

Table 12.1 Positions of G and D bands and corresponding FWHM values of carbon-doped cryogel thin films probed by two excitation laser lines.

Sample		Excitation laser line (nm)			
		532		633	
		D band	G band	D band	G band
SWCNTs doped cryogel thin films	Position (cm⁻¹)	1331.79	1591.08	1351.56	1586.44
	FWHM (cm⁻¹)	168.84	63.01	329.36	98.31
Graphene doped cryogel thin films	Position (cm⁻¹)	1340.1	1594.17	1352.45	1585.83
	FWHM (cm⁻¹)	161.76	63.61	341.91	98.31
GQDs doped cryogel thin film	Position (cm⁻¹)	1336.45	1593.32	1350.98	1583.8
	FWHM (cm⁻¹)	153	63.19	352.59	100.57

effect of used glass substrate. Listed in Table 12.1 are the positions of G and D bands and corresponding FWHM values of carbon-doped RF cryogel thin films probed by two excitation laser lines. Figure 12.3 shows Raman spectra of carbon-doped RF cryogel thin films probed by two excitation laser lines (532 and 633 nm). As can be seen from Figure 12.3 and Table 12.1, full widths at half maximum (FWHM) depend on used excitation laser line. When samples were probed by 633 nm Raman lasers, FWHMs of samples were much broader than FWHMs of samples probed by 532 nm Raman laser. Increase of FWHMs of G and D bands during probes of RF cryogel thin films by 633 nm laser indicates that used glass substrate significantly affects the shape of recorded Raman spectra.

Raman analysis has shown that doped RF cryogel thin films have amorphous structure. There is no difference between the bottom and top surface of investigated films. The FWHMs of samples are

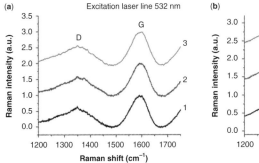
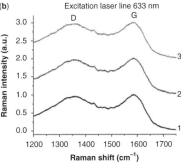

Figure 12.3 Raman spectra of doped RF cryogel thin films probed by (a) 532 nm laser and (b) 633 nm laser. Curve 1 on both graphs represents Raman spectra of doped RF cryogel thin films by SWCNTs, while curves 2 and 3 represent Raman spectra of doped RF cryogel thin films by graphene and GQDs, respectively.

much broader during probe by 633 nm laser line because of the effect of ceramic glass substrate.

12.3.3 Surface Morphology of Carbon-Doped RF Cryogel Thin Films

Atomic force microscopy and scanning electron microscopy were used to investigate the morphology of prepared RF cryogel thin films doped by carbon-based materials. Figure 12.4 shows the top view of AFM images of surface morphology of RF cryogel thin films. Figure 12.4(a,b) represent SWCNTs-doped RF cryogel thin films with different concentration of SWCNTs (0.5 and 2%). Figure 12.4(c,d) represent graphene-doped RF cryogel thin films with graphene concentration of 0.5 and 2%, respectively, while Figure 12.4(e,f) represent GQD-doped RF cryogel thin films with concentration of 0.5 and 2%. The dopant concentration increase affects the porosity of deposited RF films. As could be observed from Figure 12.4, particle size increases with dopant concentration increase. By applying special software available in the atomic force microscope we determined average particle diameter. In the case of carbon concentration of 2%, average particle diameter of SWCNTs-doped RF cryogel thin films are 158 nm (Fig. 12.4b) while particle diameters of graphene- (Fig. 12.4d) and GQDs-doped (Fig. 12.4f) RF cryogel thin films are 368 and 420 nm, respectively. The average

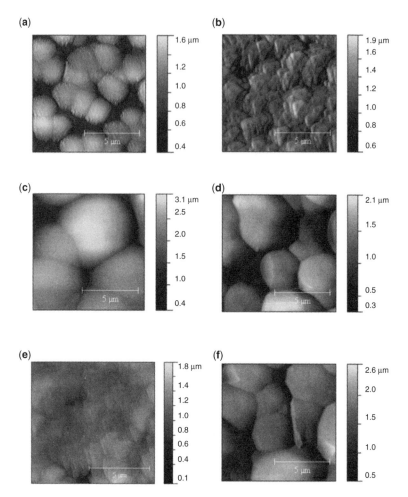

Figure 12.4 Top view AFM images of RF cryogel thin films doped by: (a) SWCNTs concentration of 0.5%; (b) SWCNTs concentration of 2%; (c) graphene concentration of 0.5%; (d) graphene concentration of 2%; (e) GQDs concentration of 0.5%; (f) GQDs concentration of 2%.

pore sizes are 100 nm for SWCNTs-doped RF cryogel thin films, 500–1000 nm for graphene-doped cryogel thin films and 100–300 nm for GQDs-doped RF cryogel thin films. Based on AFM surface analysis we could conclude that particles of RF cryogel thin films tend to aggregate and form homogeneous thin films.

Scanning electron microscopy was used to investigate surface morphology of carbon-doped RF cryogels thin films as well. In Figure 12.5, SEM micrographs of these thin films are presented. As shown in Figure 12.5(a,b,c) the typical particle cluster sizes are

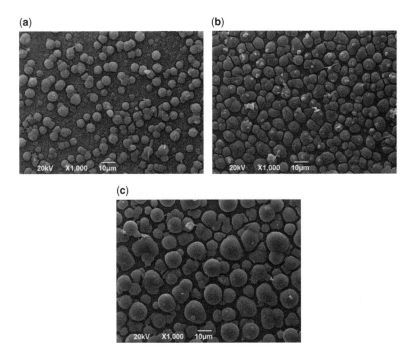

Figure 12.5 SEM micrographs of doped RF cryogel thin films by: (a) SWCNTs, (b) graphene and (c) GQDs. Concentration of used dopants was 2%.

about 5, 7 and 10 μm, respectively, while the pore diameter is up to 2–10 μm. The largest particle clusters have RF cryogel thin films which are doped by GQDs (Fig. 12.5c), while the largest pores have RF cryogel thin films doped by SWCNTs. Earlier it was reported that the RC ratio is the dominant factor influencing the cluster size. With low RC ratio, the particles derived from the clusters have small diameters and are joined together with large necks, giving the aerogel a polymeric appearance. With high RC ratio, the particles have large sizes and are interconnected to form an aerogel with a colloidal structure [22–24]. In our case with low RC ratio, carbon-based particles incorporated in RF matrix tend to aggregate and form larger clusters. SEM analyses showed that carbon-doped RF cryogel films have an island structure.

12.4   Conclusion

In this chapter a detailed structural analysis of doped RF cryogel thin films was presented. RF cryogel thin films on glass substrate were doped by SWCNTs, graphene and GQDs. Dopant concentration was changed in the range of 0.5 to 2%. Raman analysis has shown that doped RF cryogel thin films have amorphous structure (the presence of very intense and broad D and G bands). By changing the dopant concentration, particle and pore sizes of investigated films can be tuned. The largest particle cluster size and smallest pore size have GQDs-doped RF cryogel films.

Acknowledgements

The authors want to thank the Ministry of Education, Science and Technological Development of the Republic of Serbia for financial support (project number 172003).

References

1. R.W. Pekala, C.T. Alviso, in: *Mater. Res. Soc. Symp. Proc.*, edited by C.L. Renschler, J.J. Pouch, and D.M. Cox, Vol. 270, p. 3, 1992.
2. R. Petrićević, G. Reichenauer, V. Bock, A. Emmerling, J. Fricke, *J. Non-Cryst. Solids*, Vol. 225, p. 41, 1998.
3. Z.M. Marković, B.M. Babić, M.D. Dramićanin, I.D. Holclajtner Antunović, V.B. Pavlović, D.B. Peruško, B.M. Todorović Marković, *Synthetic Met.*, Vol. 162, p. 743, 2012.
4. R. Saliger, U. Fischer, C. Herta, J. Fricke, *J. Non-Cryst. Solids*, Vol. 225, p. 81, 1998.
5. R.W. Pekala, F.M. Kong, *J. Phys. (Paris) Colloq. C*, Vol. 4, p. 33, 1989.
6. Z. Marković, S. Jovanović, D. Kleut, N. Roméević, V. Jokanović, V. Trajković, B. Todorović Marković, *Appl. Surf. Sci.*, Vol. 255, p. 6359, 2009.
7. Z.M. Marković, Lj. M. Harhaji-Trajković, B.M. Todorović Marković, D.P. Kepić, K.M. Arsikin, S.P. Jovanović, A.C. Pantović, M.D. Dramićanin, V.S. Trajković, *Biomaterials*, Vol. 32, p. 1121, 2011.
8. Z.M. Marković, B.Z. Ristić, K.M. Arsikin, D.G. Klisić, L.M. Harhaji-Trajković, B.M. Todorović-Marković, D.P. Kepić, T.K. Kravić Stevović, S.P. Jovanović, M.M. Milenković, D.D. Milivojević, V.Z. Bumbaširević, M.D. Dramićanin, V.S. Trajković, *Biomaterials*, Vol. 33, p. 7084, 2012.
9. R.W. Pekala, *J. Mater. Sci.*, Vol. 24, p. 3221, 1989.
10. H. Tamon, H. Ishizaka, T. Yamamoto, T. Suzuki, *Carbon*, Vol. 37, p. 2049, 1999.
11. T. Yamamoto, T. Sugimoto, T. Suzuki, S.R. Mukai, H. Tamon, *Carbon*, Vol. 40, p. 1345, 2002.

12. H. Tamon, H. Ishizaka, T. Yamamoto, T. Suzuki, *Carbon*, Vol. 38, pp. 1099–1105, 2000.

13. T. Yamamoto, T. Nishimura, T. Suzuki, H. Tamon, *Journal of Non-Crystalline*

14. *Solids*, Vol. 288, pp. 46–55, 2001.

15. Z. Marković, D. Kepić, I. Holclajtner-Antunović, M. Nikolić, M. Dramićanin, M. Marinović-Cincović, B. Todorović-Marković, *J. Raman Spectrosc.*, Vol. 43, p. 1413, 2012.

16. D. Kepić, Z. Marković, D. Tošić, I. Holclajtner-Antunović, B. Adnadjević, J. Prekodravac, D. Kleut, M. Dramićanin, B. Todorović-Marković, *Phys. Scripta*, (2013) in press.

17. B. Todorović-Marković, S. Jovanović, V. Jokanović, M. Dramićanin, Z. Marković, *Appl. Surf. Sci.*, Vol. 255, pp. 3283–3288, 2008.

18. S. Mulik, C. Sotiriou-Leventis, N. Leventis, *Polymer*, Vol. 47, p. 364, 2006.

19. J.Y. Lee, K.N. Lee, H.J. Lee, J.H. Kim, *J. Ind. Eng. Chem.*, Vol. 8, p. 546, 2002.

20. D.R. Lide, *CRC Handbook of Chemistry and Physics*, CRC Press, pp. 9–86, 2003–2004.

21. L.M. Malard, M.A. Pimenta, G. Dresselhaus, M.S. Dresselhaus, *Phys. Rep.*, Vol. 473, p. 51, 2009.

22. G. Gouadec, P. Colomban, *Prog. Cryst. Growth Charact. Mater.*, Vol. 53, p. 1, 2007.

23. X. Lu, O. Nilsson, J. Fricke, and R.W. Pekala, *J. Appl. Phys.*, Vol. 73, p. 581, 1993.

24. R. Petričević, H. Proebstle, J. Fricke, in: *Proceedings of the 5th International Symposium on the Characterization of Porous Solids* (DECHEMA e. V., Heidelberg, 1999), p. 318.

25. R. Saliger, V. Bock, R. Petričević, T. Tillotson, S. Geis, J. Fricke, *J. Non-Cryst. Solids*, Vol. 221, p. 144, 1997.

Index

Also of Interest

Check out these published and forthcoming related titles from Scrivener Publishing

Advanced Energy Materials
Edited by Ashutosh Tiwari and Sergiy Valyukh
Forthcoming February 2014. ISBN 978-1-118-68629-4

Advanced Carbon Materials and Technology
Edited by Ashutosh Tiwari and S.K. Shukla
Published 2014. ISBN 978-1-118-68623-2

Responsive Materials and Methods
State-of-the-Art Stimuli-Responsive Materials and Their Applications
Edited by Ashutosh Tiwari and Hisatoshi Kobayashi
Published 2013. ISBN 978-1-118-68622-5

Nanomaterials in Drug Delivery, Imaging, and Tissue Engineering
Edited by Ashutosh Tiwari and Atul Tiwari
Published 2013. ISBN 978-1-118-29032-3

Biomimetics
Advancing Nanobiomaterials and Tissue Engineering
Edited by Murugan Ramalingam, Xiumei Wang, Guoping Chen, Peter Ma, and Fu-Zhai Cui
Published 2013. ISBN 978-1-118-46962-0

Atmospheric Pressure Plasma Treatment of Polymers
Edited by Michael Thomas and K.L. Mittal
Published 2013. ISBN 978-1-118-59621-0

Polymers for Energy Storage and Conversion
Edited by Vikas Mittal
Published 2013. ISBN 978-1-118-34454-5

Encapsulation Nanotechnologies
Edited by Vikas Mittal
Published 2013. ISBN 978-1-118-34455-2

Biomedical Materials and Diagnostic Devices Devices
Edited by Ashutosh Tiwari, Murugan Ramalingam, Hisatoshi
Kobayashi and Anthony P.F. Turner
Published 2012. ISBN 978-1-118-03014-1

Intelligent Nanomaterials
Processes, Properties, and Applications
Edited by Ashutosh Tiwari Ajay K. Mishra, Hisatoshi Kobayashi
and Anthony P.F. Turner
Published 2012. ISBN 978-0-470-93879-9

Integrated Biomaterials for Biomedical Technology
Edited by Murugan Ramalingam, Ashutosh Tiwari, Seeram
Ramakrishna and Hisatoshi Kobayashi
Published 2012. ISBN 978-1-118-42385-1

Integrated Biomaterials in Tissue Engineering
Edited by Murugan Ramalingam, Ziyad Haidar, Seeram
Ramakrishna, Hisatoshi Kobayashi, and Youssef Haikel
Published 2012. ISBN 978-1-118-31198-1

The Physics of Micropdroplets
Jean Berthier and Kenneth Brakke
Published 2012. ISBN 978-0-470-93880-0

Antioxidant Polymers
Synthesis, Properties and Applications
Edited by Giuseppe Cirillo and Francesca Iemma
Published 2012. ISBN 978-1-118-20854-0

Introduction to Surface Engineering and Functionally Engineered
Materials
Peter Martin
Published 2011. ISBN 978-0-470-63927-6

Handbook of Bioplastics and Biocomposites Engineering Applications.
Edited by Srikanth Pilla
Published 2011. ISBN 978-0-470-62607-8

Biopolymers: Biomedical and Environmental Applications
Edited by Susheel Kalia and Luc Avérous
Published 2011. ISBN 978-0-470-63923-8

Renewable Polymers
Synthesis, Processing, and Technology
Edited by Vikas Mittal
Published 2011. ISBN 978-0-470-93877-5

Miniemulsion Polymerization Technology edited by Vikas Mittal
Published 2010. ISBN 978-0-470-62596-5

Polymer Nanotube Nanocomposites
Synthesis, Properties, and Applications
Edited by Vikas Mittal.
Published 2010. ISBN 978-0-470-62592-7